AF566455

Oxidations in Organic Synthesis

Oxidations in Organic Synthesis

Dr. V.K. Ahluwalia
Visiting Professor
Dr. B.R. Ambedkar Center for Biomedical Research,
University of Delhi, India

CRC is an imprint of the Taylor & Francis Group,
an informa business

Ane Books Pvt. Ltd.

Oxidations in Organic Synthesis
V.K. Ahluwalia

First Published in 2012

Ane Books Pvt. Ltd.

4821 Parwana Bhawan, 1st Floor
24 Ansari Road, Darya Ganj, New Delhi -110 002, India
Tel: +91 (011) 2327 6843-44
Fax: +91 (011) 2327 6863
e-mail: kapoor@anebooks.com, anebooks@vsnl.net
Website: www.anebooks.com

For

CRC Press

Taylor & Francis Group
6000 Broken Sound Parkway, NW, Suite 300
Boca Raton, FL 33487 U.S.A.
Tel : 561 998 2541
Fax : 561 997 7249 or 561 998 2559
Web : www.taylorandfrancis.com

For distribution in rest of the world other than the Indian sub-continent

ISBN : 978-146-6506-01-05

British Library Cataloguing in Publication Data
A catalogue record for this book is available from the British Library

Printed at: Thomson Press (India) Limited, Delhi.

Preface

Oxidation plays a very important role in organic synthesis. A large number of organic syntheses involve oxidation at some stage.

The present book describes a very large number of oxidising agents and their applications in oxidations. A number of reagents of importance are described in great detail. The book also includes oxidation of specific types of organic compounds. These include various types of hydrocarbons (alkanes, alkenes and alkynes); alcohols (1°, 2°, and 3° alcohols); phenols; ethers; carbonyl compounds (aldehydes and ketones); carboxylic acids; anides; hydrazides; nitro compounds; nitroso compounds; hydroxyl amines; azo compounds; azides; hydrazo compounds; amines (1°, 2°, 3° and aromatic amines); and phosphorus, arsenic and sulphur compounds. Besides, enzymatic or microbial oxidations have also been described. A special feature of the book is the last chapter dealing with some oxidations under benign conditions.

I express my sincere gratitude to Prof. Sukhdev for the help rendered. Special thanks are due to several authors, whose work has been consulted for writing the book. Names of various authors are listed in the Suggested Readings at the end of the book.

V.K. Ahluwalia

Contents

INTRODUCTION

Oxidation plays an important role in organic chemistry. A large number of reactions described in literature involve oxidation. Before we examine details of the oxidation, it will be of interest to consider some definitions, which will be useful.

A convenient definition of oxidation of an organic substrate is the addition or gain of oxygen or removal of hydrogen. Typical examples include oxidation of an aldehyde to a carboxylic acid or conversion of ethylene to ethylene oxide. Both these examples involve addition of oxygen. On the other hand, oxidation of ethanol to acetaldehyde involve removal of hydrogen. A still other definition of oxidation is the removal of electrons from the substrate. Thus, in the oxidation of ethylalcohol to acetaldehyde, two electrons are lost.

The important criteria to decide whether a particular organic reaction is a oxidation is that the reagent used in the reaction is reduced. For example, the conversion of an olefin a glycol is affected by OsO_4.

$$>C{=}C< \; + \; OsO_4 \longrightarrow >C(O)-C(O)< \text{ (cyclic } O_2OsO_2 \text{ ester)} \xrightarrow{H_2O} >C(OH)-C(OH)< \; + \; OsO_3$$

In the above example, the reagent, OsO_4 is reduced and the organic compound is oxidised. Similarly, oxidation with potassium permanganate, Mn(VII) is converted to Mn(II) or Mn(IV).

A most important definition of oxidation is that it involves increase in the oxidation state. Thus, the conversion of a functional group in a molecule, from a lower oxidation state to a higher one will be called oxidation, while the opposite of it will be called reduction. Loss of electrons is associated with the change in oxidation state of an atom. This electron loss can be determined by finding the oxidation number. A number of rules were formulated[1] for determining the oxidation number. A simplified procedure[2, 3] for determining the oxidation number was formulated. These rules were applied only to those atoms, which are commonly encountered (like H, O, C or halogen) in simple organic molecules. The following convention for assigning oxidation number is used. Oxidation number for hydrogen is –1; oxidation number for carbon (bonded to the same atom) is 0 and oxidation number for carbon bonded to a hetro atom (oxygen or halogen) is +1. Only those atoms that are modified or changed during the reaction are examined. By comparison of the sum of oxidation numbers of the various atoms in the starting material and in the product formed, the change in oxidation number can be calculated.

As an example, oxidation of ethanol gives acetaldehyde, which is further oxidised to acetic acid.

$$\underset{-1}{CH_3CH_2OH} \xrightarrow{2e^- \text{ lost}} \underset{+1}{CH_3CHO} \xrightarrow{2e^- \text{ lost}} \underset{+3}{CH_3—C(=O)—OH}$$

In ethanol, the carbon bearing the hydroxyl group has an oxidation number of –1 (C is 0, H is –1, H is –1 and 0 is +1). The oxidation number of that carbon in acetaldehyde is +1 (C is 0, H is –1, 0 is + 1 and 0 from second bond to the carbonyl is +1). The change in the oxidation number is –1 → + 1, for a net loss of two electrons. So, this is an oxidation reaction. In a similar way, the carbon of interest in acetic acid has an oxidation number of +3 (C is 0; the three bonds to oxygen total + 3). Thus, conversion of acetaldehyde to acetic acid involves loss of two additional electrons (+1 → +3) and is an oxidation reaction.

The oxidising capacity of an oxidising reagent depends on their oxidation potential. The oxidation potential represents the ability of a metal atom to be ionized to an ion $(M \xrightarrow{-e} M^+)$ with loss of an electron. The oxidation of organic substrate is carried out by transition metal compounds containing chromium, manganese, selenium, ruthenium, silver, cerium *etc*. The pH of the reaction medium also plays an important role in the oxidising capacity of a reagent. The product obtained in an oxidation reaction depends on the nature of the substrate, the reagent and reactions conditions like temperature, pressure, solvent *etc*.

In the first chapter, a brief account of various oxidizing agents along with their applications is given.

References

1. S. Soloveichik and H. Krakauer, *J. Chem. Ed.* 1966, 43, 532.
2. J.B. Hendrickson, D.J. Cram and Y.S. Hammond, *Organic Chemistry*, 3rd ed., Mcgraw Hill, New York, 1970.
3. S.H. Pine, *Organic Chemistry*, 5th, ed; McGraw Hill, New York, 1987.

1

Oxidising Agents

Various oxidising agents used for oxidation are listed in alphabetical order. A brief description of the oxidation process for which each of the oxidising agent used is also given.

1.1 ACETYL HYPOIODITE (IODINE ACETATE), CH_3COOI

It is prepared[1,2] by the treatement of silver acetate in acetic acid with iodine at room temperature. This reagent oxidises[2] alcohols to ketones. Thus, 3-ethyl-3-pentanol is oxidised to 3-pentanone and iodoethane by acetyl hypoiodite in acetic acid and chlorobenzene under irradiation for 1 hr. At 20–25°C.

$$CH_3CH_2-\underset{\underset{CH_3}{|}}{\underset{CH_2}{|}}{\overset{OH}{\overset{|}{C}}}-CH_2CH_3 \xrightarrow[\substack{C_6H_5Cl,\ h\nu,\ 1hr \\ 20\text{–}25°C}]{CH_3COOI/HOAc} CH_3CH_2COCH_2CH_3 + CH_3CH_2I$$

3-Ethyl-3-pentanol → 3-Pentanone 90%, Iodoethane 84%

Acetyl hypoiodite brings about syn hydroxylation[3] of alkenes

$$\text{Cyclohexene} \xrightarrow[\text{(2) AcOH, } H_2O\text{, reflux, 1 hr}]{\text{(1) AcOAg, } I_2\text{, AcOH, RT, 4.5 hr}} \text{cyclohexane-1,2-diol (HO, OH)}$$

66%

The cleavage of vicinal diols with acetyl hypoiodite to give dicarbonyl compounds[1].

References

1. R. Cambic, D. Chambers, P.S. Rutledge and P.D. Woodgate, *J. Chem. Soc. Perkin Trans.* I, 1978, 1483.

2. T.R. Beebe, B.A. Barnes, K.A. Bender, A.D. Halbert, R.D. Miller, M.L. Ramsay and M.W. Ridenour, *J. Organic Chem.*, 1975, **40**, 1992.
3. F.D. Yunstone and L.J. Morris, *J. Chem. Soc.*, 1959, 487.

1.2 ALUMINIUM ISOPROPOXIDE $[(CH_3)_2CHO)]_3$ Al

It is prepared by the reaction of isopropyl alcohol with amalgamated aluminium foil in the presence of small amount of carbon tetrachloride as a catalyst. When evolution of hydrogen ceases, the molten reagent is distilled under reduced pressure. It has specific synthetic uses. Some of its applications are given below:

1. *Oxidation of alcohols (Oppenauer oxidation)*[1]*:* Alchols on refluxing with aluminium isopropoxide in acetone and benzene or tolune are oxidised to aldehydes or ketones. In this oxidation, acetone acts as a hydrogen acceptor and is converted into isopropyl alchol. The presence of excess of acetone drives the reaction to completion.

$$\underset{\text{Sec alcohol}}{R{-}\overset{OH}{\overset{|}{C}H}{-}R'} + \underset{\text{Acetone excess}}{CH_3{-}\overset{O}{\overset{\|}{C}}{-}CH_3} \xrightleftharpoons{Al(OCHMe_2)_3} \underset{\text{Ketone}}{R{-}\overset{O}{\overset{\|}{C}}{-}R'} + \underset{\text{Isopropyl alcohol}}{CH_3{-}\overset{OH}{\overset{|}{C}H}{-}CH_3}$$

The oppenauer oxidation is more suitable for high-boiling alcohols, such as hydroxysteroids[2].

Al(o-i-Pr)$_3$, cyclohexanone; PhMe, distillation; 81-93%

The reaction is carried out by refluxing secondary alcohol with acetone (or cyclohexanone) in the presence of aluminium isopropoxide. The aluminium isopropoxide serves only to form the aluminium alkoxide of the alcohol, which is then oxidised through a cyclic transition state at the expense of actone to yield a ketone and isopropyl alcohol. Various steps involved are shown ahead.

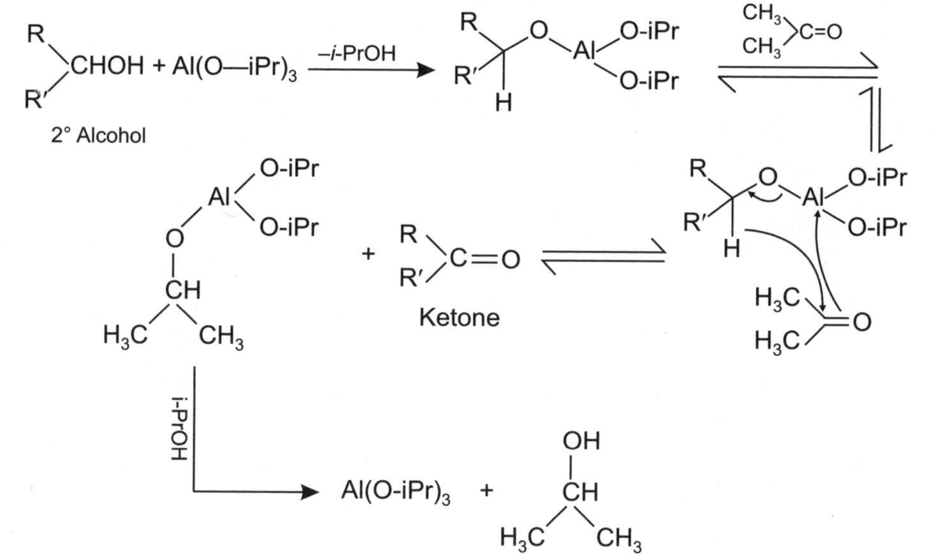

2. *Hydrolysis of oximes:* Oximes are converted into parent carbonyl compounds

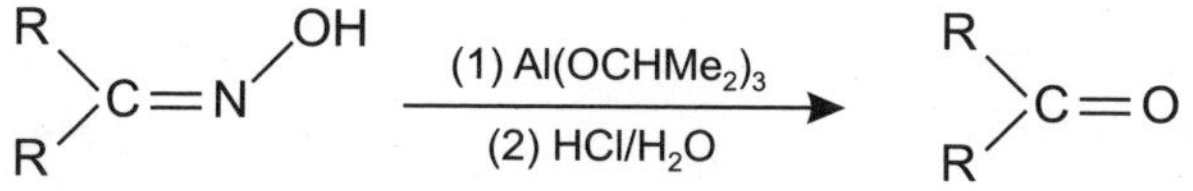

$$R_2C{=}N{-}OH \xrightarrow[(2)\ HCl/H_2O]{(1)\ Al(OCHMe_2)_3} R_2C{=}O$$

3. *Preparation of ethers:* The reaction of alkyl halides with aluminium isopropoxide in dimethyl formamide gives ethers[3]

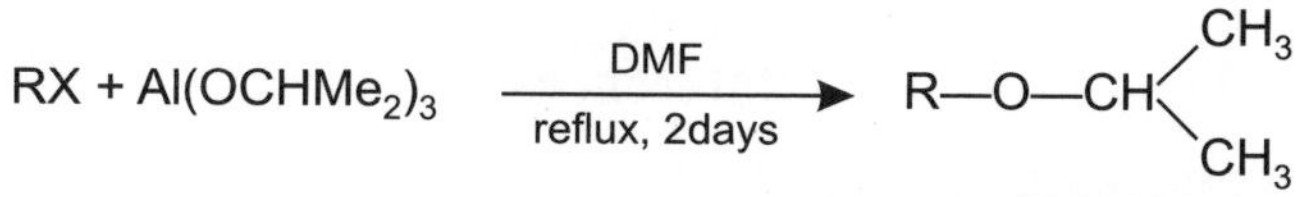

$$RX + Al(OCHMe_2)_3 \xrightarrow[\text{reflux, 2days}]{DMF} R{-}O{-}CH(CH_3)_2$$

20–80%

4. *Ring opening of epoxides*

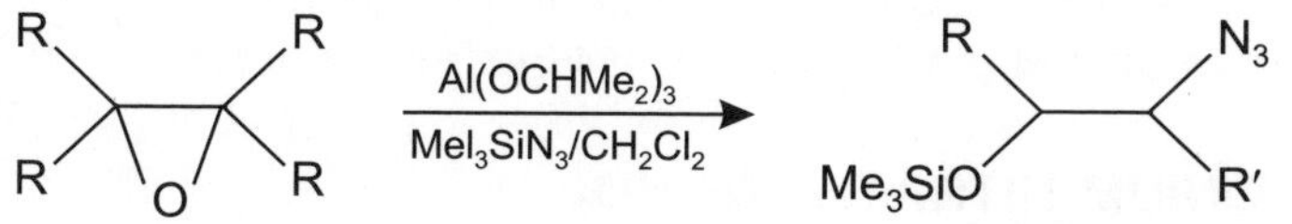

$$\xrightarrow[MeI_3SiN_3/CH_2Cl_2]{Al(OCHMe_2)_3}$$

60–90%

In some cases, rearrangement of the epoxides to allylic alcohols has been observed. Thus, α-pinene oxide rearranges to pinocarvenol at 100–200°C in presence of 1 mol% of aluminium isopropoxide. However, at higher temperature in presence of 5 mol% of the reagent pinanone is obtained.

α-Pinene epoxide → (1 mol % $Al(OCHMe_2)_3$, 100°C) → 85% Pinocarvenol

α-Pinene epoxide → (5 mol % $Al(OCHMe_2)_3$, 150°C) → Pinanone

Some other applications[4] of aluminium isopropoxide are given below :

Oppenauer Oxidation

α-Ionol ($CH{=}CHCH(OH)CH_3$) → (Oppenauer Oxidation) → α-Ionone ($CH{=}CHCOCH_3$)

References

1. C. Djerassi, *Org. React.*, 1951, **6**, 207.
2. R.V. Openauer, *Rec. Trav. Chem.* 1937, **56**, 137.
3. J.F. Eastman and R. Teranishi, *Org. Synth.* Collective Volume, 1963, **4**, 192.
4. L. Lopa-Krzymien, L.C. Leitch, *Pol. J. Chem.* 1983, **57**, 629.
5. R.B. Wood Ward, *J. Am Chem. Soc.,* 1945, **67**, 1424.

1.3 AMMONIUM NITRATE, NH_4NO_3

Acoyloins can be converted into α-diketones with ammonium nitrate in the presence of catalytic amount of cupric acetate.

$$C_6H_5COCH(OH)C_6H_5 \xrightarrow[80\%\ AcOH,\ 1.5\ hr\ reflux]{NH_4NO_3/(AcO)_2Cu} C_6H_5COCOC_6H_5$$

Acoyloin (Benzoin) → α-Diketone 90%

Reference

1. M. Weiss and M. Appel, *J. Am. Chem. Soc.*, 1947, **70**, 3666.

1.4 ARSENIC PENTOXIDE, As_2O_5

It functions as an oxidant in the Skraup synthesis of quinolines[1].

$$\text{Aniline} + \underset{\text{Glycerol}}{CH_2OH{-}CHOH{-}CH_2OH} \xrightarrow[H_2SO_4]{As_2O_5} \text{Quinoline}$$

Reference

1. M.S. Morgan and R.S. Tipsan, *J. Am. Chem. Soc.*, 1946, **68**, 1969.

1.5 ARGENTIC OXIDE, AgO

It is prepared by the electrolysis of silver nitrate in nitric acid[1]. Following are given some of the applications of argentic oxide in oxidation.

(*i*) Degredative oxidation of α-amino acids to carboxylic acids.

The oxidiation of α-amino acids to carboxylic acids with one less carbon is achieved by heating with argentic oxide at 25–70°C to give 49–100% yield of carboxylic acid[2].

$$R{-}CH(NH_2)CO_2H \xrightarrow[25\text{-}70°C]{AgO,\ H_2O} RCOOH$$

$R{=}(CH_3)_2CH$ 100% yield

$R{=}(CH_3)_2CHCH_2$ 61% yield

(*ii*) Dehydrogenation of primary amines to nitriles.

The dehydrogenation of primary amines to nitriles having one less carbon is achieved[2] by heating with argentic oxide in water at 25–70°C.

$$\underset{\text{Benzyl amine}}{C_6H_5CH_2NH_2} \xrightarrow[25\text{-}70°C]{AgO,\ H_2O} \underset{\text{Benzonitrile 77\%}}{C_6H_5CN}$$

(*iii*) Oxidation of primary alcohols to aldehydes

$$\underset{\text{Allylalcohol}}{CH_2{=}CHCH_2OH} \xrightarrow[20°C,\ 7\ min]{AgO,\ H_3PO_4} \underset{\text{Acrolein 70\%}}{CH_2{=}CHCHO}$$

$$\underset{\text{Benzylalcohol}}{C_6H_5CH_2OH} \xrightarrow[\text{3 hr.}]{\text{AgO, AcOH, 20°C}} \underset{\substack{\text{Benzaldehyde}\\ 75\%}}{C_6H_5CHO}$$

(*iv*) Oxidation of aldehydes to carboxylic acid[4]

$$\underset{\substack{\text{3-Cyclohexene}\\ \text{1-carboxaldehyde}}}{\text{3-Cyclohexene-CHO}} \xrightarrow[\text{25°C, 14 hr.}]{\text{4AgO, THF-}H_2O\text{ (9:1)}} \underset{\substack{\text{3-Cyclohexene}\\ \text{1-carboxylic acid}}}{\text{3-Cyclohexene-COOH}}$$

(*v*) Oxidation of toluene to benzaldehyde.

Methyl homologues of aromatic compounds are oxidised to aldehyde by argentic oxide[3]

$$\underset{\text{Toluene}}{C_6H_5CH_3} \xrightarrow{\text{4AgO, }H_3PO_4\text{, –10°C, 0.5 hr}} \underset{\substack{54\%\\ \text{Benzaldehyde}}}{C_6H_5CHO}$$

(*vi*) Oxidation of ethylbenzene to acetophenone.

Methylene group adjacent to aromatic rings are oxidised to keto group by argentic oxide[3]

$$\underset{\text{Ethylbenzene}}{C_6H_5\text{–}CH_2CH_3} \xrightarrow{\text{4AgO, }H_3PO_4\text{, –10°C, 1 hr.}} \underset{\substack{77\%\\ \text{Acetophenone}}}{C_6H_5\text{–}COCH_3}$$

(*vii*) Oxidation of tertiary phosphine to phosphine oxides[2].

$$\underset{\text{Triphenyl phosphine}}{(C_6H_5)_3P} \xrightarrow[\text{exothermic reaction}]{\text{AgO, }H_2O\text{, RT}} \underset{\substack{\text{Triphenyl}\\ \text{phosphine oxide}\\ (40\%)}}{(C_6H_5)_3P{=}O}$$

(*viii*) Oxidation of phosphites to phosphates[2]

$$(C_6H_5O)_3P \xrightarrow[\text{exothermic reaction}]{\text{AgO, }H_2O\text{, RT}} (C_6H_5O)_3P{=}O$$

References

1. A.A. Noyes, D. Devault, C.D. Covyell and T.J. Deahl, *J. Am. Chem. Soc.*, 1937, **59**, 1326.
2. T.G. Clarke, N.A. Hampson, J.B. Lee, J.R. Morley and B. Scanlon, *Tetrahedron Lett.*, 1968, 5685.
3. L. Syper, *Tetrahedron Lett.*, 1967, **4**, 193.
4. H. Fiesselmann, *Chem. Ber.*, 1942, **75**, 881.

1.6 ARGENTOUS OXIDE (SILVER OXIDE), Ag_2O

It is a brown-black powder and is commercially available. It can however, be prepared by the treatment of silver nitrate solution with potassium hydroxide solution. The resulting, precipitate is decanted, washed with distilled water (until free of nitrate ions). Finally the precipitate, hydrated silver oxide is filtered (suction), washed with water, alcohol, ether and dried in air.

Silver oxide is used as an oxidant in a number of reactions.

(*i*) Oxidation of aldehydes to carboxylic acids[1]

Furfural $\xrightarrow[\text{50-55°C, 20-25 min; 40°C, 15-30 min}]{Ag_2O,\ NaOH,\ H_2O,\ O_2}$ Furoic acid 73–76%

(*ii*) Cannizzaro reaction.

Aromatic aldehydes on treatment with alkali give aromatic carboxylic acids and aromatic alcohols. The reaction is catalysed by silver oxide[2-4] and also by silver (*see* silver).

3-Cyclohexene-1-carboxaldehyde $\xrightarrow[\text{ice cooling, 1 hr. RT. 2-3 hr.}]{AgNO_2,\ KOH,\ EtOH}$ 3-Cyclohexene-1-carboxylic acid 60-97%

3-Cyclohexene-1-carboxaldehyde $\xrightarrow[\text{25°, 14 hr.}]{AgO,\ THF\text{-}H_2O(9:1)}$ 3-Cyclohexene-1-carboxylic acid

(*iii*) Hydroxy aromatic compounds are oxidised to quinones[5]

OH–C$_6$H$_4$–X = X–C$_6$H$_4$–OH $\xrightarrow[\text{RT, 1hr.}]{Ag_2O,\ Na_2SO_4,\ Et_2O}$ O=C$_6$H$_4$=X–X=C$_6$H$_4$=O

X = CH or N

X = N (90% yield)
X = CH (40-45% yield)

(*iv*) Conversion of aziridine into diazirine[6]

aziridine $\xrightarrow[\text{RT, 35-65 min}]{Ag_2O,\ Et_2O}$ diazirine

(*v*) Conversion of hydroxylamines into nitroso compounds[7]

$$C_6H_5NHOH \xrightarrow[\text{2 min}]{Ag_2O,\ CH_2Cl_2} C_6H_5NO$$

N, N-Disubstituted hydroxylamines are converted[8] into free radical nitroxyls by silver oxide in the presence of anhydrous Na_2SO_4.

$$(H_3C-C_6H_4)_2NOH \xrightarrow[\text{Ice-salt cooling, 7 min}]{Ag_2O,\ Na_2SO_4,\ Et_2O} (H_3C-C_6H_4)_2N\dot{O}$$

> 60%

(*vi*) Oxidation of alcohols.

Primary alcohols are oxidised into aldehydes[9] and secondary alcohols into ketones[10]

(*vii*) Conversion of 1, 2-cyclohexanediol into adipic acid[11]

$$\text{1,2-Cyclohexane diol} \xrightarrow[\text{EtOH-KOH, 25°C}]{Ag_2O} \text{Adipic acid (90\%)}$$

Some other applications of silver oxide include conversion of hydrazo compounds to azomcompounds[6] and dehydrogenation[12] of CH – NH bonds to – C = N.

References

1. R.J. Harrisson and M. Moyle, *Org. Synth.* Collective Volume, 1963, **4**, 493.
2. I.A. Pearl, *J. Org. Chem.*, 1947, **12**, 85.
3. H. Fiesselmann, *Chem. Ber.*, 1942, **75**, 881.
4. E.J. Corey, W.N. Gilman and B.E. Ganem, *J. Am. Chem. Soc*, 1968, **90**, 5616.
5. R. Willstätter and M. Benz, *Chem. Ber.*, 1906, **39**, 3482.
6. E. Schmitz and R. Ohme, *Org. Synth.* Collective Volume, 1973, **5**, 897.
7. J.A. Maassen and T.J. DeBoer, *Rec. Trav. Chem.* Pays–Bas, 1971, **90**, 373.
8. H. Wieland and K. Roth, *Chem. Ber.*, 1920, **53**, 210.
9. L. Syper, *Tetrahedron Lett*, 1967, 4193.
10. M.Y. Sheikh and Y. Eadon, Tetrahedron Lett., 1972, 257.
11. V.K. Ahluwalia and R.K. Parashar, *Organic Reaction Mechanisms*, Narosa Publishing House, 2007, page 187.
12. R.I. Fryer, G.A. Archer, B. Brust, W. Zally and L.H. Stenbach, *J. Org. Chem.*, 1965, **30**, 1308.

1.7 BARIUM MANGANATE, $BaMnO_4$

It is obtained[1,2] from aqueous solution of barium chloride and potassium permanagate. Some of its applications are given below :

1. It oxidises alcohols, especially benzylic alcohols to carbonyl compounds[1, 2].

$$\underset{\text{Benzyl alcohol}}{C_6H_5CH_2OH} \xrightarrow[\text{reflux, 9 min}]{BaMnO_4,\ C_6H_6} \underset{\text{90\% Benzaldehyde}}{C_6H_5CHO}$$

In a similar way, benzhydrol is oxidised[2] to benzophenone

$$\underset{\text{Benzhydrol}}{C_6H_5CH(OH)\text{—}C_6H_5} \xrightarrow[\text{reflux, 0.5-2 min}]{BaMnO_4,\ C_6H_6} \underset{\text{Benzophenone}}{C_6H_5C(=O)\text{—}C_6H_5}$$

2. It oxidises hydroquinone to quinone

$$\underset{\text{Hydroquinone}}{HO\text{—}C_6H_4\text{—}OH} \xrightarrow[\text{reflux, 20 min}]{BaMnO_4,\ C_6H_6} \underset{\substack{75\% \\ \text{p-Benzoquinone}}}{O{=}C_6H_4{=}O}$$

3. It oxidises aromatic primary amines to azo compounds[2]

$$C_6H_5NH_2 \xrightarrow[\text{reflux, 30 min}]{BaMnO_4,\ C_6H_6} \underset{90\%}{C_6H_5N{=}NC_6H_5}$$

4. It oxidises phosphines to phosphine oxides[1]

$$(C_6H_5)_3P \xrightarrow[\text{reflux, 2 hr.}]{BaMnO_4,\ C_6H_6} \underset{100\%}{(C_6H_5)_3P{=}O}$$

References

1. H. Firouzabadi and E. Ghaderi, *Tetrahedron Lett.*, 1978, 839.
2. H. Firouzabadi and Z Mostafavipoor, *Bull Chem. Soc. Jpn.*, 1983, **56**, 914.

1.8 BENZENE SELENIC ANHYDRIDE, C_6H_5 Se(O) O(O) SeC_6H_5

It is prepared *in-situ* from diphenyl diselonide and tert butyl hydroperoxide. It is used for the oxidation of primary alcohols to aldehydes[1]

$$CH_3(CH_2)_8CH_2OH \xrightarrow[\text{reflux, 5 hr.}]{(PhSe)_2,\ \text{t-BuOOH},\ C_6H_6} \underset{92\%}{CH_3(CH_2)_8CHO}$$

$$C_7H_{15}CH_2OH \xrightarrow[\text{reflux, 5 hr.}]{(PhSe)_2,\ \text{t-BuOOH},\ C_6H_6} \underset{100\%}{C_7H_{15}CHO}$$

$$PhCH{=}CHCH_2OH \xrightarrow[\text{reflux, 1 hr.}]{(PhSe)_2,\ \text{t-BuOOH},\ CCl_4} \underset{87\%}{PhCH{=}CHCHO}$$

Using Benzeneselenic anhydride, secondary alcohols are oxidised to ketones[2]

$$C_6H_5CH(OH)\text{—}C_6H_5 \xrightarrow[\text{THF, RT, 3 hr.}]{(PhSeO)_2} C_6H_5COC_6H_5$$

This reagent is useful for the regeneration of ketones from their oximes, semicarbazones and phenylhydrazones.

(PhSeO)$_2$O, THF
50-60°C, 3 hr.

Oxime R = NOH
Semicarbazone R = $NNHCONH_2$
Phenylhydrazone R = $NNHC_6H_5$

89%

Benzeneselenic anhydride is a suitable dehydrogenating agent for the introduction of double bonds α to the carbonyl group[4]. One such example is conversion of cholestan-3-ol to 1, 4-cholestadiene-3 one.

(PHSeO)$_2$O
C_6H_6 reflux
24 min

Cholestan-3-ol

81%
1,4-Cholestadien-3-one

References

1. I. Kuwajima, M. Shimizu and H. Urabe, *J. Org. Chem.*, 1982, **47**, 837.
2. D.H.R. Barton, A.G. Brewster, R.A.H.F. Hui, D.J. Lester, S.V. Ley and T.G. Back, *Chem. Commun*, 1978, 952.
3. D.H.R. Barton, D.J. Lester and S.V. Ley, *Chem. Commun*, 1977, 445.
4. D.H.R. Barton, J.W. Morzycki and W.B. Motherwell, *Chem. Commun*, 1981, 1044.

1.9 BENZYLTRIETHYLAMMONIUM PERMANGANATE $C_6H_5CH_2N(C_2H_5)_3$ MnO_4

It is obtained[1] by dropwise addition of aqueous solution of benzyltriethyl ammonium chloride to a solution of potassium permanganate in water. The formed reagent is filtered and dried at 40°/02 mm pr. Like tetrabutyl ammonium permanganate, benzyltriethyl ammonium permanganate should be handled with care. Some of the important oxidation reactions are given below.

1. *Hydroxylation of Methine groups:* A frequent reaction is the oxidation at the methine group giving tertiary alcohols as the main product and secondary alcohols and ketones as minor products.[1] The hydroxylation occurs with retention of configration. Thus, the oxidation of 3-ethylpentane with benzyltriethylammonium permanganale yields[1] 25 per cent of 3-ethyl-3-pentanol, 25 per cent of 3-ethyl-2-pentanone and 25 per cent of 3-pentanone.

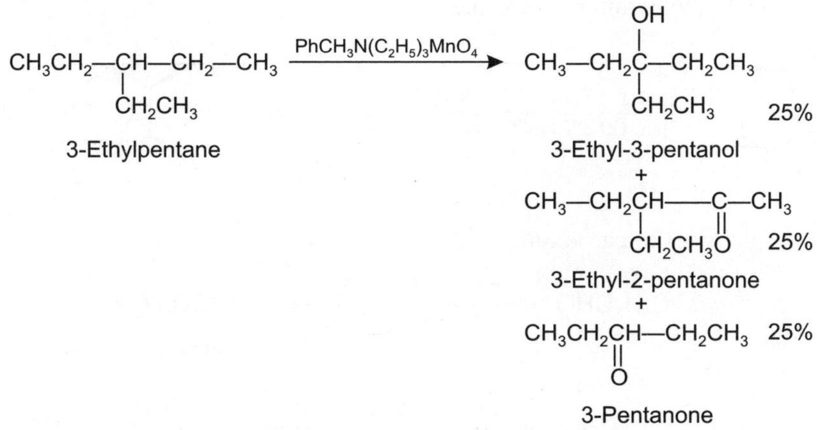

$$CH_3CH_2\text{—}\underset{\underset{CH_2CH_3}{|}}{CH}\text{—}CH_2\text{—}CH_3 \xrightarrow{PhCH_3N(C_2H_5)_3MnO_4} CH_3\text{—}CH_2\underset{\underset{CH_2CH_3}{|}}{\overset{\overset{OH}{|}}{C}}\text{—}CH_2CH_3 \quad 25\%$$

3-Ethylpentane

3-Ethyl-3-pentanol

+

$$CH_3\text{—}CH_2\underset{\underset{CH_2CH_3}{|}}{CH}\text{—}\underset{\underset{O}{||}}{C}\text{—}CH_3 \quad 25\%$$

3-Ethyl-2-pentanone

+

$$CH_3CH_2\underset{\underset{O}{||}}{CH}\text{—}CH_2CH_3 \quad 25\%$$

3-Pentanone

Similarly, methylcyclohexane yields[1] 72% of methylycohexanol and 3% of 2 methylcyclohexanone.

$$\xrightarrow{PhCH_2N(C_2H_5)_3MnO_4}$$

Methyl cyclohexene

1-Methyl cyclohexanol 72%

+

2-Methyl cyclohexanone 3%

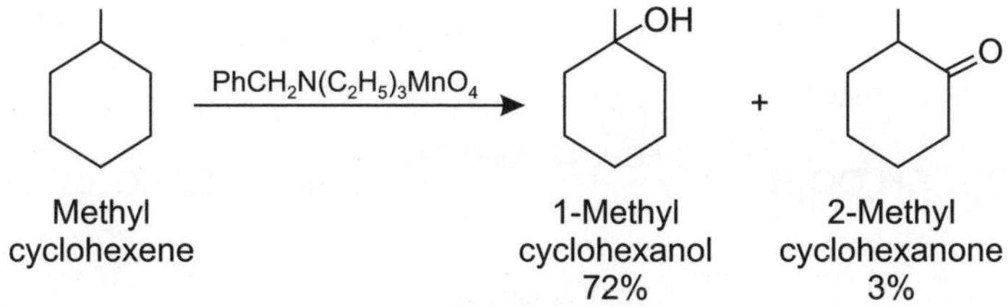

In the case of decalin, the oxidation with benzyl triethylammanium permanganate gives[1, 2] the following compounds.

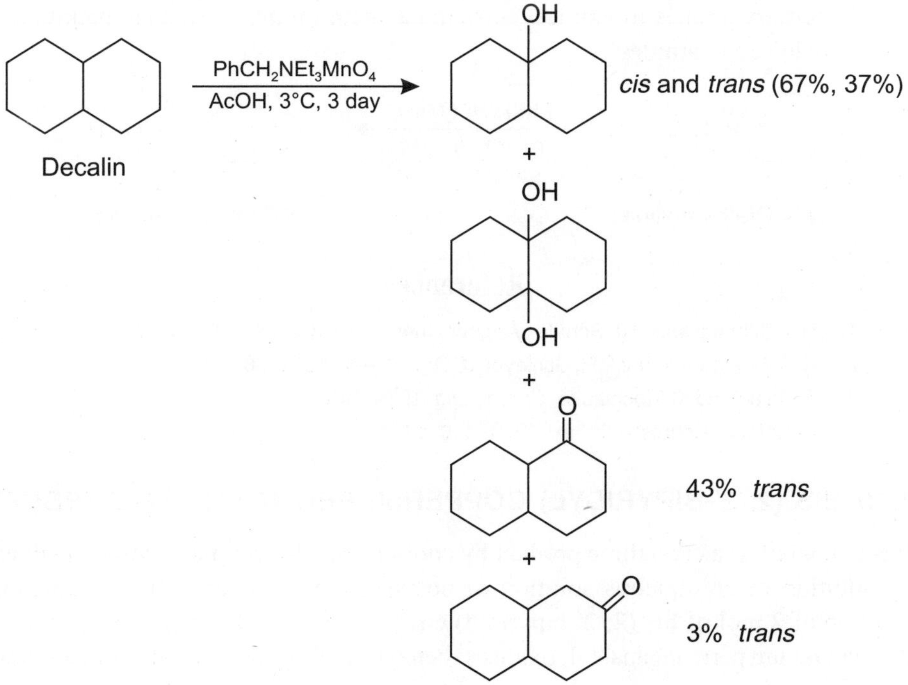

$$\xrightarrow[AcOH,\ 3°C,\ 3\ day]{PhCH_2NEt_3MnO_4}$$

Decalin

OH

cis and *trans* (67%, 37%)

+

OH

OH

+

O

43% *trans*

+

O

3% *trans*

2. Syn hydoxhylation of alkenes[3]

(1) Ph.$CH_2NEt_3MnO_4$, O-3°C, 1.5 hr
(2) 3% NaOH RT, 18 hr

HO HO 83%

3. Aldehydes to carboxylic acids[4]

$$C_6H_5CHO \xrightarrow[\text{AcOH, RT, 1.25 hr}]{PhCH_2NEt_3MnO_4} C_6H_5CO_2H$$

84%

CHO $\xrightarrow[CH_2Cl_2\text{, 1.25 hr.}]{PhCH_2NEt_3MnO_4\text{, AcOH}}$ CO_2H

O O H_2C 80%

4. Ethers to esters[1]

$$C_3H_7CH_2OC_4H_9 \xrightarrow{PhCH_2NEt_3MnO_4CH_2Cl_2} C_3H_7CO_2C_4H_9$$

80%

$$C_6H_5CH_2OC_4H_9 \xrightarrow[CH_2Cl_2]{PhCH_2NEt_3MnO_4} C_6H_5CO_2C_4H_9$$

99%

5. Tertiary amines to amides. Tert-amines with groups other than methyl are oxidised to amides[5]

$$-N(C_2H_5)_2 \xrightarrow[CH_2Cl_2\text{, 0-40°C}]{PhCH_2NEt_3MnO_4} -NCOCH_3\ (C_2H_5)$$

N,N-Diethylaniline N-Ethylacetanilide

References

1. H.J. Schmidt and H.J. Schäfer, Angew. Chem. Int. Ed. Engl., 1979, **18**, 68.
2. R.C. Bingham and V.P.R. Schleyer, *J. Org. Chem.*, 1971, **36**, 1198.
3. T. Ogino and K. Mochizuke, *Chem., Lett*, 1979, 443.
4. D. Scholz, Monatsh. *Chem.*, 1979, **110**, 1471.

1.10 BIS (2, 2′-BIPYRIDYL) COPPER(II) PERMANGANATE (BBCP)

It is obtained[1] as a crystalline product by cooling the dark-purple solution obtained by addition of an aqueous solution of potassium permanganate to an aqueous solution of 2 mol of bis (2, 2′-bipyridyl) copper (II) chloride. It is more selective than potassium permanganate. It oxidises[1] benzylic and allylic alcohols to aldehydes.

$$C_6H_5CH_2OH \xrightarrow[CH_2Cl_2,\ RT,\ 15\ min]{bis\ (2,2'\text{-bipyridyl})\ Cu(MnO_4)_2} C_6H_5CHO\ \ 95\%$$

$$PhCH = CHCH_2OH \xrightarrow[CH_2Cl_2,\ RT,\ 1\ min]{bis\ (2,2'\text{-bipyridyl})\ Cu(MnO_4)_2} C_6H_5CH = CHCHO\ \ 91\%$$

BBCP oxidises thiols to disulphides[2]

$$C_6H_5SH \xrightarrow[RT,\ 10\ min.]{BBCP,\ CH_2Cl_2} C_6H_5SSC_6H_5\ \ 100\%$$

BBCP = (2,2'-bipyridyl)$_2$ Cu(MnO$_4$)$_2$

BBCP also oxidises acyloins to α-diketones[1], and primary aromatic amines to azo cmpounds[1]. Similar results are obtained by using bis(2, 2′-bipyridyl) silver permanganate[3]

References

1. H. Fivouzabadi, A.R. Sardarian, M. Naderi and B. Vessal, *Tetrahedron*, 1984, **40**, 5001.
2. H. Fivouzabadi, M. Naderi, A. Sarddrian and B. Vessal, *Synth. Commun*, 1983, **13**, 611.
3. H. Fivouzabadi, B. Vessal and M. Naderi, *Tetrahedron Lett.*, 1982, **23**, 1847.

1.11 BISMUTH SESQUIOXIDE, Bi_2O_3

It oxidizes[1] acyloins to α-diketones in presence of acetic acid.

$$\underset{\text{Acyloin}}{C_6H_5CO\underset{OH}{CH}C_6H_5} \xrightarrow[EtOCH_2CH_2OH,\ 104°C,\ 1\ hr.]{Bi_2O_3,\ AcOH} \underset{\alpha\text{-diketone, } 40\%}{C_6H_5COCOC_6H_5}$$

Reference

1. W. Rigby, J. Chem., Soc., 1951, 793.

1.12 BENZOYL HYPOIODITE, C_6H_5COOI

It is prepared[1, 2] from iodine and 2 mol of silver benzoate. It is used in the hydroxylation of double bonds (as in the case of iodine and silver carboxylate) of vicinal diols. The reaction is stereospecific. In anhydrous medium (the Prevost reaction[1]), the reaction takes place in anti mode. However, in the presence of water[1] (the Woodwords modification), the reaction results in a syn addition. The mechanism of both these reactions has been described (*see* iodine and silver carboxylate).

References

1. P.S. Ellington, D.G. Hey and G.D. Meaken's, *J. Chem. Soc.,* 1966, 1327.
2. E. Späth, F. Kuffner and F. Kesztler. *Chem. Ber.*, 1936, **69**, 378.

1.13 BENZOYL PEROXIDE (DIBENZOYL PEROXIDE), $(C_6H_5COO)_2$

Benzoyl peroxide and iodine in carbon tetrachloride react with stilbene at 80°C to give 83 per cent yield to dibenzoyl ester of hydrobenzoin (1, 2-dibenzyloxy-1, 2-diphenyl ethane)[1].

$$\underset{\text{Stilbene}}{C_6H_5CH = CHC_6H_5} \xrightarrow[CCl_4,\ 80°C,\ 48\ hr.]{(C_5H_5COO)_2,\ I_2} \underset{\substack{\text{1,2-Dibenzoxy-1,2-diphenyl} \\ \text{ethane (83\%)}}}{C_6H_5—\underset{\substack{| \\ C_6H_5COO}}{CH}—\overset{\substack{OCOC_6H_5 \\ |}}{CH}—C_6H_5}$$

It oxideses alcohols to carbonyl compounds in presence of nickel dibromide[2].

It is commonly used as a catalyst in the anti Mavkovnikov addition of hydrogen bromide to unsaturated compounds

$$CH_3CH = CH_2 + HBr \xrightarrow{(C_6H_5COO)_2} \underset{\text{n-propyl bromide}}{CH_2—CH_2—CH_2Br}$$

It is also used as a catalyst in radical polymerisations.

References

1. A. Perret and R. Perrot, *Helv. Chim. Acta*, 1945, **28**, 588.
2. M.P. Doyle, W.J. Patrie and S.B. Williams, *J. Org. Chem.*, 1979, **44** , 2955.

1.14 BROMINE

Bromine forms crystalline molecular complexes with dioxane[1] $C_4H_8O_2{\cdot}Br_2$, pyridine, $C_5H_5N{\cdot}Br_2$ and 1,4-diazabicyclo [2, 2, 2] octane (Dabco)[3], $C_6H_{12}N_2{\cdot}2Br_2$ (Dabco·$2Br_2$).

Bromine oxidises ethers to carbonyl compounds[4]

$$\underset{\text{Benzyl methyl ether}}{C_6H_5CH_2OCH_3} \xrightarrow[h\nu,\ \Delta]{Br/CCl_4\ (1:1)} \underset{77\%}{C_6H_5CHO}$$

$$\underset{\text{Benzhydryl methyl ether}}{C_6H_5—\overset{\substack{OCH_3 \\ |}}{CH}—C_6H_5} \xrightarrow[h\nu,\ \Delta]{Br/CCl_4\ (1:1)} \underset{\substack{36\% \\ \text{Benzophenone}}}{C_6H_5COC_6H_5}$$

Bromine oxidises phenols to quinones. Thus, 4-benzyl-2, 6-dibromophenol on oxidation with bromine in a sealed tube yields 2, 6-dibromophenylchino methide[5]

$C_6H_5CH_2$–(2,6-dibromo-4-hydroxyphenyl) $\xrightarrow[\text{100°C, 4 hr sealed tube}]{Br_2/CCl_4}$ C_6H_5CH=(2,6-dibromo-cyclohexadienone)

80-85%
2,6-Dibromophenyl-chinomethide

Primary alcohols are oxidised to esters. Thus, butanol on treatment with bromine and potassium bromate in 40% HBr yields butylbutyrate.[6]

$$C_4H_9OH \xrightarrow[\text{40–100°C}]{Br_2,\ KBrO_3,\ 40\%\ HBr} C_3H_7COOC_4H_9$$

82%
Butylbutyrate

Secondary alcohols are oxidised to ketones.[7]

$$CH_3CH(OH)C_6H_{13} \xrightarrow[\text{40–100°C}]{Br_2,\ CHCl_3,\ HMPA,\ CH_2Cl_2,\ H_2O} C_6H_5COC_6H_{13}$$

95%

$$\text{Cyclohexanol} \xrightarrow[\text{0-5°C, 5 min}]{Br_2,\ CH_2Cl_2,\ HMPA,\ H_2O,\ NaHCO_3} \text{Cyclohexanone}$$

87%

$$C_6H_5CH(OH)CH_3 \xrightarrow[\text{0-5°C, 15 min}]{Br_2,\ CH_2Cl_2,\ HMPA,\ H_2O,\ NaHCO_3} C_6H_5COCH_3$$

98%

HMPA ≡ hexamethylphosphoric triamide. Oxidation of secondary alcohols to ketones[3] can also be affected with Dabco·$2Br_2$.

O-Fluorotoluene is brominated to o-fluorobenzyl bromide.[8]

$$\text{o-Fluorotoluene} \xrightarrow[\text{105-120°C, 1 hr}]{Br_2,\ h\nu} \text{o-Fluorobenzyl bromide}$$

o-Fluorotoluene

78.5%
o-Fluorobenzyl bromide

Bromine converts hydrazo compounds to azo compounds[9], oxidises sulphides to sulphoxides[10] and disulphides to sulphonic acids[11].

References

1. J.D. Billimoria and N.F. Maclagan, *J. Chem. Soc.,* 1954, 3257.
2. D.M. Williams, *J. Chem. Soc.,* 1931, 2783.
3. L.K. Blair, J. Baldwin and W.C. Smith, Jr., *J. Org. Chem.,* 1977, **42**, 1816.
4. D.G. Markees, *J. Org. Chem.,* 1958, **23**, 1490.
5. T. Zincle and W. Walter, *Justus Liebigs Ann. Chem.,* 1904, **334**, 367.
6. L. Farkas and O. Schächter, *J. Am. Chem. Soc.,* 1949, **71**, 2827.
7. M.A. Neirabeyeh, J.–C. Ziegler and B. Gross, *Synthesis,* 1976, 811.
8. M. Hudlicky, Collect. Czech. *Chem. Commun,* 1960, **25**, 1199.
9. C.G. Overberger, P.–t. Huang and M.B. Berenbaum, *Org. Synth.* Collective Volume, 1963, 4, 66.
10. S. Oal, Y. Ohnishi, S. Kozuka and W. Tagaki, *Bull. Chem. Soc. Jpn.,* 1966, **39**, 364.
11. H.T. Clarke, *Org. Synth.* Collective Volume, 1955, **3**, 226.

1.15 N-BROMOACETAMIDE

The addition of N-bromoacetamide to alkenes gives halohydrins[1], which are useful intermediates for the synthesis of epoxides.

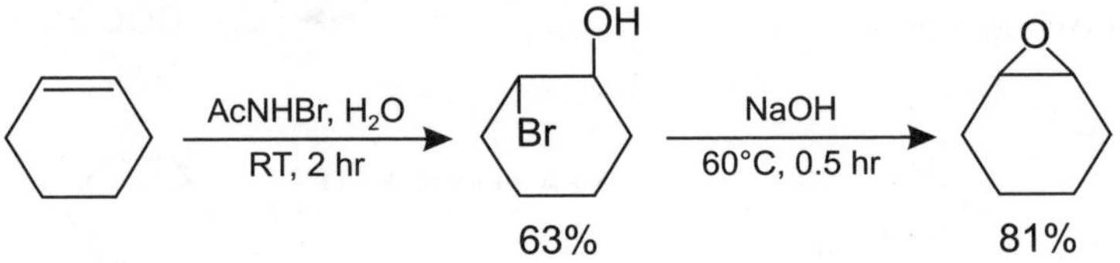

It is useful for selective oxidation[2] of secondary alcohols in preference to primary alcohols. Thus, oxidation of 20-cyano-3α, 21-dihydroxy-17-pregnen-11 one with N-bromoacetamide in pyridine-methanol gives 20-cyano-21-hydroxy-17-pregnen-3, 11-dione.

N-Bromoacetamide is also used in the selective oxidation of steroidal diols.

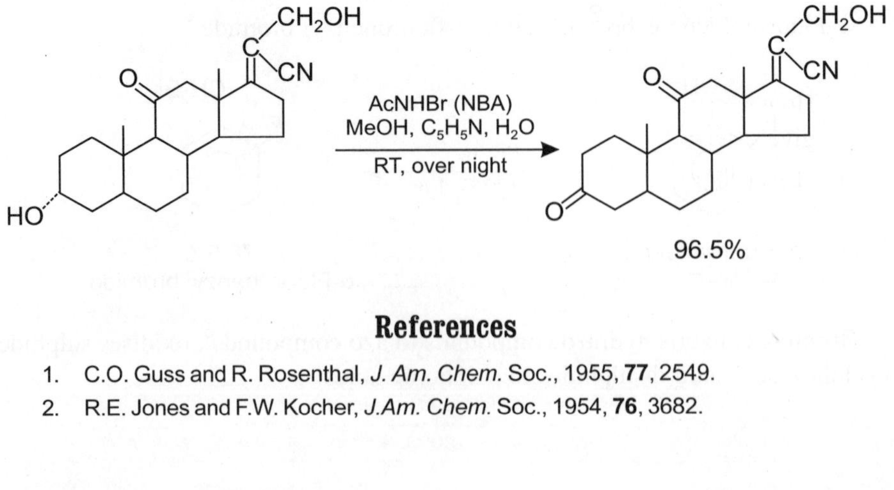

References

1. C.O. Guss and R. Rosenthal, *J. Am. Chem.* Soc., 1955, **77**, 2549.
2. R.E. Jones and F.W. Kocher, *J.Am. Chem.* Soc., 1954, **76**, 3682.

1.16 N-BROMOBENZOPHENONE IMINE, $(C_6H_5)_2$ C = NBr

It is prepared[1] from benzophenone imine, bromine and alkalies. It is used for the conversion of alcohols into carbonyl compounds.

References

1. W. Theilacker and K. Fauser, *Justus Liebigs Ann. Chem.*, 1939, **539**, 103.
2. C.G. MaCarty and C.F. Leeper, *J. Org. Chem.*, 1970, **35**, 4245.

1.17 N-BROMOSUCCINIMIDE

N.Br, NBS

It can be prepared from succinimide and bromine in presence of 3M sodium hydroxide solution. The formed NBS is washed with water, and recrystallised from hot water or acetic acid. It is stored in a refrigerator and protected from moisture to avoid decomposition. It is easier and safer to handle than bromine. Following are given some of the important reactions of NBS:

1. It oxidises alkynes to α-diketones[1].

$$C_6H_5C \equiv CC_6H_5 \xrightarrow[\text{24 hr}]{\text{DMSO, NBS (2 mol), 22°C}} C_6H_5COCOC_6H_5 \quad 98\%$$

2. Oxidation of hydrazo compounds to azo compounds.[2] Thus, di-tert-butyl hydrazodicarboxylate is dehydrogenated to di-tert-butylazodicarboxylate.

$$(CH_3)_3CO_2CNHNHCO_2C(CH_3)_2 \xrightarrow[\text{RT, 11-12 min}]{\text{NBS-}C_5H_5N,\ CH_2Cl_2} (CH_3)_3CO_2CN = NCO_2C(CH_3)_3 \quad 90\text{–}95\%$$

3. Oxidation of trimethyl silyl ethers of secondary alcohols with NBS under irradiation give ketones[3].

$$C_6H_5CH(CH_3)\text{—}OSi(CH_3)_3 \xrightarrow[\text{RT, 3.5 hr.}]{\text{NBS-}C_5H_5N,\ h\nu} C_6H_5COCH_3 \quad 76\%$$

However, trimethylsilyl ethers of primary alcohols under the above conditions give esters.[3]

4. Epoxidation of alkenes and cycloalkenes[4]

$$C_6H_5CH = CH_2 \xrightarrow[\text{60°C, 30 min}]{\text{NBS, }H_2O\text{, RT, 20\% NaOH}} C_6H_5\overset{O}{CH\text{—}CH_2} \quad 85\%$$

$$\text{cyclohexene} \xrightarrow[\text{aq. NaOH, 60°C, 30 min}]{\text{NBS, }H_2O\text{, RT, 10 min}} \text{cyclohexene oxide} \quad 81\%$$

5. Epoxidation of unsaturated carboxylic acids[4]

$$HO_2C(H)C=C(H)CO_2H \xrightarrow[\text{RT, 48 hr.}]{\text{NBS, } H_2O} HO_2C(H)C\overset{O}{-}C(H)CO_2H$$

61%

6. Epoxidation of unsaturated ketones[4]

$$(CH_3)_2C = CHCOCH_3 \xrightarrow[\text{RT, 15 min}]{\text{NBS, } H_2O} (CH_3)_2C\underset{O}{-}CHCOCH_3$$

Mesityl oxide

56%

7. Formation of halohydrins. Treatment of alkene with NBS in aqueous media give halohydrins; in this case anti addition takes place. The bromohydrins[4] on heating with alkali give the epoxides.

Cyclohexene $\xrightarrow[\text{RT, 10 min}]{\text{NBS, } H_2O}$ 2-bromocyclohexanol (OH, Br) $\xrightarrow[\text{60°C, 0.5 hr}]{\text{NaOH}}$ cyclohexene oxide

79% 81%

In fact in all reactions mentioned above (*see* 4, 5, 6) the reaction takes place via the formation of halohydrins.

8. NBS is used as a catalyst[1] in oxidation with dimethylsulfoxide (*see* dimethylsultoxide).

9. Degradative oxidation[5] of α-hydoxy acids to aldehydes and ketones.

The reaction of NBS (refluxing in water) with glycolic acid, lactic acid and mandelic acid gives formaldehyde, acetaldehyde and benzaldehyde respectively.

1. Allylic and Benzylic Bromination

Alkenes on refluxing with NBS in carbon tetrachloride solution are brominated at allylic position, the reaction is initiated by peroxide or light. Thus, cyclohexene on refluxing in ether with NBS gives 3-bromocyclohexene[6] in presence of catalytic amount of radical initiator, such as benzoyl peroxide. This type of bromination is known as **Wohl-Zigler bromination**[7]. This is the most common reaction for allylic bromination

Cyclohexene $\xrightarrow[\text{ether, 25°C benzoyl peroxide}]{\text{NBS}}$ 3-Bromo cyclohexene (50-80%) + Succinimide

Allylic methylene groups are attacked more rapidly than allylic methyl groups[8]. Thus, in 2-heptene, a secondary position is substituted more readily than a primary.

$$CH_3CH_2CH_2CH_2CH=CHCH_3 \xrightarrow[CCl_4]{NBS} CH_3CH_2CH_2CH(Br)-CH=CH-CH_3$$

2-Heptene → 3-Bromo-2-heptene (64%)

The mechanism of allylic bromination[9] is given below:

NBS → succinimidyl radical (N•) + $\dot{Br}$

Cyclohexene + $\dot{Br}$ —abstraction→ cyclohexenyl radical —Br_2→ 3-bromocyclohexene + $\dot{Br}$

The bromination of methyl or methylene group α-to an aromatic ring with NBS has also been investigated. Thus, the reaction of *p*-xylene with NBS yields α, α-dibromo-*p*-xylene. Similarly, a benzylic hydrogen atom of methylene group in diphenyl methane is substituted by NBS to give bromodiphenylmethane in 81% yield.

$$p\text{-}CH_3C_6H_4CH_3 \xrightarrow{NBS} p\text{-}BrCH_2C_6H_4CH_2Br$$

p-Xylene → α,α-Dibromo-*p*-xylene

$$C_6H_5-CH_2-C_6H_5 \xrightarrow[CCl_4/\Delta]{NBS} C_6H_5-CH(Br)-C_6H_5$$

Diphenylmethane → Bromodiphenylmethane (81%)

Some other examples are given below:

$$C_6H_5CH_2(CH_2)_2CH_3 \xrightarrow{NBS} C_6H_5CH(Br)-(CH_2)_2CH_3$$

1-Phenylbutane → 1-Bromo-1-phenylbutane

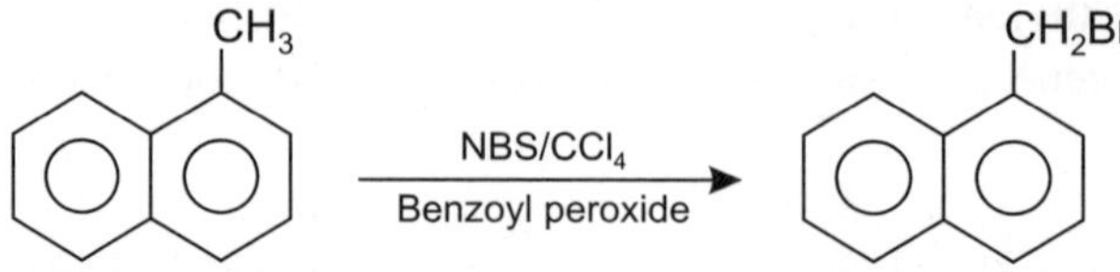

CH_3 → CH_2Br

NBS/CCl_4, Benzoyl peroxide

1-Methylnaphthalene → 1-Bromomethyl naphthalene

The reaction of 3-methylthiophene with NBS in benzene in presence of benzoyl peroxide gives 3-bromomethylthiophene in 75% yield. However, in the absence of initiator nuclear bromination takes place.[10]

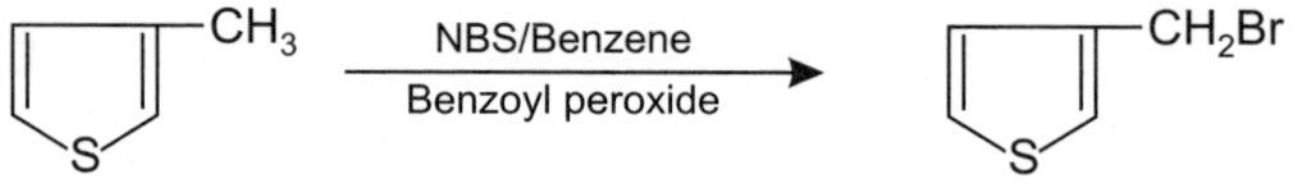

NBS/Benzene, Benzoyl peroxide

3-Methylthiophene → 3-Bromomethylthiophene (75%)

2. α-Bromination of carbonyl derivatives: Ketones having an enolisable hydrogen can be brominated in the α-position with NBS[11].

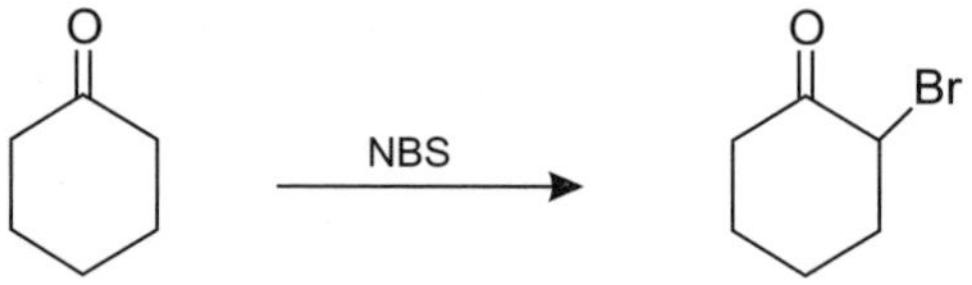

NBS

Cyclohexanone → 2-Bromocyclohexanone

3. Allylic bromination of α, β-unsaturated acids, esters, aldehydes and ketones: Unsaturated fatty acids and their deviatives, such as oleic acid[12] and methyl oleate[13] give mixture of monobromo (generally as a major product) and dibromo compounds. Ethyl crotonate reacts with NBS/CCl_4 in presence of benzoyl peroxide to give[13] 4-bromocrotonate.

$$CH_3{-}CH{=}CH{-}COOEt \xrightarrow[\text{Benzoyl peroxide}]{NBS/CCl_4} BrCH_2{-}CH{=}CH{-}COOEt$$

Ethyl crotonate → Ethyl 4-bromo crotonate

Under similar conditions, testosterone acetate is converted into 6-bromotestosterone acetate.

NBS/CCl_4, Δ

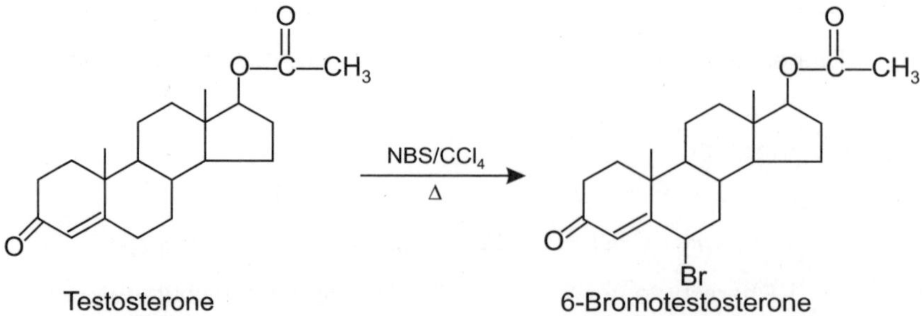

Testosterone acetate → 6-Bromotestosterone acetate

In fact, it has been found that unsaturated ketones, having methylene group in the α-position at both the carbonyl group and double bonds, are preferentially brominated in the methylene α-to the double bond.

4. Bromination of aromatic compounds[14] : The aromatic compounds react with NBS under ionic conditions to give nuclear substitution products. Thus, benzene on treatment with NBS and mixture of concentrated sulphuric acid and water (1:1, by volume) gives bromobenzene in 95% yield. Under normal Whol-zigler conditions there is virtually no reaction. It is believed that the actual brominating agent is not the free halogen, but possibly protonated N-bromo succinimide. The reaction of NBS with aromatic compounds containing highly branched chain also appear to give nuclear brominated products.

Phenols, anilines and other electron rich aromatic compounds can be monobrominated by using NBS in DMF with higher yields and higher levels of para selectivity than with bromine.

Heterocyclic compounds without alkyl substituents are generally brominated by NBS in the heterocyclic ring. Thus, thiophene is selectively brominated in position 2 by NBS in acetic acid-chloroform.

Thiophene $\xrightarrow[CH_3COOH/CHCl_3]{NBS}$ 2-Bromothiophene

N-substituted pyrroles are brominated with NBS in THF to afford 2-bromopyrroles (with one equivalent of NBS) or 2, 5-dibromopyroles (with two equivalents of NBS). However, bromination with bromine affords thermodynamically more stable 3-bromopyroles.

5. Formation of addition products: NBS is generally used as a source of electrophilic addition, thus addition predominates in polar solvents. It has been shown[11] that the addition reaction is catalysed by tetralkylammonium salts. Thus, cyclohexene reacts with NBS in presence of tetraethylammonium bromide to give 1, 2-dibromocyclohexane.

Cyclohexene $\xrightarrow[(C_2H_5)_4\overset{+}{N}\overset{-}{Br}]{NBS}$ 1,2-Dibromocyclohexane (75%)

The mechanism of bromination is represented[14] as follows:

Alkene + NBS → [Bromonium ion]$^+$ + $\bar{N}S$

$\xrightarrow{NBS}$ Br–C–C(Br)–NS + $\bar{N}S$ + NS^+ → $2\dot{N}S$ $\xrightarrow{\text{H Donor}}$ 2HNS

NS = N (succinimide ring)

References

1. S. Wolfe, W.R. Pilgrim, T.F. Garrard and P. Chamberlian, Can *J. Chem.,* 1971, **49**, 1099.
2. J.C.Kaurer, *Org. Synth,* Collective Volume, 1953, **4**, 411.
3. H.W. Pinnick and N.H. Lajis, *J. Org. Chem.,* 1978, **43**, 371.
4. C.O. Guss and R. Resenthal, *J. Am. Chem. Soc.* 1955, **77**, 2549.
5. M.Z. Barakat and M.F.A. El-Wahab, *J.Am. Chem. Soc.,* 1953 **75**, 5731.
6. H.J. Dauben, Jr. and L.L. McCoy, *J. Am. Chem. Soc.,* 1959, **81**, 4863.
7. C. Djerassi, *Chem. Rev.,* 1948, **43**, 271.
8. F.L. Grenwool, M.D. Kellert and J. Sedlak, 0S, 1958, **8**, 38.
9. Dauben and Mc Coy, *J. Am. Chem. Soc.,* 1959, **81**, 4863.
10. E. Compaigne and B.F. Tullar, 0S, 1953, **33**, 96.
11. J.R. Catch, D.H. Hey, E.R.H. Jones and W.J. Wilson, *J.Chem. Soc.,* 1948, 276.
12. L.K. Freidlin and N.I. Ziminova, Izvest. Akad. Nauk SSSR, Otdel, Khim. Nauk, 1951, 145.
13. Z. Csuros and J. Petro, Acta Chem. Acad. Sci. Hung., 1961, **29**, 321.
14. V.K. Ahluwalia and Rakesh Kumar Parashar, *Organic Reaction Mechanism*, Narosa Publishing House, 2007, pp. 521–522.

1.18 TERT-BUTYL CHROMATE, (TERT-$C_4H_9O)_2$ CrO_4

It is prepared[1] by the addition of chromium trioxide to ice-cold tert-butyl alcohol and is used in benzene or petroleum ether solutions.

This reagent oxidises primary alcohol to aldehydes.[2]

$$C_6H_5CH_2OH \xrightarrow[C_6H_6,\ 1\text{-}2°C,6\ hr]{CrO_3,\ t\text{-}BuOH,\ Cooling\ with\ ice} \underset{90\%}{C_6H_5CHO}$$

Oxidation of aliphatic alcohols with tert-butyl chromate yields mixtures of acids, aldehydes and esters[1]. Thus, butanol gives 30% of butanal, 27% butyric acid 36% butyl butyrate.

References

1. T. Suga, K. Kihara and M. Matsuura, *Bull. Chem. Soc.* Jpn. 1965, 38, 893.
2. T. Suga, K. Kihara and T. Matsuura, *Bull. Chem. Soc.,* Japn, 1965, **38**, 1141.

1.19 t-BUTYL HYDROPEROXIDE [$(CH_3)_3COOH$]

It is commercially available as a 70 or 90 per cent solution containing water and tert. butyl alcohol. Anhydrous tert. butyl hydroperoxide is obtained from the 70 per cent aqueous solution by azeotropic distillation with toluene[1]. However, it can be prepared[2] from tert-butyl hydrogen sulphate and 27 per cent hydrogen peroxide. Anhydrous (as well as highly concentrated solution) tert-butyl hydroperoxide can decompose violently in presence of strong acids and some transition metals, especially iron, cobalt and manganese[1, 3].

The conversion of alkenes into expoxides can be affected with per acids (see per acids). It has been found that a new versatile reagent, t-butyl hydroperoxide in presence of vanadium (V^{5+}) or molybdinum (Mo^{6+}) catalyst (vanadyl or molybdenyl acetylacetates are frequently used) provides an excellent method[3] for the epoxidation of alkenes[4, 5]. Epoxidation of isolated or exocyclic double bonds can be carried out by using molybdenum catalyst, such as, MoO_3 and $MoO_2(acac)_2$ and allylic alcohols can be epoxidised using vanadium catalysts. The terminal alkenes are difficult to epoxidise. However, these can be epoxised using t-butyl hydroperoxide. Thus, epoxidation of 1-decene with t-butyl hydroperoxide in presence of catalytic amount of molybdenum hexacarbonyl in boiling 1, 2-dichloroethane gives good yield of the 1-decene epoxide.

$$CH_3(CH_2)_7CH{=}CH_2 \xrightarrow[\text{1,2-dichloroethane, reflux}]{\text{t-butyl hydroperoxide, } Mo(CO)_6} CH_3(CH_2)_7\underset{\diagdown O \diagup}{CH{-}CH_2}$$

1-Decene → 1-Decene epoxide

Selective epoxidation can also be affected with t-butyl hydroperoxide. This reagent shows remarkable reactivity[4, 6] towards the double bond of allyl alcohol. Thus, the epoxidation of geraniol with t-butyl hyroperoxide in boiling benzene in the presence of catalytic amount of vanadium acetylacetone gives exclusively the 2, 3-epoxide. However, with peroxy acids, the epoxidation occurs at the other double bond.

Epoxidation with t-butyl hydroperoxide is more stereoselective than with per acids. Thus, epoxidation of the following acyclic allyl alcohol with t-butyl hydroperoxide in presence of vanadium acetylacetonate give exclusively the erythro epoxide. However, with peracids, a mixtures of ethryo and threo isomers is formed.

H_3C, H, OH, C_4H_9 — $t\text{-}C_4H_9OOH$ / $VO(acac)_2$ / toluene, 0°C → H_3C, H, OH, O, C_4H_9

> 99%
Erythro epoxide

$m\text{-}ClC_6H_4COOOH$ → Mixture of erythro anol threo epoxides

Selective epoxidation of α, β-unsaturated aldehydes[7] and ketones[8] can be achieved with t-butyl hydroperoxide without the involvement of the carbonyl function.

PhCH=CH—CHO —(t-BuOOH, MeOH, NaOH / pH 8.5, 35-40°C, 5.5 hr)→ Ph—CH—CH—CHO (O bridging)

73%

2-Cyclohexen-1-one —(t-BuOOH, C_6H_6, MeOH / Triton B($PhCH_2NMe_3OH$) / 10°C, RT 12 hr.)→ 2,3-Cyclohexen--1-one oxide

Some other applications of tert-butyl hydroperoxide include conversion of tertiary amines to amine oxides[9] and oxidation of lactams to imides[10].

Lactam —(t-BuOOH, $MnCl_2$)→ Imide

One of the most interesting and important application of tert. butyl hydroperoxide is in **enantioselective oxidations**. Normally such enantioselective oxidations are achieved by enzymes or microorganisms (see enzymes). However, such oxidations are now achieved chemically by using ciral oxidants. One such reagent is **sharpless reagent**[11], which gives the best results. It is a mixture of tetraisoproxy titanium, diethyl (R, R)-N(S, S)-tartrate, water and tertiary butyl hydro peroxide ration 1 : 2 : 1 : 1. Such oxidations are using sharpless reagent are called **sharpless**

asymmetric epoxidations. A typical example is the epoxidation of cis-2-cycloocten-1-ol with sharpless reagents yields almost exclusively *cis* epoxide, where as m-chloroperoxybenzoic acid produces exclusively the *trans* epoxide. However, in trans-2 cycloothen-1-ol, both oxidants furnish predominantly the cis epoxide[12].

cis-2-Cycloocten-1-ol —epoxidation→ *cis*-epoxide + *trans*-epoxide

	cis-epoxide	*trans*-epoxide	
t-BuOOH/VO(acac)$_2$ C_6H_6, 40°C, 24 hr.	97%	3%	(83% yield)
m-ClC$_6$H$_4$CO$_3$H, CH_2Cl_2 0°C, 24 hr.	0.2%	99.8%	(81% yield)

Enantioselective oxidation of sulphides with sharpless reagent gives 80% yield of the sulfoxides (89% ee)[11].

$C_6H_5SCH_3$ (Methyl phenyl sulphide) —Sharpless reagent, CH_2Cl_2, –20°C, 5 hr→ C_6H_5–S(=O)–CH_3 80% (ee 89%)

Another example of enantioselective oxidation is the oxidation of methyl *p*-tolyl sulfide into sulfoxide by sharpless reagent.

H_3C–C_6H_4–SCH_3 (Methyl p-toylsulphide) —Sharpless reagent, CH_2Cl_2, –20°C, 4 hr→ H_3C–C_6H_4–S(=O)–CH_3 Methyl p-tolyl sulfoxide 90% (ee 91%)

References

1. J.G. Hill, B.F. Rossiter and K.B. Sharpless, *J. Org. Chem.*, 1983, **48**, 3607.
2. N.A. Milas and D.M. Surgenor, *J. Am. Chem. Soc.*, 1946, **68**, 205.
3. K.B. Sharpless and T.R. Verhoeven, Aldrichimica Acta, 1979, **12**, 63.
4. K.B. Sharpless and R.C. Michaelson, *J. Am. Chem. Soc.*, 1973, **95**, 6136.
5. T. Katsuki and K.B. Sharpless, *J. Am. Chem. Soc.*, 1980, **102**, 5974.
6. B.E. Rossiter, T.R. Verhoeven and K.B. Sharpless, *Tet. Lett.*, 1979, 4733.
7. G..B. Payne, *J. Org. Chem.*, 1960, **25**, 275.

8. N.C. Yang and R.A. Finnegan, *J.Am. Chem., Soc.*, 1958, **80**, 5845.
9. L. Kuhnen, Chem. Ber., 1966, **99**, 3384; M.N. Sheng and J.G. Zajacek, *Org. Synth.* Collective Volume, 1988, **6**, 501.
10. A.R. Doumaux, Jr, and D.J. Tecker, *J. Org. Chem.*, 1970, **35**, 2121.
11. P. Pitchen, E. Duñach, M.N. Deshmukh and H.B. Kagan, *J. Am. Chem., Soc.*, 1984, **106**, 8188.
12. T. Itoh, K. Jitsukawa K. Kaneda and S. Teranishi *J. Am. Chem. Soc.*, 1979, **101**, 159.

1.20 Tert-BUTYL HYPOCHLORITE, Tert-C_4H_9OCl

It is prepared by passing chlorine into a solution of tert-butyl alcohol in aqueous sodium hydroxide.

It dehydrogenates amines. Thus, methyl pyrrolidine-α-carboxylate is convented into methyl 1-pyrroline-2-carboxylate in 71% yield[1].

tert-C_4H_9OCl

$COOCH_3$ N H

Methyl pyrrolidine 2-carboxylate

N $COOCH_3$

71%
Methyl 1-pyrroline--2-carboxylate

tert-Butyl hypochlorite dehydrogenate primary benzylic alcohols to the dehydes

$$C_6H_5CH_2OH \xrightarrow[CCl_4,\ 30\text{-}35°C]{t\text{-BuOCl},\ C_5H_5N} C_6H_5CHO$$

20°C, 1 hr 71%

However, primary aliphatic alcohols are converted into esters of the corresponding acids[2]. Thus, treatment of butyl alcohol in carbon tetrachloride with tert-butyl hypochlorite in presence of pyridine at 40–50° gave after 2hr. a 66% yield of butyl butyrate.

Secondary alcohols are oxidised to ketones. Thus, cyclohexanol is oxidised to cyclohexanone by tert-butyl hypochlorite in carbon tetrachloride in presence of pyridine[2]. tert-Butyl hypochlorite oxidises selenides to selenoxides.

$$C_6H_5SeCH_3 \xrightarrow[MeOH,\ -25°C,\ 40\ min]{t\text{-BuOCl},\ MeCN,\ C_5H_5N} C_6H_5\overset{O}{\overset{\|}{Se}}CH_3$$

Methyl phenyl selenide

80%
Methyl phenyl selenoxide

Sulphides are oxidised to sulfoxide[3, 4] with tert-butyl hypochlorite.

$$C_6H_5SCH_3 \xrightarrow[\text{15 min, NaHCO}_3\text{, RT 2 hr}]{\text{t-BuOCl, MeOH, }-78^\circ\text{C}} C_6H_5S(=O)CH_3$$

Methyl phenyl sulphide → Methyl phenyl sulfoxide, 90%

$(CH_3)_3C$ — 4-tert-Butyl thiocyclohexane $\xrightarrow{\text{t-BuOCl, MeOH, }-70^\circ\text{C}}$ 100% *cis* sulfoxide

References

1. H. Poisel and U. Schmidt, *Chem Ber.,* 1975, **108**, 2547.
2. C.A. Grob and H.J. Schmidt, *Helv. Chem. Acta, 1953,* **36**, 1763.
3. L. Skattebol, B. Boutette and S. Soloman, *J. Org. Chem.,* 1967, **32**, 3111.
4. C.R. Johnson, D. Mc Cants, Jr., *J. Am. Chem. Soc.,* 1965, **87**, 1109.

1.21 3-CARBOXYPYRIDINUM DICHROMATE $(3 - C_5H_4NCO_2H)_2Cr_2O_7$

It is prepared[1] by adding nicotinic acid to a solution of chromium trioxide in water at 0 – 5° followed by acetone. It oxidises[1] alcohols to carbonyl compounds, thiols to disulphides and aromatic rings to quinones.

$$C_6H_5CH_2OH \xrightarrow[C_5H_5N,\ CH_2Cl_2,\ \text{RT, 25 min}]{(3\text{-}C_5H_4NCO_2H)Cr_2O_7} C_6H_5CHO \quad 85\%$$

$$C_6H_5SH \xrightarrow[CH_2Cl_2,\ C_5H_5N,\ 0^\circ\text{C, 1 hr}]{(3\text{-}C_5H_4NCO_2H)_2Cr_2O_7} C_6H_5SSC_6H_5 \quad 80\%$$

Anthracene $\xrightarrow[\text{AcOH, reflux, 75 hr}]{(3\text{-}C_5H_4NCO_2H)_2Cr_2O_7}$ anthraquinone, 67%

References

1. C. Lóptz, A. González, F.P. Cossió and C. Paloma, *Synth. Commun,* 1985, **15**, 1197.

1.22 4-CARBOXYPYRIDINUM DICHROMATE $(4 - C_5H_5N\ CO_2H)_2\ Cr_2O_7$

It is prepared[1] by adding isonicotinic acid to aqueous solution of chromium trioxide at 0 – 5°, followed by acetone. As in the case of 3-carboxypyridinium dichromate, this reagent is also used to oxidise alcohols to carboxyl compounds, thiols to disulphides and aromatic rings to quinones.

References

1. C. López, A gonzález, F.P. Cossio and C. Palomo, *Synth. Commun,* 1985, **15**, 1197.

1.23 CALCIUM HYPOCHLORITE, $Ca(OCl)_2$

It is used for the epoxidation[1] of double bonds.

1, 4-Naphthoquinone is oxidised with calcium hypochlorite[1] to 2, 3-epoxy-dihydro-1, 4-naphthoquinone.

$Ca(OCl)_2$, H_2O, RT, 24 hr; 90%

1,4-Naphthoquinone → 2,3-Epoxy-2,3-dihydro-1, 4-naphthaquinone

Secondary alcohols are oxidsed to ketones[2]

$Ca(OCl)_2$, < 40°C, RT, 3 hr; 80%

Cyclohexanol → Cyclohexanol

References

1. T. Zincke, *Chem. Ber.,* 1892, **25**, 3599.
2. M. Schneider, J.-V. Weber and P. Faller *J. Org. Chem.,* 1982, **47**, 364.

1.24 CERIC AMMONIUM NITRATE, $(NH_4)_2Ce(NO_3)_6$

Following are given some of the oxidation reactions affected by ceric ammonium nitrate.

(*i*) Acetoxylation of aromatic side chains in benzylic positions.

Toluene on oxidation with cerium ammonium nitrate in glacial acetic acid gives benzyl acetate.

$$\underset{\text{Toluene}}{C_6H_5CH_3} \xrightarrow[\text{100\% AcOH}]{(NH_4)_2Ce(NO_3)_6} \underset{\substack{\text{Benzylacetate}\\(90\%)}}{C_6H_5CH_2OCOCH_3}$$

However, oxidation in 50% acetic acid gives benzaldehyde.

In some cases, more than one product is formed. Thus, oxidation of *p*-chlorotoluene gives a mixture of p-chlorobenzylacetate and *p*-chloro benzylalcohol.

$$\underset{\substack{\text{p-Chloro}\\\text{toluene}}}{Cl\text{—}C_6H_4\text{—}CH_3} \xrightarrow[\substack{\text{100\% HOAc}\\\text{retlux 10-40 min}}]{(NH_4)_2Ce(NO_3)_6} \underset{\substack{p\text{-Chloro-}\\\text{benzylacetate}}}{Cl\text{—}C_6H_4\text{—}CH_2OCOCH_3} + \underset{\substack{p\text{-Chloro-}\\\text{benzylalcohol}}}{Cl\text{—}C_6H_4\text{—}CH_2OH}$$

(*ii*) Oxidation of aromatic side chain.

Methyl homologues of aromatic compounds are oxidised to aldehyde by ceric ammoniumnitrate in presence of perchloric acid[2].

$$\underset{\text{Toluene}}{C_6H_5CH_3} \xrightarrow[\text{40°C, 80 min}]{(NH_4)_2Ce(NO_3)_6,\ HClO_4} \underset{\substack{\text{Benzaldehyde}\\(90\%)}}{C_6H_5CHO}$$

(*iii*) Oxidative cleavage of benzoin.

Ceric ammonium nitrate like other oxidising agents (*e.g.*, ammonium nitrate in presence of cupric acetate, bismuth sesquioxide and sodium bromate) does not convert benzoin into benzil. Instead, the bond between the alcoholic and the keto groups is cleaved to give benzaldehyde and benzoic acid[3].

$$\underset{\text{Benzoin}}{C_6H_5CO\underset{\substack{|\\OH}}{C}HC_6H_5} \xrightarrow{(NH_4)_2Ce(NO_3)_6} \underset{\substack{\text{Benzoic}\\\text{acid}}}{C_6H_5COOH} + \underset{\text{Benzaldehyde}}{C_6H_5CHO}$$

(*iv*) Oxidation of benzylic alochols.

Ceric ammonium nitrate in water or in 50% acetic acid oxidises benzylic alcohols to aldehydes at 90° in good yields[4]. In this oxidation, only catalytic amounts of the reagent and sodium bromate as a reoxidant are required to convert benzyl alcohol into benzaldehyde in 90% yield on heating in acetonitrile[5] at 80°.

$$\underset{\text{Benzylalcohol}}{C_6H_5CH_2OH} \xrightarrow[\text{CH}_3\text{CN, 80°C}]{(NH_4)_2Ce(NO_3)_6,\ NaBrO_3} \underset{\substack{90\%\\\text{Benzaldehyde}}}{C_6H_5CHO}$$

Benzylalcohol → Benzaldehyde: $(NH_4)_2Ce(NO_3)_6$, 50%AcOH, 90°C, 8 hr, 95%

(*v*) Oxidation of diols[6].

$$CH_3CH(OH)(CH_2)_8CH_2OH \xrightarrow[\text{aq. } CH_3CN, \text{ reflux 2 hr.}]{(NH_4)_2Ce(NO_3)_6,\ NaBrO_3} CH_3C(=O)(CH_2)_8CH_2OH$$

(*vi*) Oxidation of 2-adamantanone and tetracyclone.

Ceric ammonium nitrate oxidises 2-adamantanone to 2-oxahomo adamantanone in aqueous acetonitrile in 73% yield at 60° after 3hr[7]. In a similar way, tetracyclone (2, 3, 4, 5-tetraphenyl-2, 4-cyclopentadien-1-one) is oxidised by ceric ammonium nitrate in aqueous acetonitrile to tetraphenyl-2-pyrone in 77% yield[7].

(*vii*) Oxidation of methylene group to keto group.

Ceric ammonium nitrate oxidise methylene group adjacent to aromatic rings to keto group. Thus, ethyl benzene on oxidation gives acetophenone[2].

$$C_6H_5CH_2CH_3 \xrightarrow[70°C,\ 1.5\ hr.]{(NH_4)_2Ce(NO_3)_6,\ HNO_3} C_6H_5COCH_3$$

Ethylbenzene → Acetophenone 77%

The conversion of methylene to carbonyl is very easy when the methylene group is flanked by two aromatic rings. Thus, diphenylmethane yields benzophenone[2].

$$C_6H_5CH_2C_6H_5 \xrightarrow[90°C,\ 75\ min]{(NH_4)_2Ce(NO_3)_6,\ HNO_3} C_6H_5COC_6H_5$$

Diphenylmethane → 76% Benzophenone

(*viii*) Oxidation of phenols to quinones.

The oxidation of naphthols to 1, 4-naphthoquinones is affected by ceric ammonium nitrate[8].

OCH$_3$, OH, OCH$_3$, CH$_3$ (naphthol) $\xrightarrow[\text{MeCN, RT 45\% yield}]{(NH_4)_2Ce(NO_3)_6}$ OCH$_3$, O, OCH$_3$, O, CH$_3$ (1,4-naphthoquinone)

OCH$_3$, OH, OCH$_3$, Cl, CH$_3$ (naphthol) $\xrightarrow[\text{MeCN, RT}]{(NH_4)_2Ce(NO_3)_6}$ (same quinone)

Oxidations to quinones are easier if two hydroxylic groups are present in para or ortho positions is the aromatic ring. Thus, hydroquinone and its substituted derivatives are oxidised to p-benzoquinone[7].

(*ix*) Oxidation of secondary alcohols to ketones

Ceric ammonium nitrate is used to oxidise saturated and unsaturated secondary alcohols to ketones. The reagent is used only in catalytic amount with sodium bromate as a reoxidant[6].

$$C_3H_7\underset{\displaystyle OH}{\underset{|}{C}}C_8H_{17} \xrightarrow[\text{MeCN, } H_2O\text{, 80°C, 0.8 hr}]{0.2(NH_4)_2Ce(NO_3)_6\text{, 2.0 NaBrO}_3} C_3H_7COC_8H_{17}$$

(*x*) Oxidation of sulphides is sulfoxides[9]

$$C_6H_5SC_6H_5 \xrightarrow[\text{MeCN, } H_2O\text{, RT, 3 min}]{(NH_4)_2Ce(NO_3)_6 \text{ (4 mol)}} C_6H_5\overset{\overset{\displaystyle O}{||}}{S}C_6H_5$$

Diphenyl sulphide → Diphenyl sulfoxide

(*xi*) Regeneration of ketones from their derivatives.

A special advantage of ceric ammonium nitrates is the oxidative recovery of carbonyl compounds from their oximes or semicarbazones[7] and carboxylic acids from their hydrazides[10] under mild conditions.

$$C_6H_5-\overset{\overset{\displaystyle R}{||}}{C}-C_6H_5 \xrightarrow[\text{–4 to –20°C, 5 min.}]{(NH_4)_2Ce(NO_3)_6\text{, HNO}_3\text{, EtOH}} C_6H_5-CO-C_6H_5$$

Benzophenone oxime R=NOH
Benzophenone semicarbozone (R = N—NH—CO—NH$_2$)

Benzophenone

$$RCONHNH_2 \xrightarrow[\text{–4 to –20°C, 5 min}]{(NH_4)_2Ce(NO_3)_2\text{, HNO}_3\text{, EtOH}} RCOOH$$

Carboxylic acid hydrazide → Carboxylic acid

References

1. W.S. Trahanovsky and L.B. Young, *J. Org. Chem.*, 1966, **31**, 2033.
2. L. Syper, *Tetrahedron Lett.*, 1966, 4493.
3. T.–L. HO, *Synthesis* 1972, 560.
4. W.S. Trahanovsky, L.B. Young and G.L. Brown, *J. Org. Chem.*, 1967, **32**, 3865.
5. T.–L. HO, *Synthesis*, 1978, 936.
6. S. Kanemoto, H. Tomioka, K. Oshima and H. Nozaki, *Bull. Chem. Soc.*, Jpn., 1986, **59**, 105.
7. T.–L. HQ, *Synthesis*, 1973, 347.
8. H. Laatsch, *Justius Liebigs Ann. Chem.*, 1986, 1655, 1669.
9. T.–L. HO and C.M. Wong, *Synthesis*, 1972, 561.
10. T.–L. HO, H.C. HO and C.M. Wong, *Synthesis*, 1972, 562.

1.25 CERIC AMMONIUM SULPHATE $[NH_4]_2\ Ce(SO_4)_2 \cdot 2(NH_4)_2SO_4 \cdot 2H_2O$

It is used as an oxidant for converting organic polynuclear hydrocarbons into quinones[1]. Thus, naphthalene is oxidised to 1,4-naphthoquinone and phenanthrene is oxidised to 9, 10-phenanthrenequinone.

Naphthalene $\xrightarrow[4N\ H_2SO_4,\ 25°C,\ 6\ hr.]{Ce(SO_4)_2.2(NH_4)_2\ SO_4.2H_2O}$ 1,4-Napthoquinone (90–95%)

Phenanthrene $\xrightarrow[4N\ H_2SO_4,\ 50°C,\ 4\ hr.]{Ce(SO_4)_2.2(NH_4)_2\ SO_4.2H_2O}$ 9,10-Phenanthrenequinone

Reference

1. W. Periasamy and M.V. Bhatt, *Synthesis,* 1977, 303.

1.26 CHIRAL OXIDANTS

A chiral oxidant is the combination of an oxidising agent with a chiral ligand and a transition metal catalyst. The chiral oxidant is capable of controlling the stereochemistry of an oxidation reaction. This, in fact is the objective of a synthetic chemist. The preferential formation of one enantiomer is desirable when ever it is possible. Such oxidations are described as **enantioselective oxidations**. These are normally known to take place in presence of microorganisms or enzymes. However, it has now been possible to accomplish such oxidations by purely chemical means.

In a chiral oxidant, some of the metallic compounds used are vanadyl acetylacetonate, VO(acac), molybdenyl acetyl acetonate, $MoO_2(acac)_2$ and especially titanium tetraisopropoxide $[Ti[OCH(CH_3)_2]_4$. Best results are obtained by **sharpless reagent**, which is a mixture of 1 mol of titaniumtetra isopropoxide, 1-2 mol of tert-butyl hydroperoxide as the oxidant, and 2mol of (R, R) – or (S, S)-diethyl tartrate as the chiral component. Some other chiral regents, which have been used are N-ethylephedrine, quinine and quinidine acetates, N-phenyl camphoryl hydroxamic acid and (+) – (2R, 8aS) – or (–) – (2S, 8aR)-camporylsulfonyloxaziridine.

Some of the applications of chiral oxidations are given below.

1. Epoxidation of double bonds in allylic alcohols: The epoxidation of allylic alcohols was reported by Sharpless with t-butyl hydroperoxide, catalysed by the addition of tetraisopropoxytianium (IV) (Ti(OiPr)$_4$] gave the epoxy alcohol with high enantioselectively when (+) – or (–) – diethyl tartarate was added[1, 2]. Thus, the epoxidation of geraniol under these conditions in the presence of L(+)-diethyl tartrate (DET)-gave the allylic epoxide in 77% yield with 95% ee for the 2S 3S isomer[1].

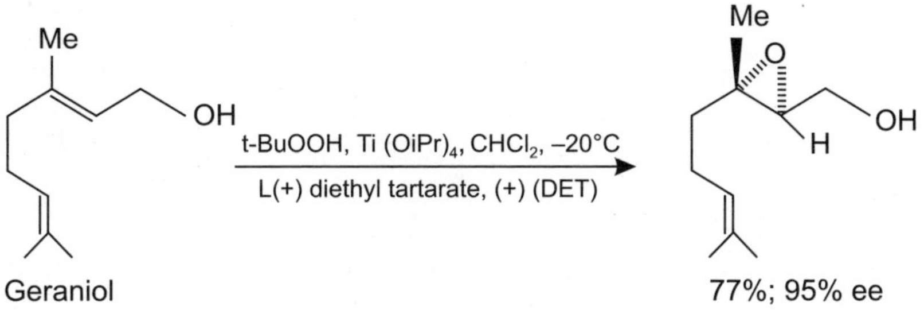

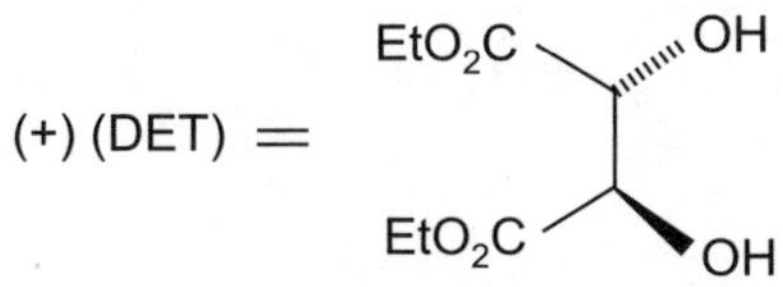

Geraniol —— t-BuOOH, Ti (OiPr)$_4$, $CHCl_2$, –20°C / L(+) diethyl tartarate, (+) (DET) ——→ 77%; 95% ee

(+) (DET) = [EtO_2C–CH(OH)–CH(OH)–CO_2Et]

The sharpless epoxidation is so dependable that a general rule has been developed to help the chemists to decide which enantiomer of DET will yield which epoxide. If the allylic alcohol is drawn in the orientation (as shown below) then (–) – DET with epoxidise it from the top face and (+) – DET from the lower face

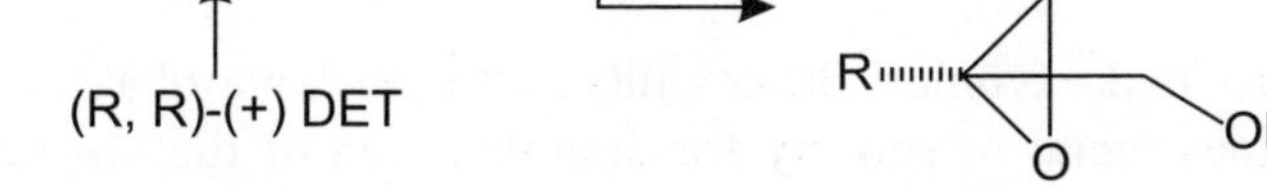

top face
(S, S)-(–) DET

(R, R)-(+) DET

Another example of sharpless asymmetric epoxidation of allylic alcohol is given below.

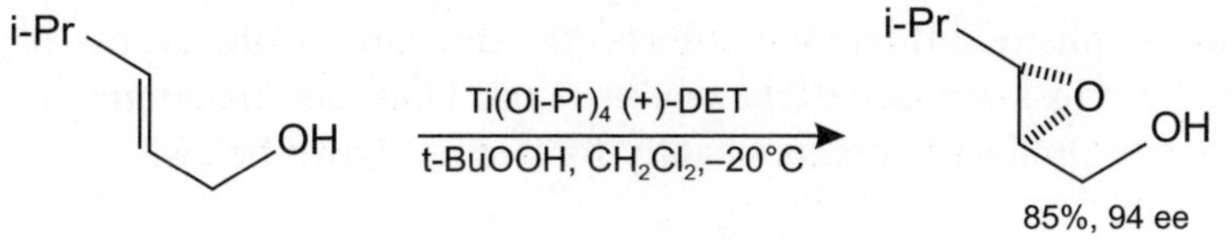

—— $Ti(Oi\text{-}Pr)_4$ (+)-DET / t-BuOOH, CH_2Cl_2,–20°C ——→ 85%, 94 ee

2. Hydroxylation of double bonds: It is well known that reaction of alkenes with osmium tetroxide (Osmylation) is a convenient method for the conversion of alkenes to 1, 2-diols. However, the reaction sometimes generates racemic products. A novel procedure for the hydroxylation of alkenes is the sharpless asymmetric hydroxylation procedure. In this method, a chiral substrate is added to the osmylation reagents with the object of generating a chiral osmate intermediate. Cinchona alkaloids, especially the esters of dihydroquinidines such as CA–1 and CA-2 were used as chiral substrates[3].

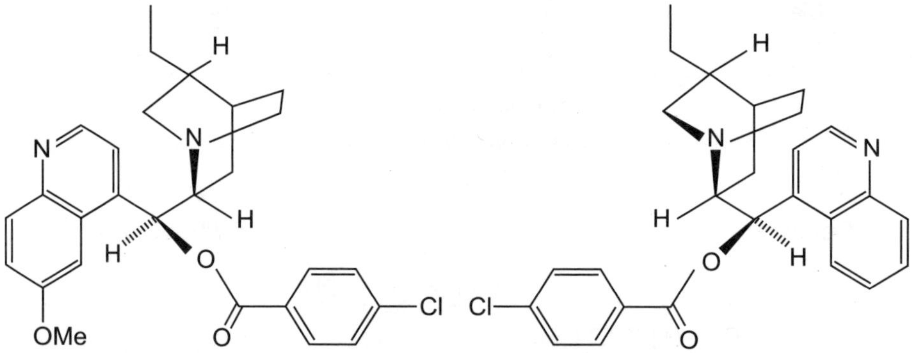

CA-1 CA-2

The chiral alkaloid is believed to form a complex with osmium and the alkene, directing attack from the opposite enantio face of the alkene[3].

The selectivity of the hydroxylation is illustrated by the reaction of *trans* stilbene with OsO_4 and NMO, which gave the diol with 99% ee in the presence of CA-1.

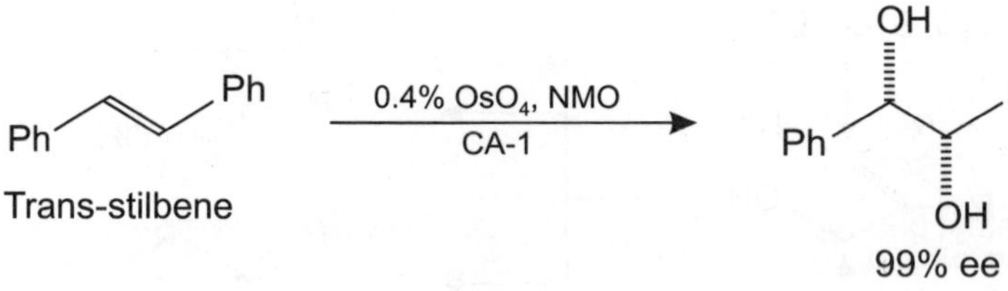

Trans-stilbene —(0.4% OsO_4, NMO / CA-1)→ diol, 99% ee

In the above hydroxylation, the chirality of the cinchona alkaloid is important, since it provides facial selectivity for complexation of the alkene as well as orientational control via the steric interactions described above.

See also hydroxylation of alkenes section 2.1.2.6.

In alkenes, the presence of chiral groups, such as phenyl sulfoxide[4] or N, S-dimethyl-S-phenylsulfoximine, affects[5] the direction of the access of osmium tetroxide from just one face of the double bond. Thus, one enantiomer of the syn-hydroxylation product is predominantly formed as shown below.

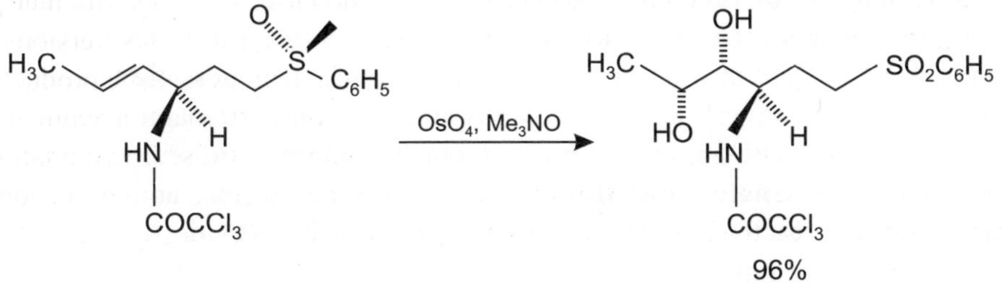

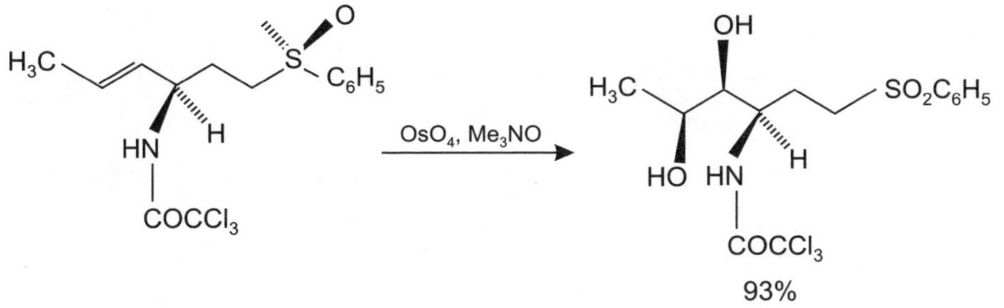

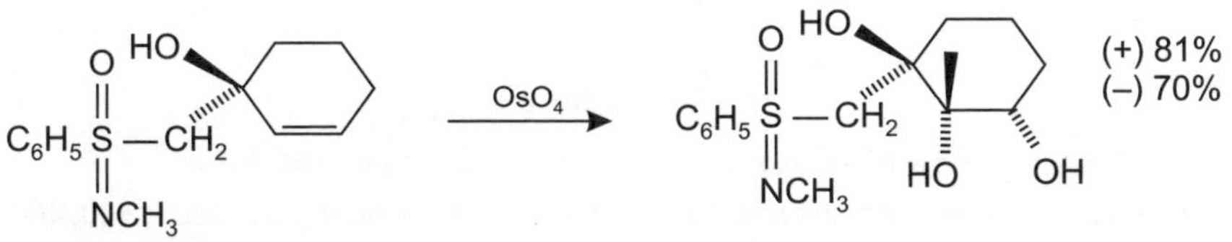

H_3C ... $S(O)C_6H_5$ ... HN–$COCCl_3$ $\xrightarrow{OsO_4,\ Me_3NO}$ H_3C–CH(OH)–CH(OH)–CH(NHCOCCl_3)–$CH_2CH_2SO_2C_6H_5$

96%

H_3C ... $S(O)C_6H_5$ ... HN–$COCCl_3$ $\xrightarrow{OsO_4,\ Me_3NO}$ H_3C–CH(OH)–CH(OH)–CH(NHCOCCl_3)–$CH_2CH_2SO_2C_6H_5$

93%

$C_6H_5S(O)(=NCH_3)$–CH_2–(1-hydroxycyclohex-2-enyl) $\xrightarrow{OsO_4}$ $C_6H_5S(O)(=NCH_3)$–CH_2–(1,2,3-trihydroxycyclohexyl)

(+) 81%
(–) 70%

3. Hydroxylation of the α-position with respect to an ester group: The hydroxylation of enolates in the α-position with respect to the ester group can be affected by chiral oxidants, such as (+) – 2R, 8aS and (–) – 2S, 8aR camphoryl sulfonylaziridines. For details see oxidation of carboxylic acids (section 2.7.1).

4. Oxidation of non-symmetrical sulphides to chiral sulfoxides: The oxidation of non-symmetrical sulphides to chiral sulfoxides canbe effected[6] by sharpless reagent.

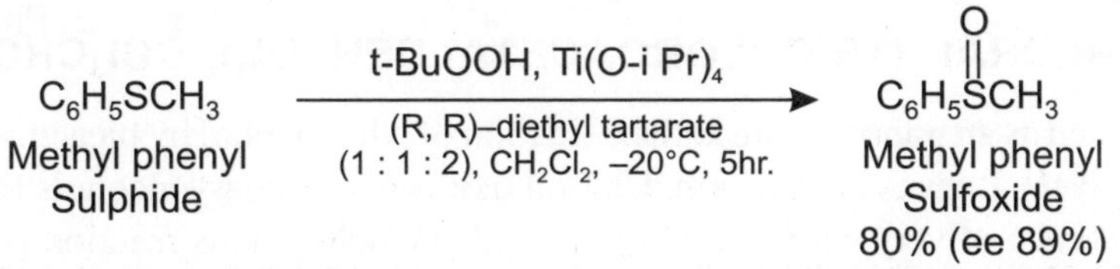

$$\underset{\text{Methyl phenyl Sulphide}}{C_6H_5SCH_3} \xrightarrow[\text{(1 : 1 : 2), } CH_2Cl_2,\ -20°C,\ 5hr.]{\text{t-BuOOH, Ti(O-i Pr)}_4 \text{ (R, R)–diethyl tartarate}} \underset{\text{Methyl phenyl Sulfoxide}}{C_6H_5S(O)CH_3}$$

80% (ee 89%)

The oxidation of methyl p-tolyl sulphide into the corresponding sulfoxide has been affected with sharpless reagent, modified by adding 1 mol of water[6, 7]. Such a reagent gives higher enantiomealtric excesses.

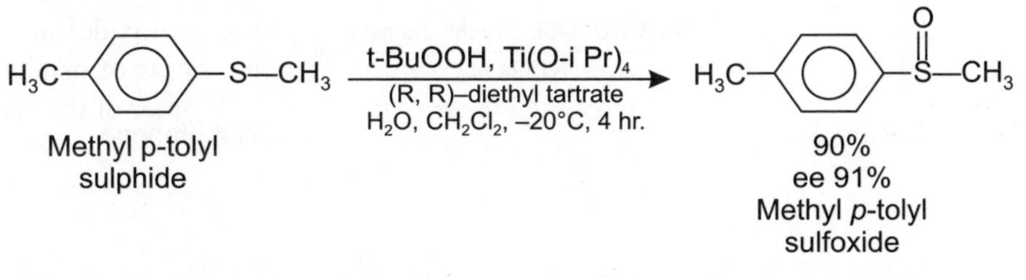

$$\underset{\text{Methyl p-tolyl sulphide}}{H_3C\text{–}C_6H_4\text{–}S\text{–}CH_3} \xrightarrow[H_2O,\ CH_2Cl_2,\ -20°C,\ 4\ hr.]{\text{t-BuOOH, Ti(O-i Pr)}_4 \text{ (R, R)–diethyl tartrate}} H_3C\text{–}C_6H_4\text{–}S(O)\text{–}CH_3$$

90%
ee 91%
Methyl *p*-tolyl sulfoxide

5. Resolution of racemic mixture: Consider the case of a substrate that is chiral (and racemic). On oxidation, one enantiomer of the substrates may react with a chiral oxidant at a different rate than the other. It is possible to stop the reaction at 50% conversion and recover a single, unreacted enantiomer of the substrate together with 50% of the product which would be oxidised. The process is referred to as **kinetic resolution**. Thus, in case of racemic amine (1), one enantiomer is unreactive, while the other is oxidised is its N-oxide.

$Ti(OiPr)_4$(2eq), t-BuOOH (0.6 eq), (t)-DIPT (1.2 eq)

(±) – 1 → 37% (95% ee) + 59% (63% ee)

(+)-DIPT

$iPrO_2C$ CO_2iPr

References

1. T. Kasturi and K.B. Sharpless, *J. Am. Chem. Soc.,* 1980, **102**, 5974.
2. B.E. Rossiter, T. Kasturi and K.B. Sharpless, *J. Am.Chem. Soc.,* 1981, **103**, 464.
3. E.N. Jacobsen, L. Markó, W.S. Mungall, G. Schröder and K.B. Sharpless. *J.Am. Chem. Soc.*, 1988, **110**, 1968; S. G. Hentges and K.B. Sharpless, *J. Am. Chem. Soc.,* 1980, **102**, 4263.
4. F.M. Hauser, S.R. Ellenberger, J.C. Clardy and L.S. Bass, *J. Am. Chem. Soc.,* 1984, **106**, 2458.
5. C.R. Johnson and M.R. Barbachyn, *J.Am. Chem. Soc.,* 1984, **106**, 2459.
6. P. Pitchen, E. Duñach, M.V. Deshmukh and H.B. Kagan, *J. Am. Soc.,* 1984, **106**, 8189.
7. P. Pitchen and H.B. Kagan, Tetrahedron, *Lett.,* 1984, **25**, 1049.

1.27 CHLORAL (TRICHLOROACETALDEHYDE), CCl_3CHO

It is obtained as an intermediate in the oxidation of ethylalcohol by bleacing powder. On hydrolysis, it gives chloroform. Chloral oxidises secondary alcohols to ketones in the presence of very active aluminium oxide (Woelm). This reaction is superior to the oppenauer oxidation, since it takes place at room temperature or at slightly elevated temperatures[1-3]. Thus, cyclobutanol is oxidised to cyclobutanone[2].

CCl_3CHO, CCl_4, Woelm aluminia, RT, 24 hr

Cyclobutanol → Cyclobutanone

Chloral is also used in the dehydrogenation of hydroxy sulphides to keto sulphides[3]. Thus secondary β-hydroxy sulphides yield β-keto sulphides on treatment with chloral on alumima.

SC_6H_5 / OH — Al_2O_3 (Woelm), CCl_3CHO, CCl_4, 25°C, 30 min, 55°C, 24 hr. → SC_6H_5 / O

References

1. G.H. Posner and M.J. Chapdelaine, *Synthesis,* 1977, 555.
2. G.H. Posner, R.B. Perfetti and A.W. Runquist, *Tetrahedron Lett.,* 1976, 3499.
3. G..H. Posner and M.J. Chapdelaine, *Tetrahedron Lett.,* 1977, 3227.

1.28 CHLORANIL (TETRACHLORO-p-BENZOQUINONE)

Cl Cl O O Cl Cl

It is useful for dehydrogenations. Thus, 1, 2, 3, 4-tetrahydrocarbazoles are transformed into carbazoles in 50 – 95% yield by refluxing with chloranil in xylene.

1,2,3,4-Tetrahydro carbazole — Chloranil, xylene, reflux → Carbazole 50-95%

Benzylic alcohols are oxidised to aldehydes[2]

$$C_6H_5CH_2OH \xrightarrow[\text{reflux, 3 hr}]{\text{Chloranil, } Et_2O} C_6H_5CHO \quad 66\%$$

References

1. B.M. Barclay and N. Campbell, *J. Chem. Soc.,* 1945, 530.
2. E.A. Braude, R.P. Linstead and K.R. Wooldridge, *J. Chem Soc.,* 1956, 3070.

1.29 CHLORINE, Cl_2

It is rearely used for oxidation purposes. However, its compounds, such as sodium, potassium and calcium hypochlorites, tert-butyl hypochlorite, sodium chlorite and sodium, potassium, silver and barium chlorates are used. Some of its applications are given below.

1. Chlorine in dimethyl sulphide oxidises[1] primary alcohols to aldehydes in high yields under mild conditions.
2. Secondary alcohols are oxidised[1, 2] to ketones

$$CH_3CH(OH)C_6H_{13} \xrightarrow[NaHCO_3,\ 0\text{-}5°C]{Cl_2,\ CHCl_3,\ HMPA,\ H_2O} CH_3COC_6H_{13}\ \ 95\%$$

$$C_6H_5CH(OH)CH_3 \xrightarrow[CH_2Cl_2,\ H_2O,\ NaH_2PO_4,\ 0\text{-}5°C,\ 15\ min]{Cl_2,\ CHCl_3,\ HMPA} C_6H_5COCH_3\ \ 98\%$$

$$\text{Cyclohexanol (H, OH)} \xrightarrow[H_2O,\ NaH_2PO_4,\ 0\text{-}5°C,\ 2\ min]{Cl_2,\ CHCl_3,\ HMPA,\ CH_2Cl_2} \text{Cyclohexanone (=O)}\ \ 81\%$$

It is found[1] that secondary alcohols are oxidised in preferance to primary alcohols.

3. Under anhydrous conditions, chlorine converts dimethyl disulphide into methanesulfinyl chloride[3]

$$\underset{\text{Dimethyl sulphide}}{CH_3SSCH_3} \xrightarrow[0\ to\ -10°C,\ vacuo\ distn.]{Cl_2/Ac_2O} \left[CH_3SCl \xrightarrow{Cl_2} CH_3SCl_3\right] \xrightarrow{Ac_2O} \underset{83\text{-}86\%}{CH_3SOCl} + 2CH_3COCl$$

However, chlorine in presence of water converts disulphides into sulfonyl chlorides[4].

$$\text{bis(2-nitrophenyl) disulphide } (O_2N\text{-}C_6H_4\text{-}S\text{-}S\text{-}C_6H_4\text{-}NO_2) \xrightarrow[70°C,\ 1.5\ hr.]{Cl_2,\ HCl,\ HNO_3,\ H_2O} \text{2-}NO_2C_6H_4SO_2Cl$$

Sulfonyl chlorides also result by the oxidation of isothiocyanates and isothioureas with chlorine in aqueousmedia[5].

$$C_6H_5CH_2SCN \xrightarrow[20\text{-}30°]{Cl_2\ H_2O} C_6H_5CH_2SO_2Cl\ \ 73\text{-}76\%$$

$$C_6H_5CH_2SC(=NH)NH_2.HCl \xrightarrow[20\text{-}30°]{Cl_2,\ H_2O} C_6H_5CH_2SO_2Cl$$

4. Mercaptans[6], alkyl thiocarboxylates[6] and sulfinic acids[7] are convered into sulfonyl chlorides with chlorine.

References

1. J. Wicha and A. Zarecki, *Tetrahedron Lett.,* 1974, 3059.
2. M.A. Neirabeyeh, J.-C. Ziegler and B. Gross, *Synthesis* 1976, 811.
3. B.S. Bal, K.S. Kochhar and H.W. Pinnick, *J. Org. Chem.,* 1981, **46**, 1492.
4. E. Wertheim, *Org. Synth.* Collective Volume, 1943, **38**, 729.
5. T.B. Johnson and I.B. Douglass, *J. Am. Chem., Soc.,* 1939, **61**, 2548.
6. I.B. Douglass and T.B. Johnson, *J. Am. Chem. Soc.,* 1938, **60**, 1486.
7. F. Ullmann and A. Lehner, *Chem. Ber.,* 1905, **38**, 729.

1.29a N-CHLOROSUCCINIMIDE, $C_4H_4ClNO_2$ (NCS)

It oxidises primary alcohols[1] to aldehydes and secondary alcohols to ketones. Thus, 1-octanol is converted into octanal in 96% yield on treatment with dimethyl sulphide and NCS in toluene at –25°C. In a similar way, cyclohexanol is converted into cyclohexanone (90% yield) by heating on a steam bath for 4 minutes with NCS in pyidine and benzene[2].

References

1. E.J. Corey and C.U. Kim, *J. Am. Chem.Soc.,* 1972, **94**, 7586.
2. C.A. Grob and H.J. Schmid, *Hel. Chim. Acta,* 1953, **36**, 1763.

1.29b CLAYFEN

Clayfen is montmorillonite k 10 clay supported iron (III) nitrate and is commercially available. It is used as a novel oxidant for the oxidation[1] of alcohols to carbonyl compounds by heating in a MW oven for 10–60 seconds in the absence of solvent.

$$R_2R_1CH{-}OH \xrightarrow[\text{MW, 15-60 sec.}]{\text{Clayfen}} R_2R_1C{=}O \quad \text{87-96\%}$$

R_1 = Ph, p-MeC_6H_4, p-$MeOC_6H_4$, R_2 = H
R_1 = p-$MeOC_6H_4$; R_2 = p-MeO C_6H_4CO
R_1 = Ph; R_2 = Et, PhCO.

Reference

1. R.S. Varma and R. Dahiya, *Tetrahedron, Lett.,* 1997, **38**, 2043.

1.30 CHROMATES

Acetyl chromate $(CH_3CO)_2\ CrO_4$
Trychloroacetyl chromate $(CCl_3CO_2)_2\ CrO_4$
Trifluoroacetyl chromate $(CF_3CO_2)_2\ CrO_4$

Benzoyl chromate $(C_6H_5CO)_2\ CrO_4$

All these reagents are prepared from chromium trioxide and the respective anhydrides[2]. Out of all these reagents, trifluoroacetyl chromate reacts with many solvents and even with silicone stopcock grease.

The remaining acyl chromates are used for epoxidation[1] and for the conversion of alkenes into aldehydes[2].

References

1. N. Miyaura and J.K. Kochi, *J. Am. Chem. Soc.*, 1983, **105**, 2368.
2. H. Schildknecht and W. Föttinger, Justus Liebigs *Ann. Chem.* 1962, **659**, 20.

1.31 CHROMIUM TRIOXIDE, CrO_3

It is used for oxidation of various functional groups in acetic acid, dilute sulphuric acid, a mixture of acetic acid and dilute sulphuric acid, dilute sulphuric acid and acetone (**Jones reagent**), acetic anhydride and acetic acid (**Fieser reagent**), Water, Water and ether[1–3], dichloromethane[4], benzene[5], pyridine[6], dimethyl formamide[7] and hexamethyl phosphoric triamide (HMPA)[8].

It is also used as **Collins reagent** $(CrO_3 \cdot 2C_5H_5N)$, a chromium trioxide-pyridine complex or dipyridine chromium (VI).

Besides the above, chromium trioxide is also used as oxidant in the following reagent.

- Chromium troxide-3, 5-dimethylpyrazole complex.
- Pyridimium chlorochromate (PCC).
- 4-Dimethylaminopyridinium chlorochromate.
- Poly (vinylpyridinium) chlorochromate (PVPCC).

All these reagents are discussed in separate sections.

Following are given some of the typical oxidations of chromium trioxide:

1. Allylic oxidations of methylene to carbonyl groups are carried out by chromic oxide[9, 10] or its complexes with pyridine[11, 12]. Such oxidations are performed in steroidal esters or ketones.

RO_2C (steroid) → RO_2C (steroid enone, O)

$CrO_3.2C_5H_5N$, CH_2Cl_2, RT, 17 hr, yield 74% OR $CrO_3.C_5H_5N.HCl.CH_2Cl_2$ reflux 48 hr. yield 54%

R = CH_3 or C_6H_5

2. Chromium trioxide in conc. nitric acid is used for the epoxidation of alkenes and cycloalkenes. Some examples[13, 13a] are given below.

$$CH_2{=}CHC_6H_{13} \xrightarrow[\text{CH}_2\text{Cl}_2,\ \text{CCl}_4,\ -78^\circ\text{C},\ 30\ \text{min}]{\text{CrO}_2(\text{NO}_3)_2,\ \text{DMF}} \underset{O}{CH_2{-}CHC_6H_{13}}\quad 43\%$$

$$C_6H_5CH{=}CH_2 \xrightarrow[\text{CH}_2\text{Cl}_2,\ -78^\circ\text{C},\ 30\ \text{min}]{\text{CrO}_2(\text{NO}_3)_2,\ \text{DMF}} \underset{O}{C_6H_5CH{-}CH_2}\quad 83\%$$

$$(C_6H_5)HC{=}CH(C_6H_5) \xrightarrow[\text{CH}_2\text{Cl}_2,\ \text{CCl}_4,\ -78^\circ\text{C},\ 30\ \text{min}]{\text{CrO}_2(\text{NO}_3)_2,\ \text{DMF}} \text{trans-stilbene oxide}\quad 30\%$$

$$\text{Cyclohexene} \xrightarrow[\text{Et}_2\text{O, RT, }0^\circ\text{C}]{\text{CrO}_2(\text{NO}_3)_2,\ \text{C}_5\text{H}_5\text{N},\text{CH}_2\text{Cl}_2} \text{Cyclohexene oxide}\quad 81\%$$

Epoxidation of alkenes can also be brought about by tert-butyl hydroperoxide in presence of chromium compounds.

3. Cycloalkenes are oxidised to unsaturated ketones by chromium trioxide in acetic acid.

$$\xrightarrow[\text{77-78}^\circ\text{C, 25 min}]{\text{CrO}_3,\ \text{AcOH}}$$

4,4,14a-Trimethyl-Δ^8-pregnene-20-one 3,21-diacetate → 3β,21-Diacetoxy-4,4,14a-Trimethyl-Δ^8-pregnene-7,11,20-trione (75%)

4. Oxidation of alkenes and cycloalkenes to ketones can be brought[14] about by chromium trioxide in sulphuric acid in presence af Hg(OCOEt).

$$CH_2 = CHC_6H_{13} \xrightarrow[\text{Hg (OCOEt)}_2,\ \text{Me}_2\text{CO},\ 25^\circ\text{C, 4 hr}]{\text{CrO}_3,\ \text{H}_2\text{SO}_4,\ \text{H}_3\text{O}} CH_3COC_6H_{13}\quad 82\%$$

5. Aromatic compounds are oxidised to quinones[13a,15] by chromium trioxide in acetic acid.

Naphthalene → (CrO_3, AcOH; 0°C, 2-3 hr, RT 3.5 day) → Naphthoquinone 32-35%

2,3-Dimethyl naphthalene → (CrO_3, AcOH; 20-30°C, 15 min; RT, 3 days) → 2,3-Dimethyl naphthoquinone

Phenanthrene → (CrO_3, AcOH; reflux, 8-10 hr) → 9,10-Phenanthraquinone

The oxidation[15a] of chrysene (1,2-benzophenanthrene) by refluxing for 8-10 hr. with chromic oxide in acetic acid gives a 96-97% yield of 5,6-chrysenequinone.

Chrysene → (CrO_3, AcOH; reflux, 8-10 hr) → 5,6-Chrysenequinone

The oxidation[16] of anthracene to anthraquinone takes place in position 9 and 10 by sodium dichromate in sulphuric acid in presence of a PTC. This has industrial applications in the synthesis of lake and vat dyes.

Anthracene $\xrightarrow[Bu_4NHSO_4,\ 70°C,\ 2\ min]{Na_2Cr_2O_7,\ H_2SO_4}$ Anthraquinone (82-92%)

6. Oxidation of primary alcohols to aldehydes is brought about by compounds of hexavalent chromium. Some typical examples are given below:

(*a*) $C_7H_{15}CH_2OH \longrightarrow C_7H_{15}CHO$

Reaction Conditions	***Yield (%)***	***Reference***
$Na_2Cr_2O_7$, 3M H_2SO_4 Bu_4NHSO_4, CH_2Cl_2 RT·1 min	95	16
CrO_3, Bu_4 NCl(catalyst) CH_2Cl_2, 90 min	60	17
$CrO_3 \cdot 2C_5H_5N$, CH_2Cl_2 RT, 15 min	90	18

(*b*) $CH_2=CHCH_2OH \longrightarrow CH_2=CH\,CHO$

Reaction conditions	***Yield (%)***	***Reference***
$CrO_3 \cdot 2C_5H_5N$, C_5H_5N, Cooling with ice, RT·15-22 hr.	63	19
$CrO_3 \cdot Me_2CO$, 0°C, 15-20 min	91	20
CrO_3, HMPA, RT, 12 hr.	85	21
$CrO_3 \cdot 2C_5H_5N$, cooling with ice RT·15-22 hr.	63	19

(*c*) $C_6H_5CH_2OH \longrightarrow C_6H_5CHO$

Reaction conditions	***Yield (%)***	***Reference***
$Na_2Cr_2O_7$, H_2O, 98°C, 3 hr.	85	22
$Na_2Cr_2O_7$, Bu_4NHSO_4 3MH_2SO_4, CH_2Cl_2, RT, 1 min	100	16
CrO_3, Bu_4NCl, H_2O CCl_4, 60°, 1 hr.	81	23
CrO_3, Me_2CO, 0°C, 20 min	76	20
$CrO_3 \cdot Et_2O$, CH_2Cl_2; SiO_2, 35 min	75	24
$2CrO_3$, HMPA, RT, 3 Hr.	85	8
$CrO_3 \cdot 2C_5H_5N$, CH_2Cl_2, RT, 15 min	89	18

$6CrO_3{\cdot}2C_5H_5N$, CH_2Cl_2, 25°C, 5-15 min (Collins reagent)	95	25
CrO_3, HCl, N⟨◯⟩—NMe_2, CH_2Cl_2, RT, 3hr	64	26

(*d*) $PhCH{=}CHCH_2OH \longrightarrow PhCH{=}CHCHO$

Reaction conditions	***Yield (%)***	***Reference***
$CrO_3{\cdot}2C_5H_5N$, cooling with ice, RT, 15-23 hr	81	19
H_2CrO_4, C_5H_5N, SiO_2, AcOH, CH_2Cl_2, 30-55°C, 4 hr	87	27
CrO_3, Me_2CO, 0°C, 20 min	84	20
CrO_3, HMPA, RT, 1 hr	92	8
CrO_3, $2C_5H_5N$, CH_2Cl_2, RT, 15 min	96	18

7. Oxidation of secondary alcohols to ketones.

The oxidation of secondary alcohols to ketones by compounds of hexavalent chromium is not only the most common, but also the best known reaction as far as the mechanism is concerned.

$$3RR'CHOH + 2CrO_3 + 3H_2SO_4 \longrightarrow 3RR'CO + Cr_2(SO_4)_3 + 6H_2O$$

The mechanism of the reaction is given below.

$$RR'CH{-}OH + CrO_3 \longrightarrow RR'CH{-}O{-}Cr(=O)_2OH \longrightarrow RR'C{=}O + (HO)_2Cr{=}O$$

2° Alcohol — Ester of alchol and chromic acid

The tetravalent chromium (IV) acid $[(HO)_2Cr{=}O]$ reats with chromium trioxide to form chromium (V) acid, using oxides the alcohol via its ester.

$$\overset{IV}{(HO)_2Cr{=}O} + \overset{VI}{CrO_3} \longrightarrow 2HO\overset{V}{Cr}(=O)_2$$

Chrominum (IV) acid

$$RR'CH{-}OH + 2HO{-}\overset{V}{Cr}(=O)_2 \longrightarrow RR'CH{-}O{-}Cr(=O){=}O \longrightarrow RR'C{=}O + HO\overset{III}{Cr}O$$

The presence of three oxidising species, chromium (VI), chromium (V) and chromium (IV) acids, accounts for the difference in the oxidation products of some alcohols. Following are given some typical examples:

(*a*) $CH_3CH(OH)-C_6H_{13} \longrightarrow CH_3COC_6H_{13}$

Reaction conditions	***Yield (%)***	***Reference***
CrO_3, HMPA, RT, 15hr.	40	8
CrO_3, $CH_2Cl_2Et_2O$, Cellite, 35 min	80	3
CrO_3, Amberlyst A26, C_6H_6, reflux, 9 hr	97	28
$3(2C_5H_5N \cdot CrO_3)$, C_5H_5N ice bath, 30 min RT, 15-22 Hr.	18	16
3,5-dimethylpyrazole (N–H) $N.CrO_3$, CH_2Cl_2, RT 30 min	93	29
CrO_3, Bu_4NO_3 SCF_3, CH_2Cl_2 RT. 1 hr	80	17

(*b*) Cyclohexanol (H, OH) $\longrightarrow$ Cyclohexanone (=O)

Reaction conditions	***Yield (%)***	***Reference***
CrO_3, HMPA, RT, 15hr.	49	8
CrO_3, Amberlyst A26, C_6H_{14}, 3hr	77	30
CrO_3Bu_4NCl, CH_2Cl_2, RT, 2hr	70	17
$2(C_5H_5N)_2CrO_4$ on SiO_2, AcOH, 30-35°C, 9 hr	64	27
$2(C_5H_5N)_2CrO_4$ on SiO_2, AcOH, C_6H_6, reflux 3 hr	76	27

(*c*) $C_6H_5CH(OH)CH_3 \longrightarrow C_6H_5COCH_3$

Reaction conditions	***Yield (%)***	***Reference***
CrO_3, HMPA, RT, 2hr.	100	8
CrO_3,CH_2Cl_2,Et_2O, cellite, 35 min	93	3
CrO_3, Amerlst A26, C_6H_6, 1hr.	77	28
$3(2C_5H_5N \cdot CrO_3)$, C_5H_5N, cooling with ice, 30 min, RT, 15-22 hr	71	29
$6(2C_5H_5N \cdot CrO_3)$, CH_2Cl_2, 25°C, 5-10 min	96	25

In certain cases, secondary alcohols are directly oxidise[31] to carboxylic acid. Thus, 2-methylcyclohexanol on oxidation with chromium trioxide gives δ-acetylvaleric or heptanon-6-oic acid

$$\text{2-Methyl cyclohexanol} \xrightarrow[\text{30°C, 1 hr. RT}]{CrO_3,\ H_2SO_4,\ H_2O} \underset{\substack{46\text{-}55\% \\ \delta\text{-Acetyl valeric acid}}}{CH_3CO(CH_2)_4CO_2H}$$

8. Crhomium oxide is used in dehydrogenatic hydrogenations.

A large scale application of intramolecular coupling combined with concomitant aromatisation is the industrial conversion[32] of heptane into toluence in 72 % yield on heating at 500°C in the presence of chromium and aluminium oxides.

9. Oxidation of methyl groups into aldehydes.

Chromium trioxide in the presence of acetic anhydride, acetic acid and sulphuric acid convert o-and p-nitrotoluenes into the corresponding nitrobenzaldehyde diacetates, from which the aldehydes are obtained by hydrolysis.

$$O_2N\text{-}C_6H_4\text{-}CH_3 \longrightarrow O_2N\text{-}C_6H_4\text{-}C(OCOCH_3)_2 \xrightarrow[\text{HCl/EtOH, 45 min reflux or } H_2SO_4\text{/EtOH, reflux 30 min}]{\text{hydrolysis}} O_2N\text{-}C_6H_4\text{-}CHO$$

Nitrotoluene	*Reaction conditions*	*Yield (%)*	*Reference*
Ortho	CrO_3, Ac_2O, H_2SO_4(36-37%) 10°C, 3hr	74	32, 33
Para	CrO_3, Ac_2O, AcOH(48-54%) H_2SO_4, <10°C	89-94	34

10. Cleavage of alkenes. Unsaturated hydrocarbons with tetrasubstituted double bonds are cleaved[32] to ketones by chromium trioxide in sulphuric acid.

$$(CH_3)_2C = C(C_6H_5)_2 \xrightarrow[-78°C,\ 15\ min,\ 3\ hr.]{CrO_3,\ 65\%\ H_2SO_4} \underset{22\%}{(CH_3)_2C = O} + \underset{66\%}{(C_6H_5)_2CO}$$

11. Cleavage of diols.

The treatment of 17 β, 17*a*, β-dimethyl-D-homoandrostane-3β, 17α, 17aα, triol 3-β-acetate with chromium trioxide yields[32] 3β-17, 17α-seco homoandrostane-17, 17*a*-dione

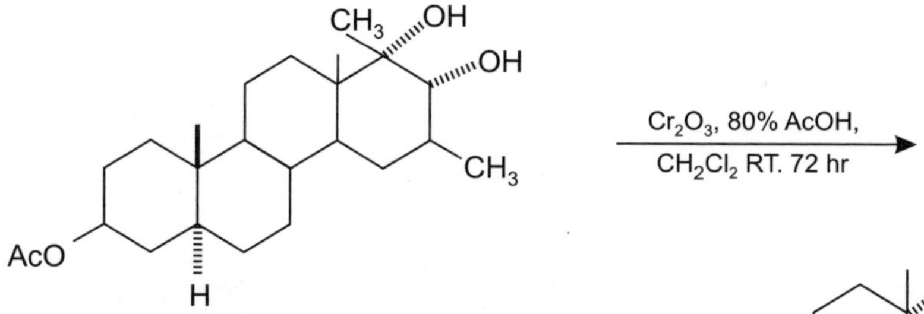

CH3 OH OH CH3 AcO H

Cr_2O_3, 80% AcOH, CH_2Cl_2 RT. 72 hr

17β-, 17aβ-Dimethyl -D-homoandrostane- -3β, 17α, 17aα,-triol 3β. acetate

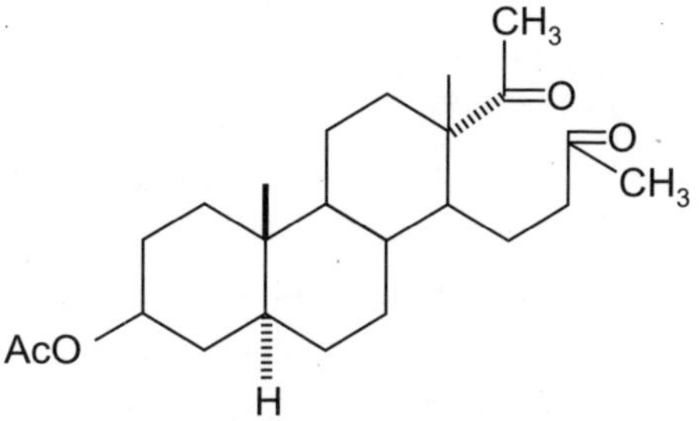

CH3 O O CH3 AcO H

3β-Acetoxy-17, 17α-dimethyl -17,17a-secohomoandrostane- -17, 17a dione

In a similar way, a mixture of *cis* and *trans*-1, 2-dimethylacenaphthene is oxidised[33] with chromium trioxide in acetic acid to 1, 8-diacetylnaphthalene in 89% yield.

Cyclic diols (*e.g.*, cyclopentene 3, 5-diol) on oxidation[33a] with chromium trioxide in sulphuric acid gives the corresponding cyclic diketones in 65-80% yield.

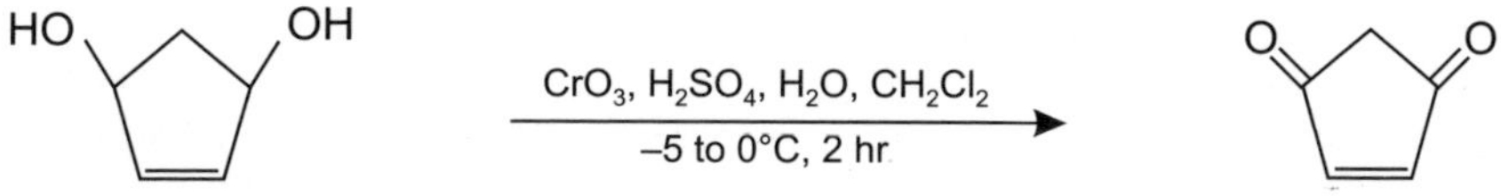

HO OH

CrO_3, H_2SO_4, H_2O, CH_2Cl_2
–5 to 0°C, 2 hr

O O

Cyclopentene 3,5 diol Cyclopentene 3,5-dione

12. Hydryxylation at methine groups.

The oxidation[35] of adamantane occurs almost exclusively at the tetiary carbon, C-1.

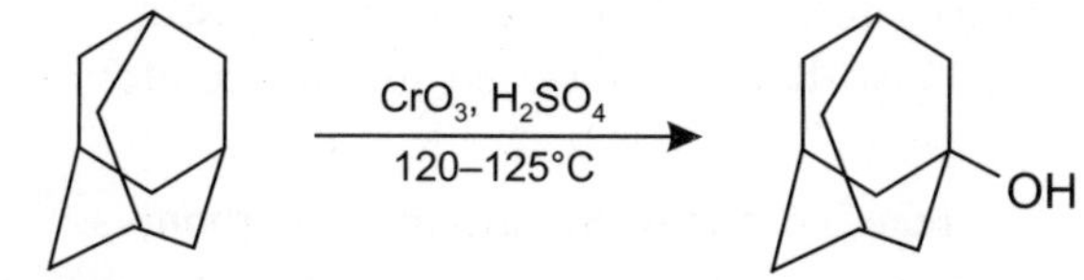

CrO_3, H_2SO_4
120–125°C

OH

Adamantane 1-Adamantol 52%

13. Oxidation of unsaturated alcohols at multiple bonds.

Oxidation of 1-(buten-3-yl) cyclohexanol with chromium trioxide in acetic acid and acetic anhydride results[36] in the formation of carboxylic acid, which cyclise to form γ-lactone in 50% yield.

$$\text{1-(Buten-3yl) cyclohexanol (HO, } CH_2CH_2CH{=}CH_2\text{ on cyclohexane)} \xrightarrow[10°C,\ 74\ hr.]{CrO_3,\ AcOH,\ Ac_2O} \text{Cyclohexane } \gamma\text{-lactone (50\%)}$$

1–(Buten-3yl) cyclohexanol

50%
Cyclohexane γ-lactone

In a similar way, 1-(penten-4-yl) cyclohexanol gives the corresponding δ-lactone.

14. Oxidation of ethers to carboxylic acids and esters.

Chromium trioxide in acetic acid converts[4] hexadecyl ethyl ether into palmitic acid and hexadecyl acetate.

$$\underset{\substack{33\% \\ \text{Hexadecyl} \\ \text{ethyl ether}}}{C_{16}H_{33}OC_2H_5} \xrightarrow[20°C,\ 1\ hr]{CrO_3,\ AcOH,\ CH_2Cl_2} \underset{\substack{33\% \\ \text{Palmitic} \\ \text{acid}}}{C_{15}H_{33}CO_2H} + \underset{\substack{9\% \\ \text{Hexadecyl} \\ \text{acetate}}}{C_{16}H_{33}OCOCH_3}$$

In a similar way, hexadecyl methyl ether is converted into palimitic acid and hexadecyl formate. In secondary benzyl ethers cleavage to the ketones usually predoaminates[31] over the formation of benzoates.

$$C_6H_{13}CH(CH_3)OCH_2C_6H_5 \xrightarrow[0°C,\ 12\ hr]{CrO_3, H_2SO_4,\ Me_2CO} \underset{79\%}{C_6H_{13}COCH_3} + \underset{21\%}{C_6H_5CO_2CH(CH_3)C_6H_{13}}$$

15. Oxidation of hydroxy ketones to diketones.

It frequently occurs in steroidal alcohols. In case the alcoholic group, usually secondary, is remote enough from the keto group, its oxidation takes place independently. Thus, a solution of chromium trioxide in aqueus sulphuric acid oxidises 5-pregnen-3-β-ol-20-one in acetone solution at room temperature within 2-5 min gives 5-pregnen-3, 20-dione in 90% yield[37a]. Similarly, 11β-hydroxytestosterone-17-acetate is converted into 11-ketotestosterone 17-acetate in 92% yield[38].

16. Oxidation of reactive methylene groups to carbonlys.

If the methylene group is flanked by two carbonyl groups (as in barbituric acid), it is oxidised[39] to the keto group by chromium trioxide.

$$\underset{\text{Barbituric acid}}{CO(NH{-}CO)_2CH_2} \xrightarrow[25 \to 50°C,\ >30\ mm]{CrO_3,\ H_2O,\ AcOH} CO(NH{-}CO)_2CO$$

Barbituric
acid

17. Oxidation of phenols to quinones Oxidation of 2, 4, 6-tribromo-*m*-cresol by chromium trioxide in 70% acetic acid at 70–75°C within 10 min gives dibromo-*m*-toluquinone (3, 5-dibromo-2-methyl-*p*-benzoquinone) in 77% yield[40].

$$\text{2,4,6-Tribromo } m\text{-cresol} \xrightarrow[\text{70–75°C, 10 min}]{CrO_3\text{, 70\% HOAc}} \text{Dibromo-}m\text{-toluquinone}$$

2,4,6-Tribromo *m*-cresol — Dibromo-*m*-toluquinone

It is thus seen that oxidation to p-benzoquinone occurs even if the para position to OH group is occupied by a substituent. The substitutent in the para position is either eliminated or modified.

18. Oxidation of unsaturated esters to unsaturated keto esters is accomplished by chromium trioxide in acetic acid and acetic anhydride. Thus, methyl 2-nonenoate is converted in benzene solution at 0 – 20°C into methyl 4-keto-2-nonenoate in 86% yield[10]:

$$\underset{\text{Methyl-2-nonenoate}}{CH_3CH_2CH_2CH_2CH_2CH_2\,CH{=}CH-COOCH_3} \xrightarrow[\text{0–20°C}]{CrO_3/HOAc} \underset{\substack{80\% \\ \text{4-Keto-2-nonenoate}}}{CH_3CH_2CH_2CH_2CH_2\,\underset{\|}{\overset{}{C}}{}\,CH{=}COOCH_3}$$

(C=O at the keto carbon) 80% 4-Keto-2-nonenoate

19. Chromium trioxide impregnated pre-moistered alumina in used as an efficient oxidising system[41] for the oxidation of benzyl alcohols to carbonyl compounds by simply mixing the substrate and the reagent at room temperature, and heating in a MW over for 40 seconds

$$\begin{matrix}R \\ R_1\end{matrix}\!\!>CH{-}OH \xrightarrow[\text{MW, 40 sec.}]{\text{Wet } CrO_3\text{-}Al_2O_3} \begin{matrix}R \\ R_1\end{matrix}\!\!>C{=}O$$

68-90%

R = Ph, p-MeC_6H_4, p-$MeOC_6H_4$, p-$NO_2C_6H_4$; R_1 = H
R = Ph, R_1 = Me,Ph, PhCO.

In the above oxididation reaction there is no over oxidation to carboxylic acids.

References

1. H.C. Brown, C.P. Garg and K. –T. Liu, *J. Org. Chem.*, 1971, **36**, 387.
2. E. Santaniello, F. Ponti and a. Manzocchi, *Synthesis*, 1978, 534.
3. S.J. Flatt, G.W. J. Fleet and BJ. Taylor, *Synthesis*, 1979, 815.

4. I.T. Harrison and S. Harrison, *Chem. Commun.*, 966, 752.
5. G. Cardillo, M. Orena and S. Sandri, *Tetrahedron Lett.*, 1976, 3983.
6. D. Edwards and J.B. Stenlake, *J. Chem. Soc.*, 1954, 3272.
7. G. Snatzke, *Chem. Ber.*, 1961, **94**, 729.
8. G. Cardillo, M. Orena and S. Sandri, *Synthesis*, 1976, 394.
9. D. Rosenthal, P. Grabowich, E.F. Sabo and J. Fried, *J. Am. Chem. Soc.*, 1963, **85**, 3971.
10. M. Nakayama, S. Shinke, Y. Matsushita, S. Ohina and S. Hayashi, *Bull. Chem. Soc. Jpn.*, 1979, **52** 184.
11. W.G. Durben, M. Lorber and D.S. Fullerton, *J. Org., Chem.*, 1969, **34**, 3587.
12. R.W. Ratcliffe, *Org. Synth.* Collective Volume, 1988, **6**, 373.
13. N. Miyaura and J.K. Kochi, *J. Am. Chem. Sec.*, 1983, **105**, 2368.
13*a*. R. Wendband and J. Lalonde, *Org. Synth.* Collective Volume, 1963, **4**, 713.
14. H.R. Rogers, J.X. Mc Dermott and G.M. Whitesides, *J. Org. Chem.*, 1975, **40**, 3577.
15. L.I. Smith and I.M. Webster, *J. Am. Chem. Soc.*, 1937, **59**, 662.
15*a*. E. Bamberger and C. Burgdorf, *Chem. Ber.*, 1890, **23**, 2433.
16. D. Pletcher and S.J.D. Tait, *J. Chem. Soc.*, Perkin Trans. 2, 1979, 788.
17. T. Brunelet, C. Jouitteau and G. Gelbard, *J. Org. Chem.*, 1986, **51**, 4016.
18. R. Rateliffe and R. Rodehorst, *J. Org. Chem.*, 1970, **35**, 4000.
19. J.R. Holum, *J. Org. Chem.*, 1961, **26**, 4814.
20. K.E. Harding, L.M. May and K.F. Dick, *J. Org. Chem.*, 1975, **40**, 1664.
21. J. Kuwajima, M. Shimizu and H. Urabe, *J. Org. Chem.*, 1982, **47**, 837.
22. D.G. Lee and U.A. Spitzer, *J. Org. Chem.*, 1970, **35**, 3589.
23. S. Cacchi, *Synthesis*, 1979, 356.
24. S.J. Flatt, G.W.J. Fleet and B.J. Taylar, *Synthesis*, 1979, 815.
25. J.C. Collins, W.W. Hess and F.J. Frank, *Tetrahedron Lett*, 1968, 3363.
26. F.S. Guziec, Jr, and F.A. Luzzio, *J. Org. Chem.*, 1982, **47**, 1787.
27. R.P. Singh, H.N. Subbarao and S. Dev, *Tetrahedron*, 1979, **35**, 1789.
28. G. Cainelli, G. Cardillo, M. Orena and S. Sandri, *J. Am. Chem. Soc.*, 1976, **98**, 6737.
29. E.J. Coery and G.W.J. Fleet, *Tetrahedron Lett.*, 1973, 4499.
30. G. Cairnelli, G. Cardillo, M. Orena and S. Sandri, *J. Am. Chem. Soc.*, 1976, 98.
31. J.R. Schaeffer and A.D. Snoddy, *Org. Synth,* Collective Volume, 1963, **4**, 11.
32. S.M. Tsang, E.H. Wood and J.R. Johnson, *Org. Synth.* Collective Volume, 1955, **3**, 641.
33. T. Nishimuna, *Org. Synth.* Collective Volume, 1963, **4**, 713.
34. S. Lieberman and R. Connor, *Org. Synth.* Collective Volume, 1943, **2**, 441.
35. S. Landa, J. Vais and J. Burkhand, *J. Chem.,* 1967, **7**, 233.
36. M.F. Schlecht and H.–J. Kim, *Tetrahedron Lett,* 1985, **26**, 127.
37. B.S. Bal, K.S. Kohhar and H.W. Pinnick, *J. Org. Chem.,* 1981, **46**, 1492.
37*a*. C. Djerasi, R. R. Engle and A. Bowers, *J. Org. Chem.,* 1956, **21**, 1547.
38. O. Manceva, G. Rosenkranz and F. Sondheimer, *J. Chem. Soc.,* 1953, 2189.
39. A.V. Holmgren and W. Wenner, Org. Synth. Collective Volume, 1963, **4**, 23.
40. L.I. Smith and D.V. Byers, *J. Am. Chem. Soc.,* 1941, **63**, 612.
41. R. S. Varma and R. K. Saini, *Tetrahedron Lett.*, 1998, **39**, 1481.

1.32 CHROMIUM TRIOXIDE-3,5-DIMETHYLPYRAZOLE COMPLEX, $CrO_3 \cdot C_5H_8N_2$

It is prepared[1] *in situ* by adding one mol of 3, 5-dimethylpyrazole to a suspension of one mol of chromium trioxide in dichloromethane at room temperature. The scope and limitations of this oxidant are similar to those of chromium trioxide-pyridine complex. Thus, it oxidises Primary alcohols to aldehydes. Some examples[1] are given below:

$$C_5H_{11}C \equiv C\,CH_2OH \xrightarrow[\text{RT. 0.5 hr.}]{C_5H_8N_2,\ CrO_3,\ CH_2Cl_2} C_5H_{11}C \equiv CCHO \quad 75\%$$

$$C_6H_5\,CH_2OH \xrightarrow[\text{RT. 45 min.}]{C_5H_8N_2,\ CrO_3,\ CH_2Cl_2} C_6H_5CHO \quad 83\%$$

$$PhCH{=}\,CHCH_2OH \xrightarrow[\text{RT. 45 min.}]{C_5H_8N_2,\ CrO_3,\ CH_2Cl_2} PhCH{=}\,CHCHO \quad 90\ \%$$

It also oxidises secondary alcohols to ketones. Some examples are given below:

$$CH_3CH(OH)\,C_6H_{13} \xrightarrow[\text{RT. 30 min}]{C_5H_8N_2.\ CrO_3,\ CH_2Cl_2} CH_3COC_6H_{13} \quad 93\ \%$$

$$C_6H_5CH(OH)\,CH_3 \xrightarrow[\text{RT. 30 min}]{C_5H_8N_2.\ CrO_3,\ CH_2Cl_2} C_6H_5\,CO\,CH_3 \quad 100\ \%$$

$$C_6H_5\,CH(OH)\,C_6H_5 \xrightarrow[\text{RT. 30 min}]{C_5H_5N_2.\ CrO_3,\ CH_2Cl_2} C_6H_5\,CO\,C_6H_5 \quad 98\ \%$$

$C_5H_8N_2 \equiv$ 3,5-dimethylpyrazole (ring with CH_3, H_3C, N, N–H)

Reference

1. E.J. Corey and G..W.J. Fleet, *Tetrahedron Lett.*, 1973, 4499.

1.33 CHROMIUM TRIOXIDE PYRDINE COMPLEX (COLLINS REAGENT), $CrO_3{\cdot}2C_5H_5N$

It is commercially available. It is prepared by adding one mole of chromum oxide to a solution of 2 mol of pyrdine in dichloromethane[1]. The complex may also be formed on the surface of silica gel by adding pyridine to silica gel impregnated with chromic oxide[2].

It oxidises primary alcohols to aldehydes. Some examples are given below:

$$C_7H_{15}CH_2OH \xrightarrow[\text{RT, 15, min}]{CrO_3.2C_5H_5N,\ CH_2Cl_2} C_7H_{15}CHO$$
90%

$$CH_2{=}CHCH_2OH \xrightarrow[\text{Cooling with ice, RT, 15-22 hr}]{CrO_3.2C_5H_5N,\ C_5H_5N} CH_2{=}CHCHO$$
63%

$$C_6H_5CH_2OH \xrightarrow[\text{5-15 min (collins reagent)}]{6CrO_3.2C_5H_5N,\ CH_2Cl_2.\ 25°C} C_6H_5CHO$$
95%

$$PhCH{=}CHCH_2OH \xrightarrow[\text{with ice, RT, 15-23 hr}]{CrO_3.2C_5H_5N,\ \text{Cooling}} PhCH{=}CHCHO$$
81%

It oxidises secondary alcohols to ketones. Some examples are given below:

$$CH_3CH(OH)C_6H_{13} \xrightarrow[\text{25°C, 5-15 min}]{6(2C_5H_5N.CrO_3),\ CH_2Cl_2} CH_3COC_6H_{13}$$
97%

$$\text{Cyclohexanol (C(H)(OH))} \xrightarrow[\text{25°C, 5-15 min}]{6(2C_5H_5N\ CrO_3),\ CH_2Cl_2} \text{Cyclohexanone (C=O)}$$
98%

$$C_6H_5CH(OH)C_6H_5 \xrightarrow[\text{25°C, 5-15 min}]{6(2C_5H_5N.CrO_3),\ CH_2Cl_2} C_6H_5COC_6H_5$$
96%

Collins reagent also oxidises methylene group next to double bond[5] or triple bonds[6] into carbonyl groups to yield α, β-unsaturated ketones.

References

1. R. Rateliffe and R. Rodehorst, *J. Org. Chem.*, 1970, **35**, 4000.
2. R.P. Singh, H.N. Subbarao and S. Dev, *Tetrahedron*, 1979, **35** , 1789.
3. J.R. Halum, *J. Org. Chem.*, 1961, **26**, 4814.
4. J.C. Collins, W.W. Hess and F.J. Frank, *Tetrahedron Lett.*, 1968, 3363.
5. W.G. Dauben, M. Lorber and D.S. Fullerton, *J. Org. Chem.*, 1969, 34, 3587.
6. J.E. Shaw and J.J. Sherry, *Tetrahedron Lett.*, 1971, 4379.

1.34 CHROMYL CHLORIDE, $Cr_2O_2Cl_2$·(b·p·117°C)

It is prepared[1] from chromium trioxide or sodium dichromate, hydrochloric acid and sulphuric acid.

It oxidises alkenes to aldehydes[2].

$$CH_2=C(CH_3)-CH_2C(CH_3)(CH_3)CH_2 \xrightarrow[(2)\ Zn]{(1)\ CrO_2Cl_2} OHC-CH(CH_3)-CH_2-C(CH_3)(CH_3)-CH_3$$

2,4,4-Trimethylpentene → 2,4,4-Trimethylpentanal

Cycloalkenes are oxidised to ketones. Thus, oxidation of *trans*-cyclododecene with chromyl chloride followed by treatment with zinc dust in acetic acid gives[3] cyclododecanone in 75% yield.

Chromyl chloride oxidises methyl groups in aromatic compounds to the corresponding aldehydes (**Etard reaction**).

$$O_2N\text{-}C_6H_4\text{-}CH_3 \xrightarrow[\text{RT, few days}]{CrO_2Cl_2,\ CS_2} O_2N\text{-}C_6H_4\text{-}CHO$$

p-Nitrotoluene → 60-70% *p*-Nitrobenzaldehyde

Oxidation of aromatic compounds with longer chains are not always suitable, since the reaction results in more than are product[4].

Chromyl chloride adsorbed on alumina-silica gel in dichloromethane oxidises[5] primary alcohols to aldehydes.

$$C_7H_{17}CH_2OH \xrightarrow[\substack{\text{adsorbed on } Al_2O_3\text{-}SiO_2 \\ \text{RT, 5 hr}}]{CrO_2Cl_2 \text{ in } CH_2Cl_2} \underset{94\%}{C_7H_{15}CHO}$$

In a similar way, secondary alcohols are oxidised to ketones[5]

$$CH_3\underset{\large OH}{\underset{|}{C}}HC_6H_{13} \xrightarrow[CH_2Cl_2,\ 25°C,\ 24\ hr]{CrO_2Cl_2/SiO_2\text{-}Al_2O_3} \underset{94\%}{CH_3COC_6H_{13}}$$

$$C_6H_5\underset{\large OH}{\underset{|}{C}}HCH_3 \xrightarrow[CH_2Cl_2,\ 25°C,\ 5\ hr]{CrO_2Cl_2/SiO_2\text{-}Al_2O_3} \underset{100\%}{C_6H_5COCH_3}$$

Trimethyl silyl enol ethers on treatment with chromyl chloride in dichloromethane at –78°C yields α-hydroxy ketones in 60 – 82% yields[6]

$$(CH_3)_3CCOCH_3 \xrightarrow[\text{MeCN, Reflux}]{Me_3SiCl,\ Et_3N} (CH_3)_3\underset{\large OSi(CH_3)_3}{\underset{|}{C}}{=}CH_2 \xrightarrow[-78°C,\ 30\ min]{CrO_2Cl_2/CH_2Cl_2} \underset{82\%}{(CH_3)_3CCOCH_2OH}$$

$$C_6H_5COCH_3 \xrightarrow[-78°C,\ 30°C\text{ to RT}]{\substack{\text{LDA, THF} \\ Me_3SiCl}} C_6H_5\overset{\large OSi(CH_3)_3}{\overset{|}{C}}{=}CH_2 \xrightarrow[-78°C,\ 30\ min]{CrO_2Cl_2/CH_2Cl_2} \underset{60\%}{C_6H_5COCH_2OH}$$

References

1. H.H. Sisler, *Inorg. Synth.*, 1946, **2**, 205.
2. F. Freeman, R.H. Dubois and T.G. McLaughlin, *Org. Synth.* Collective Volume, 1988, **6**, 1028.
3. K.B.Sharpless and A.Y. Teranishe, *J. Org. Chem.,* 1973, **38**, 185.
4. V.W. Miller and G. Rohde, *Chem. Ber.*, 1890, **23**, 1070.
5. J. San Filippo, Jr. and C.-I. Chern, *J. Org. Chem.,* 1977, **42**, 2182.
6. T. V. Lee and J. Toczek, *Tetrahedron Lett.,* 1982, **23**, 2917.

1.35 COPPER ACETATE, $Cu(OCOCH_3)_2$ or $Cu(OCOCH_3)_2 \cdot H_2O$

As in the case of copper sulphate (*see* copper sulphate), copper acetate is used for the oxidative coupling of acetylenes[1].

$$(CH_3)_2\underset{\large OH}{\underset{|}{C}}\,C \equiv CH \xrightarrow[H_2O,\ AcOH]{Cu(OCOCH_3)_2} (CH_3)_2\underset{\large OH}{\underset{|}{C}}C \equiv C\underset{\large OH}{\underset{|}{C}}(CH_3)_2$$

It is also used for the conversion of acyloins into α-diketones[2]. Its presence favourably affects the acetoxylation of toluene to benzyl acetates by sodium persulphate.[3]

References

1. G.. Eglintion and A.R. Galbraith, *J. Chem. Soc.*, 1959, 889; I.D. Campbell and G. Eglinton, *Org. Synth. Collective* Volume, 1973, **5**, 517.
2. P. Ruggli and P. Zeller, *Helv. Chem. Acta*, 1945, **28**, 741; A.T. Blomquist and A. Goldstein, *Org. Synth*. Collective Volume, 1963, **4**, 838.
3. A. Belli and C. Giordano, *Synthesis*, 1980, 477.

1.36 COPPER CHROMITE, $CuCr_2O_4$

It is used for the dehydrogenation of primary alcohols to aldehydes[1] and secondary alcohols to ketones[1, 2].

References

1. R.E. Dunbar and M.R. Arnold, *J. Org. Chem.,* 1945, **10**, 501; R.I. Longley, Jr., and W.S. Emerson, *Org. Synth,* Collective Volume, 1963, **4**, 677.
2. W.R. Nes, *J. Org. Chem.,* 1958, **23**, 899.

1.37 COPPER OXIDE, CuO

Primary alcohols are catalytically dehyrogenated by passing vapours of alcohols at 275 – 350°C over copper oxide catalyst supported on asbestos. Thus, ethyl alcohol is converted into acetaldehyde in 88% yield at 93% conversion by passing its vapours at 275°C over mixtures of oxides of copper, cobalt and chromium on asbestos[1].

Vapour phase dehydrogenation of primary alcohols to aldehydes takes place at 250 – 300°C over cupric oxide in a current of helium[2].

$$CH_2 = C(CH_3)CH_2OH \xrightarrow{Cu,\ 350°C,\ air} CH_2 = C(CH_3)CHO$$

Secondary alcohols are oxidised[2] to ketones in vapour phase of 250 – 300°C

$$\text{Cyclohexanol} \xrightarrow{CuO,\ He,\ 250\text{-}300°C} \text{Cyclohexanone (90\%)}$$

Cyclohexanol

Cyclohexanone
90%

Copper oxide also transforms hydrazo compounds into azo compounds[3]:

$$C_6H_5NHNHC_6H_5 \xrightarrow[250\text{-}300°C]{CuO} C_6H_5N = NC_6H_5$$

References

1. J.M. Church and H.K. Joshi, *Ind. Eng. Chem.*, 1951, **43**, 1804.
2. M.Y. Sheikh and G. Eadon, *Tetrahedron Lett.*, 1972, 257.
3. H.L. Lochte, W.A. Noyes and J.R. Bailey *J. Am. Chem. Soc.*, 1922, **44**, 2556.

1.38 COPPER CORBONATE, BASIC, $CuCO_3 \cdot Cu(OH)_2$

It is useful for the hydroxlation of benzoic acids exclusively in the ortho position to produce salicylic acids[1].

Reference

1. W.W. Kaeding and G.R. Collins, *J. Org. Chem.*, 1965, **30**, 3750.

1.38A COPPER NITRATE

Copper (II) nitrate impregnated on K 10 clay (claycop)-hydrogen peroxide is an effective reagent for the oxidation[1] of a variety of substrates. In these cases, the maintenance of pH is not required.

$$(R_2)(R_1)CH-R_3 \xrightarrow[Claycop\text{-}H_2O_2]{MW} (R_2)(R_1)C=O \quad 91\text{-}95\%$$

R_1 = H, p-$NO_2C_6H_4$
R_2 = H, Ph
R_3 = H, Br, COOH, NH_2

Reference

1. R. S. Varma and R. Dahiya, Tetrahedron Lett., 1998, **39**, 1307.

1.39 COPPER SULPHATE, $CuSO_4 \cdot 5H_2O$

It is used along with $K_2S_2O_8$ for regiospecific oxidation of dimethylanisoles to methoxymethyl benzaldehydes; this reagent selectively oxidises only the methyls in the ortho or para position with respect to methoxy group[1].

$$CH_3O\text{-}C_6H_3(CH_3)\text{-}CH_3 \xrightarrow[\text{MeCN/H}_2\text{O (1:1), reflux, 15-30 min}]{CuSO_4.5H_2O,\ K_2S_2O_8} CH_3O\text{-}C_6H_3(CH_3)\text{-}CHO$$

3,4-Dimethyl anisole → 2-Methyl-4-methoxy benzaldehyde 92%

Oxidative coupling of acetylenic alcohols to diols takes place with copper sulphate[2].

$$(CH_3)_2\underset{\displaystyle OH}{\underset{|}{C}}C{\equiv}CH \xrightarrow[H_2O,\ AcOH]{CuSO_4.5H_2O,\ AcOH} (CH_3)_2\underset{\displaystyle OH}{\underset{|}{C}}C{\equiv}C{-}C{\equiv}C{-}\underset{\displaystyle OH}{\underset{|}{C}}(CH_3)_2$$

58%

Copper sulphate is also used for the conversion of α-hydroxy ketones (acyloins) to α-diketones[3].

References

1. F.M. Hauser and S.R. Ellenberger, *Synthesis*, 1987, 723.
2. G. Eglinton and A.R. Galbraith, *J. Chem. Soc.*, 1959, 889.
3. W.W. Hartman and J.B. Dickey, *J. Am. Chem. Soc.*, 1933, **55**, 1228; T.S. Gardner, E. Wenis and J. Lee, *J. Org. Chem.*, 1958, **23**, 823.

1.40 CUPRIC PERMANGANATE, $Cu(MnO_4)_2.8H_2O$

It is commercially available. It oxidises primary alcohols to carboxylic acids. Thus, decanol and benzyl alcohol are oxidised[1] with cupric permanganate to yield decanoic acid and benzoic acid, respectively at room temperature after 24 hrs (yields 91 and 84%, respectively).

In a similar way, aldehydes are oxidised[1] to carboxylic acids and secondary alcohols to ketones.

Reference

1. N.A. Nouveldin and D.G. Lee, *J. Org. Chem.*, 1982, **47**, 2790.

1.41 CUPROUS CHLORIDE, CuCl

Alcohols are oxidised[1] to ketones by passing oxygen at room temperature under irradiation through a solution of catalytic amount of chloroplatanic acid and cuprous chloride, the yield is 97%.

The oxidation[2] of o-phenylene diamine with oxygen in presence of cupric chloride in pyridine at 20°C yields mueononitile (1, 4-dicyano-1, 3-butadiene).

$$o\text{-}C_6H_4(NH_2)_2 \xrightarrow[\text{pyridine, 20°C}]{O_2,\ CuCl} CNCH{=}CH{-}CH{=}CHCN$$

1,4-Dicyano-1,3-butadiene

References

1. R.P.A. Sneeden and R.B. Turner, *J. Am. Chem. Soc.*, 1955, **77**, 190.
2. J. Tsuji and H. Kakayangi, *Org. Synth.*, Collective Volume, 1988, **6**, 662.

1.42 DIMETHYLAMINOPYRIDINIUM CHLOROCHROMATE, 4-$(CH_3)_2NC_5H_4NH \cdot CrO_3Cl$

It is prepared[1] by the addition of one mol of 4-dimethylaminopyridine to one mol of chromium trioxide in dilute hydrochloric acid. It is used to oxidise[1] allylic and benzylic alcohols to aldehydes in dichloromethane at room temperature.

$$(CH_3)_2C{=}CHCH_2CH_2C(CH_3){=}CHCH_2OH \xrightarrow[\text{[O]}]{Me_2N\text{-}C_5H_4N.HCl,\ CrO_3,\ CH_2Cl_2,\ RT.15\ hr} (CH_3)_2C{=}CHCH_2CH_2C(CH_3){=}CHCHO$$

$$C_6H_5CH_2OH \xrightarrow[CH_2Cl_2,\ RT,\ 3\ hr.]{3CrO_3,\ HCl,\ NC_5H_4\text{-}NMe_2} \underset{64\%}{C_6H_5CHO}$$

Saturated alcohols are not oxidised by this reagent.

Reference

1. F.S. Guzier, Jr., and F.A. Luzzio, *Synthesis*, 1976, 809.

1.43 DESS-MARTIN "PERIODINANE"

It is prepared[1] by the oxidation of o-iodobenzoic acid with potassium bromate-sulphuric acid followed by acetylation of the product (*see* potassium bromate).

$$\underset{\text{o-Iodobenzoic acid}}{o\text{-}IC_6H_4COOH} \xrightarrow[H_2SO_4]{KBrO_3} \underset{93\%}{C_6H_4(CO)O\text{-}I(O)OH} \xrightarrow[100°C,\ 40\ min]{Ac_2O,\ AcOH} \underset{\text{Dess-Martin reagent}}{C_6H_4(CO)O\text{-}I(OCOCH_3)_3}$$

It is used for the oxidation of alcohols under mild conditions and in high yields.[1]

$$C_7H_{15}CH_2OH \xrightarrow[AcOH,\ CH_2Cl_2,\ 25°C,\ 1\ hr.]{\text{Dess-Martin reagent}} \underset{93\%}{C_7H_{15}CHO}$$

$$\text{Cyclohexanol} \xrightarrow[CH_2Cl_2,\ RT,\ 30\ min;\ NaOH\ 10\ min]{\text{Dess-Martin reagent}} \underset{90\%}{\text{Cyclohexanone}}$$

Reference

1. D. B. Dess and J.C. Martin, *J. Org. Chem.*, 1983, **48**, 4155; R. J. Linderman and D. M. Graves, Tetrahedron Lett., 1987, **28**, 4259.

1.44 DIAZOMETHANE, CH_2N_2

Diazomethane, a toxic and highly reactive gas is used in ether solution. It is prepared[1] by warming N-methyl-N-nitrosourethane (which is preaped by the reaction of methylamine with ethylchloroformate followed by treatment with HNO_2) with methanolic potassium hydroxide.

$$\underset{\text{Methyl amine}}{CH_3NH_2} + \underset{\text{Ethyl chloro formate}}{ClCOOC_2H_5} \longrightarrow CH_3NHCOOC_2H_5 \xrightarrow{HNO_2} \underset{\text{M-methyl-N-nitroso-urethane}}{CH_3\text{—}N(NO)\text{—}COOC_2H_5}$$

$$\underset{\text{N-methyl-N-nitroso-urethane}}{CH_3\text{—}N(NO)\text{—}COOC_2H_5} \xrightarrow{KOH} \underset{\text{Diazo methane}}{CH_2N_2} + CO_2 + C_2H_5OH$$

Diazomethane is represented by several resonance hybrid structures.

$$CH_2{=}\overset{+}{N}{=}\overset{-}{N} \longleftrightarrow \overset{-}{C}H_2\text{—}N{=}\overset{+}{N} \longleftrightarrow \overset{-}{C}H_2\text{—}\overset{+}{N}{\equiv}N \longleftrightarrow \overset{+}{C}H_2\text{—}N{=}\overset{-}{N}$$

It is extensively used for methylation of compounds containing acidic hydrogen *i.e.*, carboxylic acids, phenols and enols. It has following two important applications.

1. Homologations: Diazomethane reacts with aldehydes to give methyl ketones[2].

$$\underset{\text{Benzaldehyde}}{C_6H_5\text{—}C({=}O)\text{—}H} \xrightarrow{CH_2N_2} \underset{\text{Acetophenone}}{C_6H_5\text{—}C({=}O)\text{—}CH_3} \longrightarrow \underset{\text{Epoxide}}{C_6H_5\text{—}\underset{\diagdown O \diagup}{CH\text{—}CH_2}}$$

In a similar way, ketones are converted with their higher homologues.

$$R\text{—}C({=}O)\text{—}R' \xrightarrow{CH_2N_2} R\text{—}C({=}O)\text{—}CH_2R' \longrightarrow R\text{—}\underset{\diagdown O \diagup}{C(R')\text{—}CH_2}$$

In both the above cases, the yields are low due a formation of epoxides as predominant products.

Certain cyclic ketones undergo ring enlargement on treatment with diazomethane. Thus, cyclohexanone is converted into cycloheptanone[3].

$$\text{Cyclohexanone} \xrightarrow[BF_3]{CH_2N_2} \underset{65\%}{\text{Cycloheptanone}} + \underset{\text{epoxide}}{15\%}$$

The mechanism of homologation is as given below.

$$R{-}\overset{O}{\overset{\|}{C}}{-}R' \xrightarrow{CH_2{=}\overset{+}{N}{=}\overset{-}{N}} R{-}\underset{R'}{\overset{O^-}{C}}{-}CH_2{-}\overset{+}{N}{\equiv}N \ \text{(betanin)}$$

$$\downarrow -N_2$$

$$RCH_2{-}\overset{O}{\overset{\|}{C}}{-}R' \xleftarrow{\text{rearrangement}} R{-}\underset{R'}{\overset{\bar{O}}{C}}{-}\overset{+}{C}H_2 \xrightarrow[\text{formation}]{\text{epoxide}} R{-}\underset{R'}{C}\overset{O}{\frown}CH_2$$

2. Arndt-Eistert Reaction: In this reaction, a carboxylic acid is converted[4] into its next higher homologue or to a derivative of the homologous acid like an amide or ester. The various steps involved are:

$$RCOOH \xrightarrow{SOCl_2} RCOCl \xrightarrow{CH_2N_2} \underset{(\alpha\text{-Diazo ketone})}{RCOCHN_2} \xrightarrow{Ag_2O} RCH_2COOH$$

$$RCOCHN_2 \begin{cases} \xrightarrow{ROH/Ag_2O} \underset{\text{Ester}}{RCH_2COOR} \\ \xrightarrow{NH_3/Ag_2O} \underset{\text{Amide}}{RCH_2CONH_2} \end{cases}$$

The last step, *i.e.*, conversion of α-diazoketone into carboxylic acid ester or amide is called **Wolff rearrangement**.

The mechanism of Arndt Eistert reaction is given below:

$$R{-}\overset{O}{\overset{\|}{C}}{-}Cl + \overset{-}{C}H_2{-}\overset{+}{N}{\equiv}N \longrightarrow R{-}\overset{O}{\overset{\|}{C}}{-}CH_2{-}\overset{+}{N}{\equiv}N$$

$$R{-}\overset{O}{\overset{\|}{C}}{-}CH_2{-}\overset{+}{N}{\equiv}N \xrightarrow{\overset{-}{C}H_2{-}\overset{+}{N}{\equiv}N} \underset{\alpha\text{-diazoketone}}{R{-}\overset{O}{\overset{\|}{C}}{-}\overset{-}{C}H{-}\overset{+}{N}{\equiv}N} + CH_3{-}\overset{+}{N}{\equiv}N$$

$$R{-}\overset{O}{\overset{||}{C}}{-}\overset{-}{C}H{-}\overset{+}{N}{\equiv}N \xrightarrow{Ag_2O} \left[R{-}\overset{O}{\overset{||}{C}}{-}\ddot{C}H\right] \longrightarrow R{-}\overset{H}{\overset{|}{C}}{=}C{=}O$$

carbene ketone

$$R{-}\overset{H}{\overset{|}{C}}{=}C{=}O \xrightarrow{H_2O} RCH_2COOH$$

ketone — Carboxylic acid

$$\xrightarrow{R'OH} RCH_2COOR'$$

Ester

$$\xrightarrow{NH_3} RCH_2CONH_2$$

Amide

References

1. H. Von. Pechamann. Bev., 1894, **27**, 1888.
2. C.D. Gulsche, *Org. React.*, 1954, **8** , 364.
3. H.O. House, E.J. Grubbs and W.F. Cannon, *J. Am. Chem. Soc.*, 1960, **82**, 4099.
4. F. Arndt and B. Eistert, Ber., 1935, **68**, 200.

1.45 N N–DICYCLOHEXYLCARBODIMIDE (DCC), $H_{11}C_6 - N = C = N - C_6H_{11}$

It is commercially available and can be prepared by the oxidation of dicyclo hexylurea with *p*-toluenesulfonyl chloride in hot pyridine.

$$\underset{\text{Dicyclohexylurea}}{C_{11}C_6NHCONHC_6H_{11}} \xrightarrow[\text{hot } C_5H_5N]{p\text{-}CH_3C_6H_4SO_2Cl} \underset{\text{DCC}}{H_{11}C_6N{=}C{=}NC_6H_{11}} + H_2O$$

DCC is a powerful dehydrating agent. Some of the applications of DCC are given below:

1. Synthesis of Peptides[1] : DCC is used for the coupling of amino acids for the formation of amide bonds. For example, for the synthesis of a dipeptide, from two amino acids, the amino group of one of the amino acid and the carboxyl group of the other carboxylic group must be protected before the two are brought together in presence of DCC.

$$P{-}NH{-}CH(R'){-}COOH + H_2N{-}CH(R){-}COOR \xrightarrow{DCC} P{-}NH{-}CH(R'){-}CO{-}NH{-}CH(R){-}COOR$$

P = protecting group

α peptide

2. Synthesis of Esters[2]**:** Carboxylic acids and alcohols react in presence of DCC to give esters. The reaction proceeds by the reaction of carboxylic acid with DCC to form a strong acetylating agent, which reacts with alcohols as shown below.

$$R{-}\overset{O}{\overset{\|}{C}}{-}OH \ (\text{Carboxylic acid}) + H_{11}H_6N=C=NC_6H_{11} \ (\text{DCC}) \rightleftharpoons R{-}\overset{O}{\overset{\|}{C}}{-}O{-}C(=N{-}C_6H_{11})NHC_6H_{11} \ (\text{Strong acetylating reagent}) \xrightleftharpoons{H^+} R{-}\overset{O}{\overset{\|}{C}}{-}O{-}C(={}^{+}NH{-}C_6H_{11})NHC_6H_{11} \ (A)$$

$$(A) + R'{-}\ddot{O}{-}H \ (\text{Alcohol}) \longrightarrow R(R'O)C(O^-){-}O{-}C(={}^{+}NH{-}C_6H_{11})NHC_6H_{11} \longrightarrow R{-}\overset{O}{\overset{\|}{C}}{-}OR' \ (\text{Ester}) + H_{11}C_6NH{-}\overset{O}{\overset{\|}{C}}{-}NHC_6H_{11}$$

The above reaction can also be used for the synthesis of amides. Thus, the intermediate (A) obtained above rects with an amine to give an amide.

$$R{-}\overset{O}{\overset{\|}{C}}{-}O{-}C(={}^{+}NH)NHC_6H_5 \ (A) + R'{-}\ddot{N}H_2 \longrightarrow R(R'N)C(O^-){-}O{-}C(={}^{+}NH{-}C_6H_5)NH{-}C_6H_5 \xrightarrow{-H^+} H_{11}C_6NH{-}\overset{O}{\overset{\|}{C}}{-}NHC_6H_{11} + R{-}\overset{O}{\overset{\|}{C}}{-}NR' \ (\text{amide})$$

3. Synthesis of Peroxides[3] by reaction of carboxylic and with hydrogen peroxide in the presence of DCC.

$$2ArCOOH \ (\text{Carboxylic acid}) + H_2O_2 \xrightarrow{DCC} Ar{-}\overset{O}{\overset{\|}{C}}{-}O{-}O{-}\overset{O}{\overset{\|}{C}}{-}Ar \ (\text{Peroxide})$$

4. Dehydroxylation of Alcohols[4]: The O-acylurea adduct, formed from the reaction of alcohol or phenol with DCC can be hydrogenated to yield the corresponding hydrocarbon.

$$\text{R—OH} + H_{11}C_6N=C=NC_6H_{11} \longrightarrow \text{R–O–C}(=N\text{–}C_6H_{11})\text{NHC}_6H_{11} \xrightarrow[\text{Pd/C}]{H_2} \text{R—H}$$

O-acylurea

5. Synthesis of α, β Unsaturated Ketones[5] and Esters[6]: Dehydration of β-hydroxyketones and β-hydroxyesters can conveniently brought about with DCC to form α, β unsaturated ketones and esters, respectively.

$$\beta\text{-Hydroxyketone} \xrightarrow[-H_2O]{\text{DCC}} \alpha,\beta\text{-Unsaturated ketone}$$

$$\beta\text{-Hydroxyester} \xrightarrow[-H_2O]{\text{DCC}} \alpha,\beta\text{-Unsaturated ester}$$

6. Synthesis of Ethers[7] and Thioethers[8]: Ethers and thioethers can be prepared from alcohols and thioalcohols by reaction with an alcohol in presence of DCC.

$$\text{Ar—OH} + \text{MeOH} \xrightarrow{\text{DCC}} \text{Ar—O—Me}$$

$$\text{Ar—SH} + \text{MeOH} \xrightarrow{\text{DCC}} \text{Ar—S—Me}$$

7. Synthesis of α-diazoketones: α-Diazoketone can be synthesised by the coupling of diazomethane with carboxylic acid in presence of DCC (*see* also diazomethane).

$$\text{RCOOH} + CH_2N_2 \xrightarrow{\text{DCC}} \text{RCOCHN}_2$$

References

1. M. Bodanszky, Peptide Chemistry, *A Practical Textbook*, Springer, New York 1988.
2. B. Neises and W Steglich, OS, 1985, **63**, 183.
3. F.D. Greene and J. Kazan, *J. Org. Chem.*, 1963, **28**, 2168.

4. E. Vowinkel and I. Butha, CB, 1974, **107**, 1353.
5. C. Alaxandre and F. Rouesse, CR(C), 1972, **274**, 1837.
6. C. Alaxandre and F. Rouesse, BSF (2), 1971, 1837.
7. E. Vowinkel, C.B., 1966, **99**, 42, 1499.
8. E. Vowinkel, C. Wolff, C.B., 1974, 107, 496.
9. D. Hodson, G. Holl and D.K. Wall, *J. Chem., Soc.*, 1970, 971.

1.46 DDQ (2, 3-DICHLORO-5, 6-DICYANO-P-BENZOQUINONE)

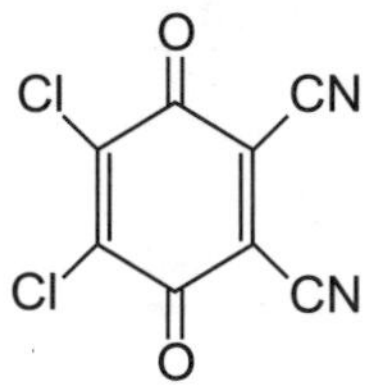

DDQ is commonly used for dehydrogenation, particularly in cases where substrate is difficult to dehydrogenate. It is used under anhydrous conditions, since it decomposes in presence of water. The mechanism of dehydrogenation with DDQ involves transfer of hydride ions from hydrocarbon to the quinone oxygen, followed by the transfer of a proton to the phenolate ion. Thus, DDQ is reduced to hydroquinone.

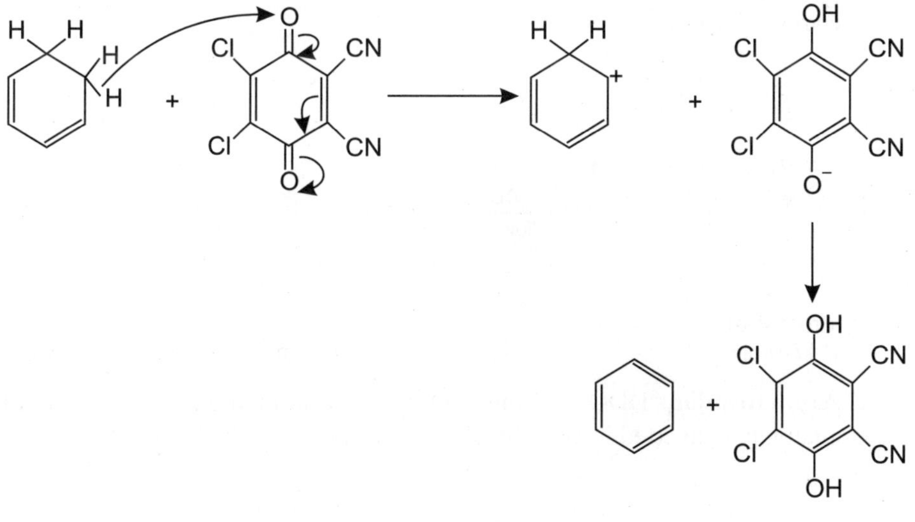

DDQ is easily available. However, it is synthesised as follows:

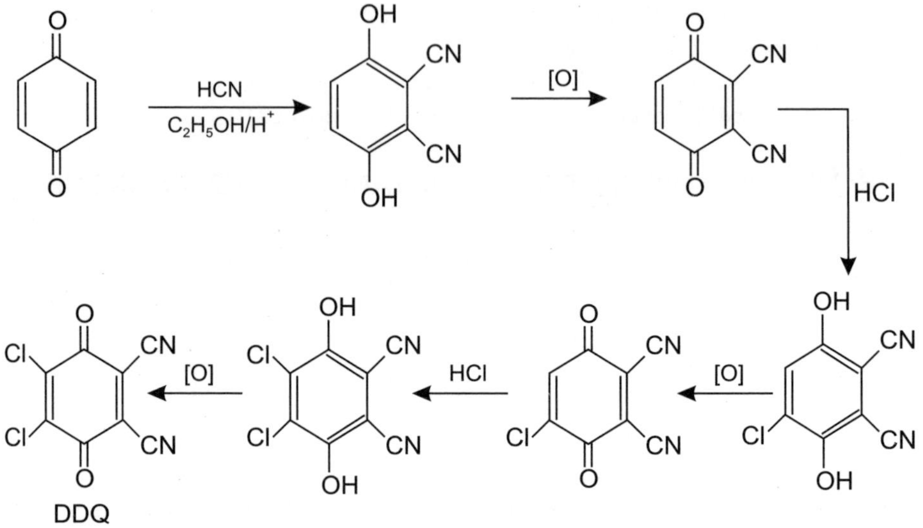

HCN
C_2H_5OH/H^+
[O]
HCl
[O]
HCl
[O]
DDQ

Some of the important applications of DDQ are given below:

1. Dehydrogenation: Generally dehydrogenations result in the formation of carbon-carbon double bonds. Thus, androst-4-en-3, 17-dione on heating with DDQ in dioxane in presence of gaseous hydrogen chloride gives a 72% yield of androsta-4, 6-diene-3, 17-dione.[2]

DDQ, HCl
dioxane, 30 min

Androst-4-en-
-3, 17-dione

72%
Androsta-4, 6-diene-3, 17-dione

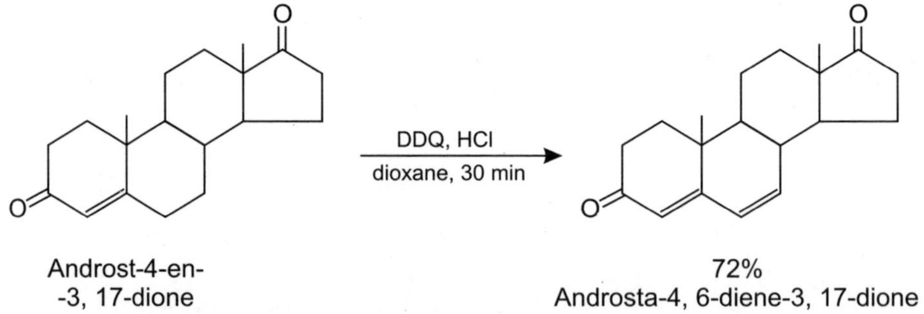

2. Aromatisation: DDQ is commonly used for aromatisation. Thus, tetralin is quantitatively converted into naphthalene by refluxing with DDQ in benzene[3].

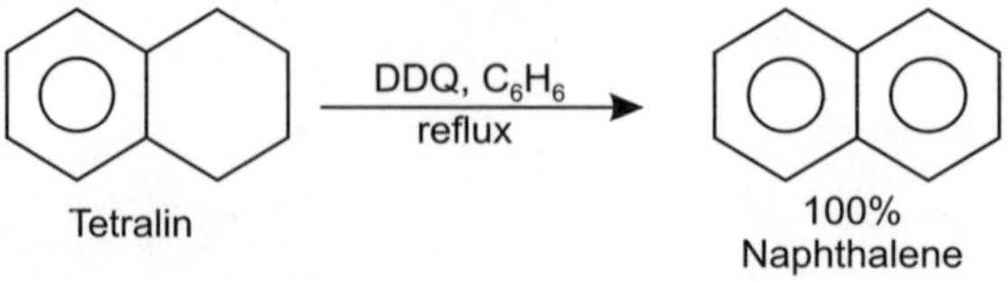

DDQ, C_6H_6
reflux

Tetralin

100%
Naphthalene

In a similar way acenaphthene is converted into acenaphthylene.

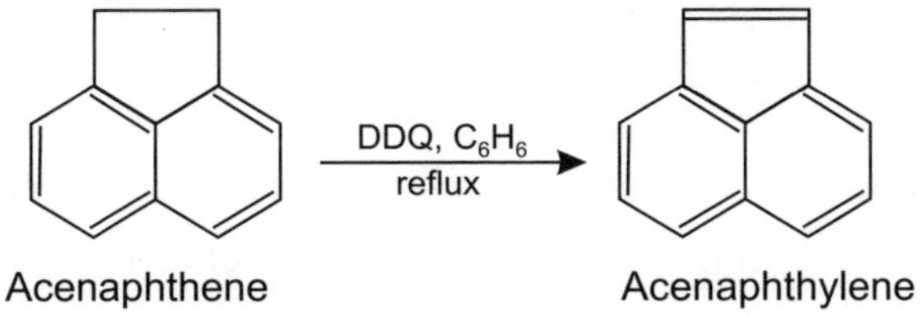

Acenaphthene $\xrightarrow[\text{reflux}]{\text{DDQ, } C_6H_6}$ Acenaphthylene

3. Oxidation of allylic alcohols[4]

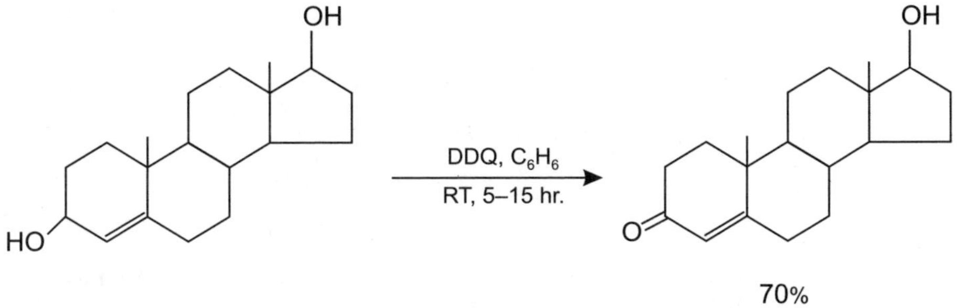

$\xrightarrow[\text{RT, 5–15 hr.}]{\text{DDQ, } C_6H_6}$

70%

Allylic alchols are more readily oxidised than saturated secondary alcohols, which in turn are more readily[4] oxidised than saturated primary alcohols.

4. Oxidation of phenols to quninones: DDQ is a powerful oxidizing agent for phenols, which are converted into ortho and para quniones. Oxidation to *p*-benzoquinones occur even if the para position is occupied by a substituent and even if an orthoposition is free.[5]

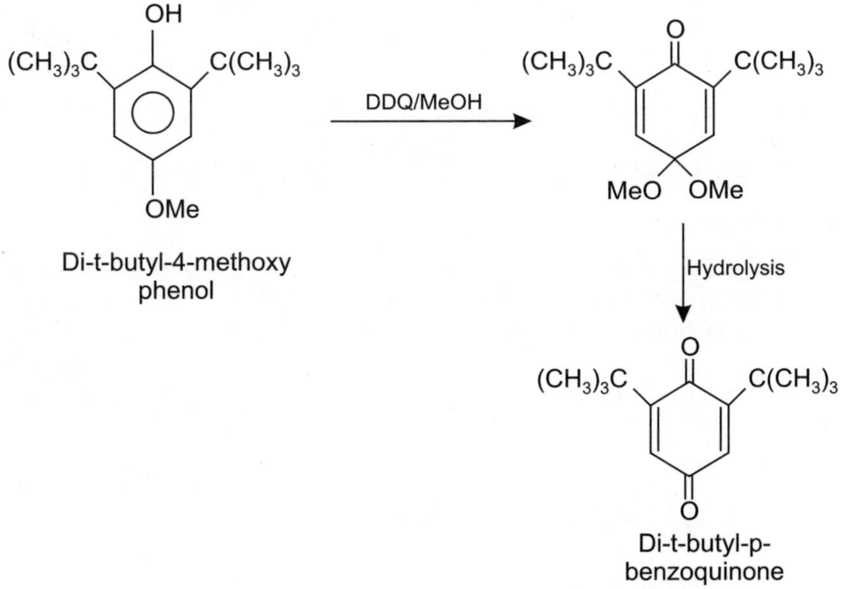

OH, $(CH_3)_3C$, $C(CH_3)_3$, OMe

Di-t-butyl-4-methoxy phenol

$\xrightarrow{\text{DDQ/MeOH}}$

O, $(CH_3)_3C$, $C(CH_3)_3$, MeO OMe

Hydrolysis

O, $(CH_3)_3C$, $C(CH_3)_3$, O

Di-t-butyl-p-benzoquinone

Mesitol on oxidation with DDQ gives 3, 5-dimethyl-4-hydroxy benzaldehyde.

Mesitol

DDQ

CH_3OH

H_2O

3,5-Dimethyl -4-hydroxybenzaldehyde

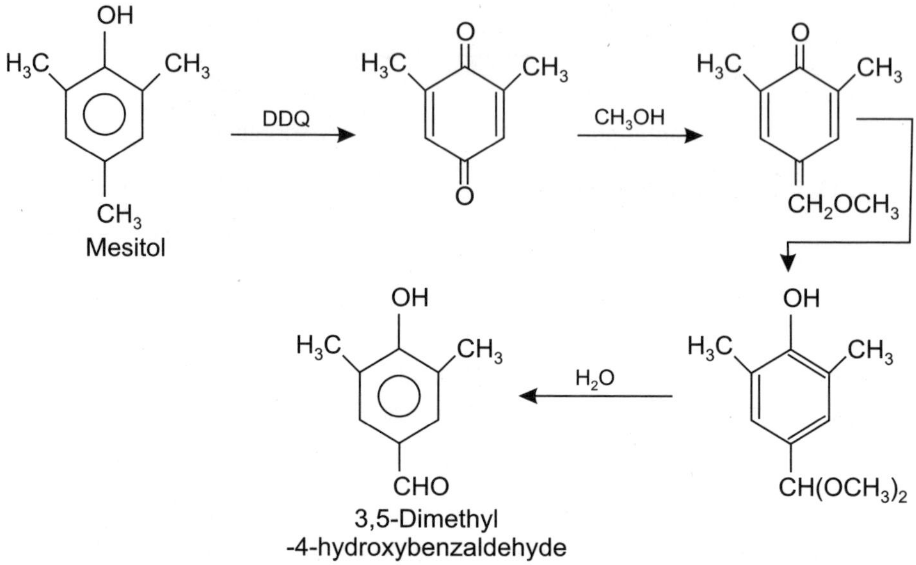

Sometimes, the oxidation to quinones is accompnied by doubling of the molecule[6].

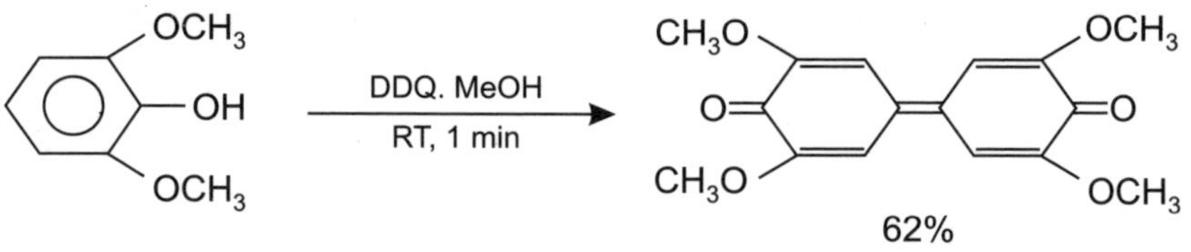

DDQ. MeOH RT, 1 min

62%

5. Oxidation of primary alcohols to aldehydes[4]

$$C_6H_5CH{=}CH\,CH_2OH \xrightarrow[15\text{ hr.}]{DDQ.CCl_4} C_6H_5CH{=}CHCHO$$

6. Oxidative cyclisation[7]

An important use of DDQ is found in cyclodehydrogenation reactions. Thus, 8-diphenylmetyl-1-naphthoic acid on refluxing with DDQ in benzene undergoes oxidative cyclisation to δ-lactone.

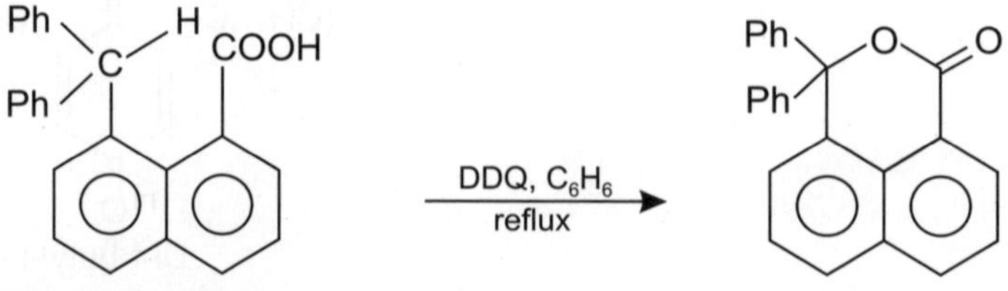

DDQ, C_6H_6 reflux

8-Diphenylmethyl -1-naphthoic acid

δ-lactone

Some other examples of cyclodehydrogenation are given below:

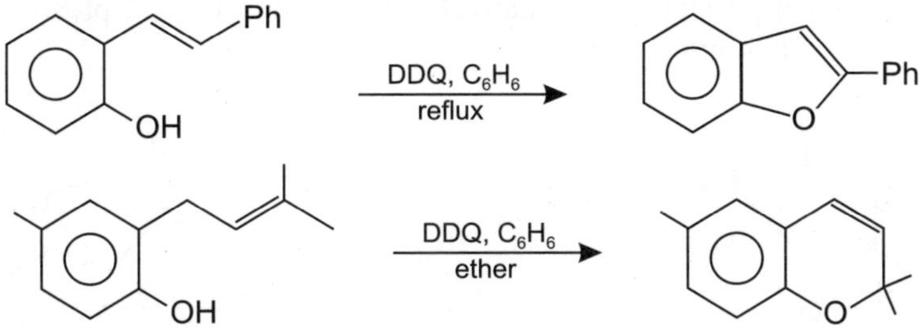

DDQ, C_6H_6
reflux

Ph

DDQ, C_6H_6
ether

OH

7. Isomerisation: The hydroaromatic compounds with a blocking group, *e.g.*, 1, 1-dimethyl-1, 2, 3, 4-tetrahyaphthalene undergoes aromatisation[8] with a 1, 2-shift of a methyl group to give 1, 2-dimethylnaphthalene. The mechanism of the reaction is given below:

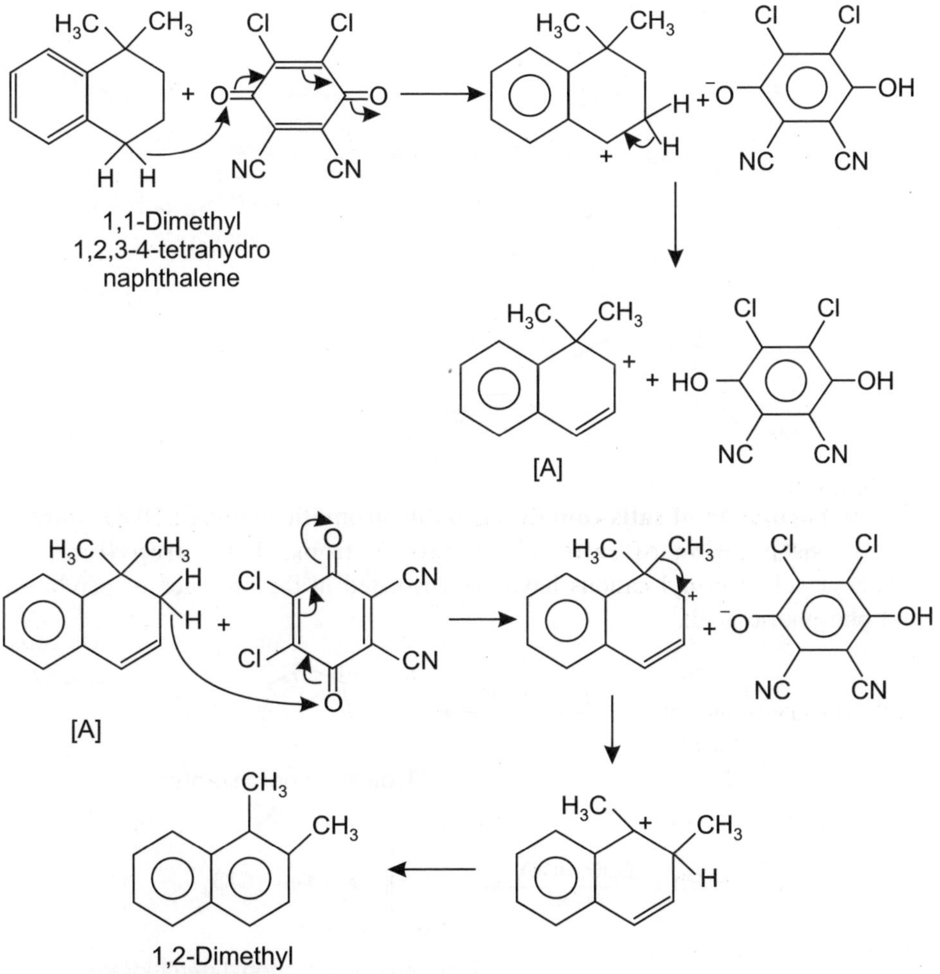

1,1-Dimethyl
1,2,3-4-tetrahydro
naphthalene

[A]

[A]

1,2-Dimethyl
naphthalene

8. Oxidative Dimerisation: Enols and enolisable ketones in which α, β-unsaturation is not possible are oxidatively dimerised[6]. Thus, 2-phenylindane-1, 3-dione is converted into tetrone.

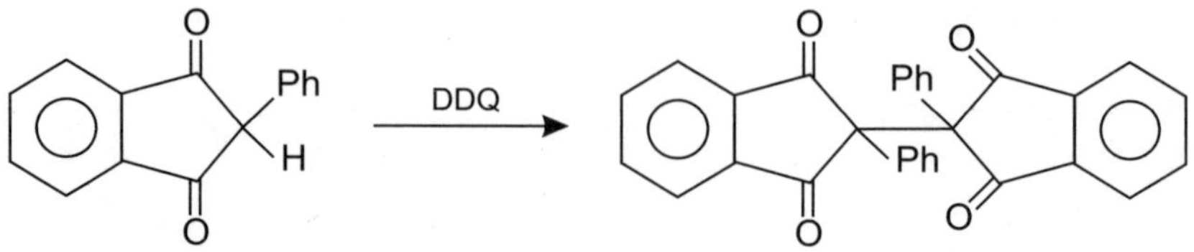

2-Phenylindane –1,3-dione —DDQ→ Tetrone

9. Oxidation[9] of benzylic and allylic alkyl groups: The benzylic and allylic alkyl groups are oxidised by DDQ to carbonyl group. Some examples are given below:

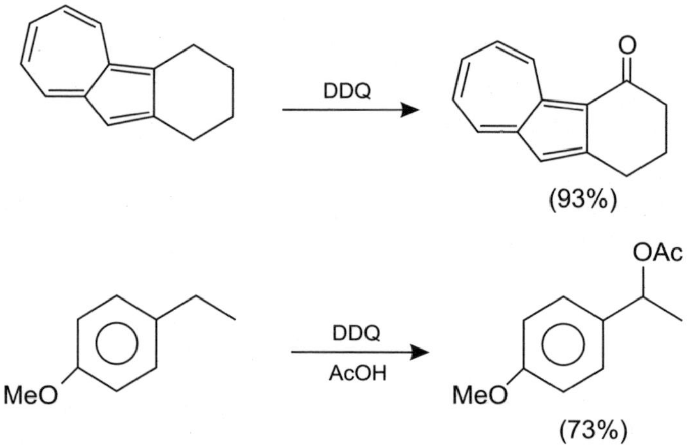

DDQ (93%)

DDQ, AcOH (73%)

10. Formation of salts containing stable aromatic cations[7]: DDQ is useful for the preparation of salts of aromatic cations. Thus, tropylium and triphenylcyclopropenyl cations have been isolated in the presence of perchloric and phosphoric acid.

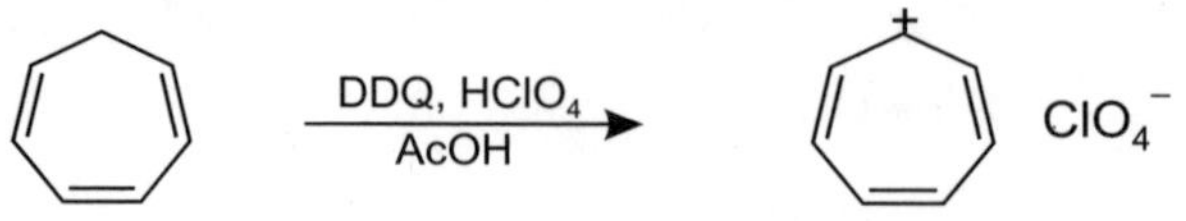

$\xrightarrow[\text{AcOH}]{\text{DDQ, HClO}_4}$ ClO_4^-

Tropylium perchlorate

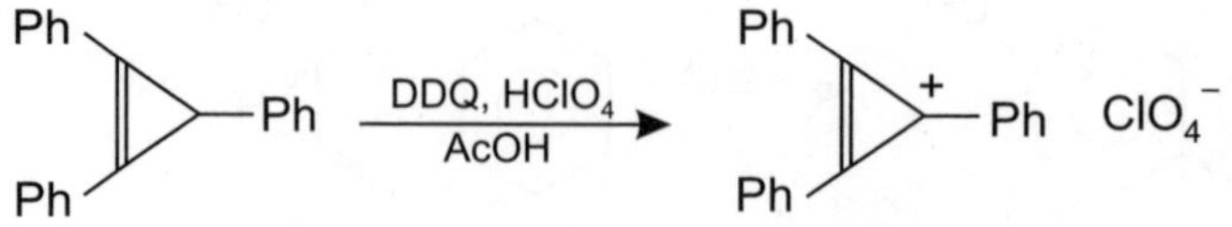

$\xrightarrow[\text{AcOH}]{\text{DDQ, HClO}_4}$ ClO_4^-

Triphenylcyclopropenyl perchlorate

References

1. Thiele and Gunther, *Ann.* 1906, **349**, 45.
2. A.B. Turner and H.J. Ringold, *J. Chem. Soc.,* 1967, 1720.
3. E.A. Braude, A.G. Brook and R.P. Linstead, *J. Chem. Soc.,* 1954, 3569.
4. D. Burn, V. Petrow and G.O. Weston, *Tetrahedron Lett.,* 1960, 114.
5. A. Mckillop and D.W. Young, *Synth. Commun.,* 1977, **7**, 467.
6. H.D. Becker, *J. Org. Chem.,* 1965, **30**, 982.
7. V.K. Ahluwlia and R.K. Parashar, *Organic Reaction Mechanisms*, Narosa Publishing House, 2007, pp. 530–537 and the reference cited therein.
8. E.J. Braude, R.P. Jackman, G. Linstead and G. Lowe, *J. Chem. Soc.,* 1960, 3123.
9. E.J. Corey and Y.B. Xiang, *Tet. Lett,* 1987, **28**, 5403.

1.47 DIETHYL AZODICARBOXYLATE, $C_2H_5O_2CN = NCO_2C_2H_5$

It is a strong hydrogen acceptor in dehydrogenations and oxidations. Some of its important applications are given below:

1. Oxidation of benzylic alcohols to aldehydes[1]

$$C_6H_5CH_2OH \xrightarrow[C_6H_6,\text{ reflux 4 hr.}]{EtO_2CN{=}NCO_2Et} \underset{61\%}{C_6H_5CHO}$$

2. Dehydrogenation of hydrazo compounds to azo compounds[1]

$$\underset{\text{Hydrazobenzene}}{C_6H_5NHNHC_6H_5} \xrightarrow[C_6H_6,\text{ reflux 30 min.}]{EtO_2CN{=}NCO_2Et} \underset{\substack{100\% \\ \text{azobenzene}}}{C_6H_5N{=}NC_6H_5}$$

3. Oxidation of hydroxylamines to nitroso compounds[2]

$$\underset{\substack{\text{Phenyl} \\ \text{hydroxylamine}}}{C_6H_5NHOH} \xrightarrow[Et_2O,\ 0°C,\text{ few hours}]{EtO_2CN{=}NCO_2Et} \underset{\substack{89\% \\ \text{Nitrosobenzene}}}{C_6H_5NO}$$

4. Oxidation of mercaplans (thiols) to disulphides[1]

$$CH_2{=}CHCH_2SH \xrightarrow[\substack{\text{exothermic reaction} \\ \text{RT, few hours}}]{EtO_2CN{=}NCO_2Et} \underset{93\%}{CH_2{=}CH\ CH_2SSCH_2CH{=}CH_2}$$

5. Oxidation of secondary alcohols to ketones[1]

Both saturated and benzylic alochols are dehydrogenated by refluxing with diethyl azodicarboxylate in benzene for 10 hr.

$$\underset{\substack{| \\ OH}}{C_6H_5CHC_6H_5} \xrightarrow[C_6H_6,\text{ reflux 10 hr.}]{EtO_2CN{=}NCO_2Et} \underset{71\%}{C_6H_5COC_6H_5}$$

6. Oxidation of sulphides to sulfoxides[3]

$$CH_3SCH_2CH_2CH(NH_2)COOH \xrightarrow[\text{aq. EtOH, RT.}]{EtO_2CN{=}NCO_2Et} CH_2\overset{O}{\overset{\|}{S}}CH_2CH_2CH(NH_2)COOH$$

Methionine → 97.5% Methionine S-oxide

7. Demethylation of tetiary amines to secondary amines[4]

N-Methyl isopelletierine (2-(CH_2COCH_3)-piperidine, NCH_3) $\xrightarrow[\text{Spontaneous boiling}]{EtO_2CN{=}NCO_2Et,\ MeOH}$ ($NCHO$, CH_2COCH_3) $\xrightarrow{HCl,\ H_2O,\ \text{reflux 3 hr.}}$ (NH, CH_2COCH_3)

80-82%
Isopelletierine

References

1. F. Yoneda, K. Suzuki and N. Nitta, *J. Org. Chem.*, 1967, **32**, 727.
2. E.C. Taylor and F. Yoneda, *Chem. Commun.*, 1967, 199.
3. R. Axen, M. Chaykovsky and B. Witkop, *J. Org. Chem.*, 1967, **32**, 4117.
4. K. Hess, *Chem. Ber.*, 1919, **52**, 964.

1.48 DIMETHYL SULPHIDE AND CHLORINE

Dimethyl sulphide and chlorine or N-chlorosuccinimide (NCS) oxidise primary alcohol to aldehydes in high yields under mild conditions. Thus, 1-octonol is converted into octanal in 96% yield on treatment with dimethyl sulphide[1] and N-chlorosuccinimide in toluene at –25°C. The mechanism[1] of the reaction is given below:

NCS–Me_2S (succinimido–$N{-}S(CH_3)_2Cl^-$) $\xrightarrow[-C_4H_5NO_2]{RCH_2OH}$ $\xrightarrow[-HCl]{Et_3N}$

$$\longrightarrow (CH_3)_3\overset{+}{S}{-}O{-}CHR \ (H \cdots \ddot{N}Et_3) \longrightarrow (CH_3)_2S + O{=}CHR + Et_3\overset{+}{N}H\overset{-}{Cl}$$

In case there are two hydroxyl groups (one secondary and the other tertiary) only the secondary hydroxyl is oxidised[2].

$$\xrightarrow[-78^\circ C,\ 2\ hr;\ Et_3N,\ 25^\circ C,\ 10\ min]{Me_2S,\ Cl_2,\ CH_2Cl_2}$$

72%

References

1. E.J. Corey and C.U. Kim, *J.Am. Chem. Soc.*, 1972, **94**, 7586.
2. E.J. Corey and C.U. Kim, *Tetrahedron Lett.*, 1973, 919.

1.49 DIMETHYL SULFOXIDE (DMSO), $(CH_3)_2$ SO

It is a versatile reagent. Its main use is the conversion of primary alcohols into aldehydes and secondary alcohols into ketones. These oxidations are accomplished under mild conditions, at times at temperatures below 0°C. However, these oxidations require the presence of an activator for the formation of a complex with the alcohol and a base is required for the abstraction of a proton from the complex to initiate its collapse to an aldehyde or a ketone. The course of the reaction is given below:

$$RR'CHOH + ClCOCl \xrightarrow{-HCl} RR'CH-O-CO(Cl) \xrightarrow{-Cl^-}$$

$$(CH_3)_2S{=}O$$

$$RR'CH-O-CO-O-\overset{+}{S}(CH_3)_2 \xrightarrow{-CO_2} \xrightarrow{Et_3N} RR'C(H)-O-\overset{+}{S}(CH_3)_2 \quad Et_3N$$

$$\longrightarrow RR'C{=}O + S(CH_3)_2 + (C_2H_5)_3\overset{+}{N}H$$

In an alternative[2] mechanism, the activator converts dimethyl sulfoxide into a reactive sulfonium intermediate, which in turn reacts with the alcohol to form a complex (A). This is followed by abstraction of a proton from the complex (A) with a base usually triethylamine; the proton is abstracted from one of the methyl groups linked to sulphur. Finally, the disintegration of the formed anions takes place due to intramolecular shift of electrons; the product obtained is dimethyl sulphide and the carbonyl compared. At times, a side reaction resulting from a competitive shift of electrons along the dotted arrows to give a sulphur ylide (B), which recombines with the alkoxide thus, generated to form a methylthiomethyl alkyl ether (C). The side reaction takes place at relatively higher temperature (above –60°C) and is favoured by some DMSO activators like dicyclohexylcarbodiimide and trifluoro acetic anhydride. The side reaction can be suppressed by using bulkier bases like triisopropylamine[3,4]. The best activator is oxalyl chloride[5]. Other activators like chlorine, phosgine, thionyl chloride, sulphur trioxide pyridine, phosphoric acid, pyridinium trifluoroacetate give very good yields[2]. The above mechanism is represented as shown below:

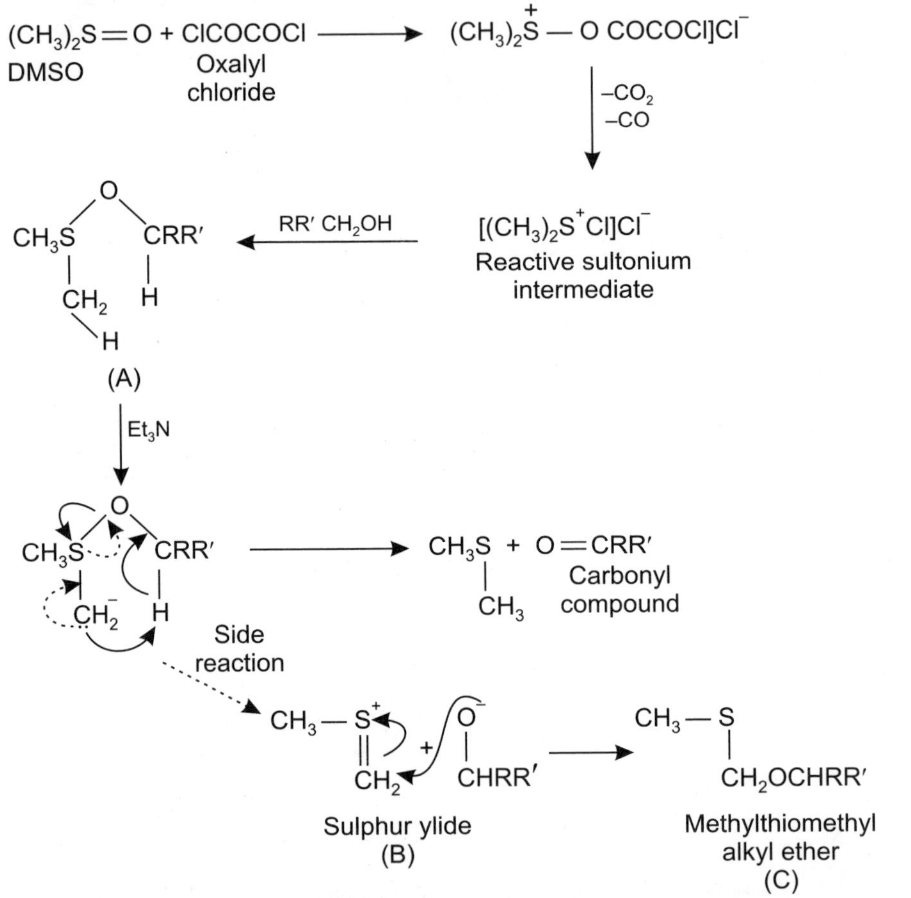

$$(CH_3)_2S{=}O + ClCOCOCl \longrightarrow (CH_3)_2\overset{+}{S}{-}O\,COCOCl]Cl^-$$

DMSO; Oxalyl chloride

$$\xrightarrow{-CO_2,\ -CO} [(CH_3)_2S^+Cl]Cl^-$$

Reactive sultonium intermediate

$$[(CH_3)_2S^+Cl]Cl^- \xrightarrow{RR'\,CH_2OH} \text{(A)}$$

(A) $\xrightarrow{Et_3N}$ CH_3S + $O{=}CRR'$ (Carbonyl compound)

Side reaction:

$CH_3{-}\overset{+}{S}({=}CH_2)$ + $^-O{-}CHRR'$ ⟶ $CH_3{-}S{-}CH_2OCHRR'$

Sulphur ylide (B); Methylthiomethyl alkyl ether (C)

The oxidation of alcohols with DMSO and oxalyl chloride is known as **swern oxidation**.

Some examples of the oxidation of alcohols to carbonyl compounds are given below:

$$CH{\equiv}C(CH_2)_2CH_2OH \xrightarrow[\text{-60°C to -50°C, 22 min; } Et_3N\text{, 5 min} \rightarrow \text{RT}]{(COCl)_2,\ CH_2Cl_2,\ DMSO} CH{\equiv}C(CH_2)_2CHO \quad 99.6\%$$

$$H_3CC{\equiv}CC{\equiv}CCH_2CH_2CH_2OH \xrightarrow[C_6H_{11}N\,CNC_6H_{11}\ (DCC)]{DMSO,\ CH_2Cl_2\ (COCl)_2} H_3CC{\equiv}CC{\equiv}C\,CH_2CH_2CHO \quad 73\%$$

$$C_6H_5CH_2OH \xrightarrow[-50°C,\ 40\ min,\ Et_3N,\ 10\ min.,\ RT.\ 40\ min]{DMSO,\ (CF_3CO)_2,\ CH_2Cl_2} C_6H_5CHO \quad 80\%$$

$$C_6H_5CH_2OH \xrightarrow[-60°C,\ 35\ min,\ Et_3N,\ 5\ min.]{DMSO,\ (COCl)_2,\ CH_2Cl_2} C_6H_5CHO \quad 98\%$$

$$CH_3CH(OH)C_6H_{11} \xrightarrow[35\ min,\ Et_3N,\ -60°C,\ 20\ min]{DMSO,\ (COCl)_2,\ CH_2Cl_2,\ -60°} CH_3COC_6H_{11} \quad 98\%$$

$$\text{Cyclohexanol} \xrightarrow[Et_3N,\ -60°C,\ 5\ min]{DMSO,\ (COCl)_2,\ CH_2Cl_2,\ -60°C,\ 35\ min} \text{Cyclohexanone} \quad 94\%$$

$$C_6H_5CH(OH)C_6H_5 \xrightarrow[Et_3N,\ 20\ min,]{DMSO,\ (COCl)_2,\ CH_2Cl_2,\ -60°C,\ 35\ min} C_6H_5COC_6H_5 \quad 98\%$$

$$\text{Cyclohex-2-en-1-ol} \xrightarrow[-60°C,\ Et_3N,\ -60°C]{2.4\ DMSO,\ 1.1\ (COCl)_2,\ CH_2Cl_2} \text{Cyclohex-2-en-1-one} \quad 80\%$$

Some other applications of DMSO are given below:

1. **Oxidation of aliphatic alkyl halides to aldehydes:** Aliphatic primary halides (chlorides, bromides, iodides), especially the iodides are oxidised to aldehydes by treatment with DMSO[8]. The reactivity of alkyl chloride and bromides can be increased by converting them *in situ* to the corresponding iodides by the addition of sodium iodide[8] to the raction mixture.

$$RCH_2X \longrightarrow RCHO$$

R	*X*	*Reaction conditions*	*% Yield*
C_3H_7	Br	NaI, DMSO, $NaHCO_3$, 115°C, 2 hr.	75
C_5H_{11}	I	DMSO, $NaHCO_3$, 150°C, 3 min	86
C_7H_{15}	Cl	NaI, DMSO, $NaHCO_3$, 115°C, 2 hr	73
C_7H_{15}	Br	NaI, DMSO, $NaHCO_3$, 115°C, 2 hr	74
C_7H_{15}	I	DMSO, $NaHCO_3$, 148°C, 4 min	74

Note: *Data taken reference 8*

A modification of the above procedure is to convert the halides into alkyl tosylates on treatment with silver tosylate in acetonitrile at 0 – 5°C followed by heating with DMSO and sodium bicarbonate[9].

$$C_7H_{15}CH_2X \xrightarrow[0\text{–}5°C]{\text{Tos OAg, MeCN}} C_7H_{15}CH_2OTos \xrightarrow[150°C\ 5\ min]{\text{DMSO, NaHCO}_3} C_7H_{15}CHO\ \ 70\text{–}74\%$$

X=Cl, Br, I

Oxidation of secondary alkyl bromide by DMSO in presence of sodium iodide and sodium bicarbonate at 115°C for 2 hr. give the corresponding ketones[8] in 55–65%.

$$CH_3CH(Br)R \xrightarrow[NaHCO_3,\ 115°C,\ 2\ hr.]{DMSO,\ NaI} CH_3COR$$

2–Bromobutane, R = C_2H_5 — 2–Butanone, (R = C_2H_5), 65%

2–Bromooctane, R = C_6H_{13} — 2–Octanone, (R = C_6H_{13}), 56%

2. **Oxidation of benzylic halides to aromatic ketones:** The oxidation of benzylic aldehydes to aromatic ketones can be accomplished with DMSO[8, 9a].

$$O_2N\text{–}C_6H_4\text{–}CH_2X \longrightarrow O_2N\text{–}C_6H_4\text{–}CHO$$

X	Reaction conditions	% yield
Br	DMSO, MeCN, RT, 8 hr.	48
Cl	DMSO, NaI, 120°C, 4hr.	62%

3. **Oxidation of aldehydes to carboxylic acids:** An elegant and gentle method[10] for the conversion of aldehydes to carboxylic acids is to convert the aldehyde first into bisulphite addition compound followed by oxidation with DMSO

$$H_3C\text{–}CH(CHO)\text{–}C_6H_4\text{–}C_6H_5 \xrightarrow{NaHSO_3} H_3C\text{–}CH(CH(OH)SO_3Na)\text{–}C_6H_4\text{–}C_6H_5 \xrightarrow[RT\ 24\ hr.]{DMSO,\ Ac_2O} [H_3C\text{–}CH(COSO_3Na)\text{–}C_6H_4\text{–}C_6H_5] \xrightarrow{K_2CO_3,\ H_2O,\ H^+} H_3C\text{–}CH(CO_2H)\text{–}C_6H_4\text{–}C_6H_5\ \ 84\%$$

4. **Oxidation of alkynes to α-diketones:** Diarylacetylenes (internal alkyne) are oxidised into α-diketones by DMSO and N-bromosuccinimide[11].

$$C_6H_5C{\equiv}CC_6H_5 \xrightarrow{\text{DMSO, NBS (2 mol), 22°C, 24 hr}} \underset{98\%}{C_6H_5COCOC_6H_5}$$

5. **Oxidation of α-bromo ketones to α-dicarbonyl compounds[9a]:** α-Bromo ketones are converted into α-dicarbonyl compounds by DMSO.

$$R\text{-}C_6H_4\text{-}COCH_2Br \xrightarrow[\text{RT, 9 hr.}]{\text{DMSO}} R\text{-}C_6H_4\text{-}COCHO$$

R = H, yield, 71%
R = Br, yield 84%

6. **Oxidation of diods:** The oxidation of diods having alcoholic groups of the same nature as in 2, 3-dihydroxy naphthalene is converted into dialdehyde (2, 3-diformylnaphthalene) by DMSO activated by acetic anhydride[4].
7. **Oxidation of epoxides to α-hydroxy ketones:** Treatment of epoxides with DMSO in the presence of trifluoromethanesulfonic acid and disopropylethylamine give α-hydroxy ketones[12].

$$\text{Cyclohexane epoxide} \xrightarrow[\text{23°C, 30–45 min; } CH_2Cl_2\text{, –78°C, } iPr_2NEt]{\text{DMSO, } CF_3SO_3H} \text{α-Hydroxycyclohexanone (61\%)}$$

Cyclohexane epoxide

61%
α-Hydroxycyclohexanone

8. **Oxidation of hydroxy ketones to diketones:** DMSO in presence of various activators is a suitable reagent for the oxidation of steroidal hydroxy ketones[13] to diketones.

$$\text{Testosterone} \xrightarrow[C_6H_6]{0.5\ H_3PO_4\text{, 5DCC, 35°, DMSO}} \text{4-Androstene-3, 17-dione (99\%)}$$

Testosterone

99%
4-Androstene-3, 17-dione

9. **Oxidation of ketones to α-dicarbonyl compounds:** Acetophenones on oxidation with DMSO in presence of hydrogen bromide gives 86%. α-dicarbonyl compounds.

Br—C6H4—$COCH_3$ —(DMSO, 48% HBr; 55°C, 24 hr)→ Br—C6H4—COCHO

p-Bromoacetophenone 86% *p*-Bromophenyl glyoxal

References

1. D.H.R. Barton, B.J. Ganer and R.H. Wightman, *J. Chem. Soc.*, 1964, 1855.
2. K. Omura, D. Swern, *Tetrahedron*, 1977, **34**, 1651.
3. S.L. Huang, K. Omura and D. Swern, *Synthesis*, 1978, 297.
4. C.M. Amon, M.G. Banwell and G.L. Gravatt., *J. Org. Chem.*, 1987, **52**, 4851.
5. A.J. Mancuso, S.-L. Huang and D. Swern, *J. Org. Chem.*, 1978, **43**, 2480.
6. F. Bohlmann and R. Miethe, *Chem. Ber.*, 1967, **100**, 3861.
7. K. Omura, A.K. Sharma and D. Swern, *J. Org. Chem.*, 1976, **41**, 957.
8. P. Dave and H.-S. Engel, *Synth. Commun*, 1986, **16**, 1343.
9. K. Kornblum, W.J. Jones and G.L. Anderson, *J. Am. Chem. Soc.*, 1959, **81**, 4113.

9a. N. Kornblum, J.W. Powers, G.L. Aderson, W.J. Jones, H.O. Larson, O. Levand and W.M. Weaver, *J. Am. Chem. Soc.*, 1957, **79**, 6562.

10. P.G. M. Wuts and C.L. Bergh, *Tetrahedron Lett*; 1986, **27**, 3995.
11. R.P. Linstead, K.O.A. Michaelis and S.L.S. Thomas, *J. Chem. Soc.*, 1940, 1139.
12. B.M. Trost and M.J. Fray, *Tetrahedron Lett.*, 1988, **29**, 2163.
13. K.E. Pfitzer and J.G. Moffatt, *J. Am. Chem. Soc.*, 1965, 87, 5661.

1.50 DIOXIRANES

These are prepared[1] from acetone and other aliphatic ketones by the reaction with oxone (*see* oxone).

R_2CO —(Oxone; $NaHCO_3$, H_2O)→ $R_2C(O_2)$

Ketones Dioxiranes

Dioxiranes are used for the epoxidations[1, 2].

$C_6H_5CH{=}CHC_6H_5$ —($(H_3C)_2C(O_2)$)→ $C_6H_5CH(O)CHC_6H_5$

Olefin Epoxide

The epoxidation of α, β-unsaturated esters can also be effected by dimethyl dioxirane.[1]

$C_6H_5CH{=}CHCOOEt$ —($(H_3C)_2C(O_2)$)→ $C_6H_5CH(O)CHCOOC_2H_5$

Ethyl *trans*-cinnamate Ethyl 2, 3-epoxydihydro cinnamate (63%)

Primary amines are oxidised to nitro compounds by dimethldioxirane[3].

$$C_2H_5CH(CH_3)NH_2 \xrightarrow{Me_2C(O_2),\ Me_2CO} C_2H_5-CH(CH_3)-NO_2$$

2-Aminobutane → 2-Nitrobutane

Tertiary amines (like pyridine) are converted into amine oxide by dimethyldioxirane[1].

$$\text{Pyridine} \xrightarrow[\text{KOH (pH 75–80), RT, 2 hr.}]{Me_2C(O_2)} \text{Pyridine oxide } (N^+-O^-)$$

Dimethyldioxirane also oxidises sulphides to sulfoxides[1]

$$C_6H_5SCH_3 \xrightarrow[\text{RT}]{Me_2C(O_2),\ Me_2CO} C_6H_5S(=O)CH_3$$

Methyl phenyl sulphide → Methyl phenyl sulfoxide (65%)

References

1. R.W. Murray and R. Jeyaraman, *J. Org. Chem.*, 1985, **50**, 2847.
2. D.J. Reif and H.O. House, *Org. Synth.* Collective Volume, 1963, **4**, 860.
3. R.W. Murray, R. Jeyaraman and L. Mohan, *Tetrahedron Lett.*, 1986, 27.

1.51 DIPHENYL PICRYLHYDRAZYL

It is prepared[1] from N, N-diphenyl-N′-(2, 4, 6 trinitrophenyl) hydrazine by oxidation with lead dioxide.

$$(C_6H_5)_2N\,NH-C_6H_2(NO_2)_3 \xrightarrow{PbO_2} (C_6H_5)_2N\,\dot{N}-C_6H_2(NO_2)_3$$

N, N-Diphenyl-N′-(2,4, 6-trinitrophenyl) hydrazine → Diphenylpicrylhydrazyl

It converts[2] hydrazobenzene into azobenzene.

$$\underset{\text{Hydroazobenzene}}{C_6H_5NHNHC_6H_5} \xrightarrow[\text{CHCl}_3]{\text{Diphenylpicryl}} \underset{\substack{100\% \\ \text{Azobenzene}}}{C_6H_5N=NC_6H_5}$$

It is also used for the dehydrogenation of 1, 4-dihydronaphthalene to naphthalene[2].

References

1. S. Goldschmidt and K. Renn, *Chem. Ber.*, 1922, **55**, 628.
2. E.A. Braude, A.G. Brook and R.P. Linstead, *J. Chem. Soc.*, 1954, 3574.

1.52 FERRIC CHLORIDE $FeCl_3$ OR $FeCl_3 \cdot 6H_2O$

It is used for the oxidation of amines[1], hydroquinones[2] and phenols[3] to quinones.

Diaminodurene (in the form of its hexachlorostannate) $\cdot H_2SnCl_6$ $\xrightarrow[30°C,\ 24\ hr.]{FeCl_3,\ HCl,\ H_2O}$ Duroquinone, 90%

2-Mercapto hydroquinone $\xrightarrow{2N\ FeCl_3\ \text{in alcohol}}$ Disulphide of p-benzoquinone, 100%

Tetrafluoro hydroquinone $\xrightarrow[RT]{FeCl_3,\ H_2O,\ Et_2O}$ Tetrafluoro p-benzoquinone, 100%

Ferric chloride is used in dehydrogenative coupling of phenols[4].

β-Naphthol $\xrightarrow[\text{reflux}]{FeCl_3\text{, (anhy.)/ether}}$ α, α′–Bi–β–Naphthol

References

1. L.I. Smith, *Org. Synth.* Collective Volume, 1943, **2**, 254.
2. W. Alcalay, *Helv. Chem. Acta,* 1947, **30**, 578.
3. M. Hudlicky and H.M. Bell, *J. Fluorine Chem.,* 1975, **6**, 201.
4. H. Walder, *Chem. Ber.,* 1882, **15**, 2166.

1.53 FERRIC SULPHATE, $Fe_2(SO_2)_3$

It is used either as such or is generated *in situ* from ferric oxide and sulphuric acid[1]. It oxidises hydroquinones to quinones. Thus, 2-bromo-3, 5-dimethyl hydroquinone, an oxidation with ferric sulphate and dilute hydrochloric acid gives 84% yield of 2-bromo-3, 5-dimethyl p-benzoquinone[2], which is isolated by steam distillation.

2-Bromo-3, 5-dimethyl hydroquinone $\xrightarrow[\text{2. Steam distillation}]{\text{1. } Fe_2(SO_4)_3\text{, HCl}}$ 2-Bromo-3-5, dimethyl *p*-benzoquinone (84%)

A solution of hydrogen peroxide in presence of ferrous or ferric sulphate is known as **Fenton reagent** and is used[3] for degradation of carbohydrates (**Ruff degradation**) by the following sequence of reactions:

D(–)–Arabinose $\xrightarrow[Br_2|H_2O]{[O]}$ CO_2H…CH_2OH $\xrightarrow{CaCl_2}$ $CO_2\frac{Ca}{2}$…CH_2OH $\xrightarrow[\text{(2) } H_2O_2\text{, 40–60°C, RT. Overnight}]{\text{(1) } Ba(OAc)_2\text{, } H_2O_2\text{, } Fe_2(SO_4)_3\text{, } H_2O\text{, reflux}}$ D(–)–Ethythrose 46%

(CHO, H–C–OH, H–C–OH, CH_2OH)

References

1. H.E. Fierz–David, L. Blangey and H. Dübendorfer, *Helv. Chem. Acla,* 1946, **29**, 1661.
2. L.I. Smith and P.F. Wiley, *J. Am. Chem. Soc.,* 1946, **68**, 894.
3. W.G. Overend, M. Stacey and L.F. Wiggins, *J. Chem. Soc.,* 1949, 1358.

1.54 FREMY SALT (POTASSIUM NITROSODISULFONATE) $\dot{O}N(SO_3K)_2$

It is a free radical and is capable of converting aromatic amines into quinones.[1]

$(KO_3S)_2NO^{\bullet}$, Me_2CO, KH_2PO_4, H_2O, 75 min,

2,6-Dimethyl aniline → 2,6-Dimethyl *p*-benzoquinone (82%)

It also converts phenols into quinones[2,3]

$\bullet ON(SO_3K)_2$, H_2O, Et_2O, 25°C, 20 min

3,4-Dimethylphenol → 3,4-dimethyl o-benzoquinone 49-50%

$\dot{O}N(SO_3Na)_2$, H_2O, C_7H_{16}, 12°C, 4 hr.

2,3,6-Trimethyl phenol → 2,3,6-Trimethyl *p*-benzoquinone 77-79%

References

1. H.-J. Teuber and M. Hasselbach, *Chem. Ber.,* 1959, **92**, 674.
2. H.-J. Teuber, *Org. Synth.* Collective Volume, 1988, **6**, 480.
3. P.A. Wehrli and F. Pigott, *Org. Synth.* Collective Volume, 1988, **6**, 1010.

1.55 HEXAMINE (HEXAMETHYLENE TETRAMINE), $C_6H_{12}N_4$

Hexamine converts primary amines (in which NH_2 group is linked to CH_2 group) to aldehydes[1].

$$C_6H_5CH_2NH_2 \xrightarrow[C_6H_{12}N_4,\ H_2O]{40\%\ CH_2O,\ HCl,\ reflux;} C_6H_5CHO$$

Benzyl amine → 66%

Allylic halides are converted into α, β-unsaturated aldehydes or ketones by the **Sommelet reaction**[2, 3, 4] by treatment with hexamine followed by refluxing with water or by steam distillation.

$$RCH{=}CHCH_2X \xrightarrow{C_6H_{12}N_4} RCH{=}CH\ CH_2 \ .\ C_6H_{12}N_4 \ .\ X$$

reflux with water or steam distillation

$$R\ CH{=}CHCHO$$

2-Chloromethyl thiophene → ($C_6H_{12}N_4$, $CHCl_3$, reflux 1 hr.; steam distillation) → Thiophene 2-aldehyde, 51%

Another interesting example[5] is given below:

3-Chloromethyl-4-methylbenzyl chloride → (1. $C_6H_{12}N_4$, $CHCl_3$, reflux 1 hr.; 2. H_2O, reflux 4 hr.) → Toluene 2, 4-dialdehyde, 70%

Phenols on refluxing with hexamine in glacial acetic acid followed by hydrolysis by refluxing with water give hydroxyaldehydes, the formyl group preferably enters in the para position with respect to phenolic group.

2, 6-Dimethyl phenol → (1. $C_6H_{12}N_4$/glacial HOAc reflux 4–6 hr.; 2. reflux with water) → 3,5-Dimethyl-4-hydroxy benzaldehyde

References

1. J. Graymore and D.D. Davies, *J. Chem. Soc.*, 1945, 293.
2. S.J. Angyal and R.C. Rassack, *J. Chem. Soc.*, 1949, 2700.
3. F.W. Dunn, T.D. Waugh and K. Dittmer, *J. Am. Chem. Soc.*, 1946, **68**, 2118.
4. M. Sommetet, *Bull. Soc. Chem.* Fr., 1913, **13**, 1054.
5. J.H. Wood, C.C. Tung, M.A. Perry and R.E. Gibson, *J. Am. Chem. Soc.*, 1950, **72**, 2992.

1.56 HYDROGEN PEROXIDE

Hydrogen peroxide (H_2O_2) is commercially available in aqueous solution of 30% concentration. The 30% hydrogen peroxide is a colourless liquid (d. 1.110) and is stabilized against decomposition, which normally occurs in presence of traces of iron, copper, aluminium and transition metals.

Following are given some of the reactions in which hydrogen peroxide is used as an oxidising agent:

(*i*) **Oxidation of Aldehydes and Ketones:** Hydrogen peroxide is used as an oxidising agent in presence of alkali for the oxidation of aldehydes and ketones into hydroxyl group. In this case, a phenolic group must be present in ortho or para position to the carbonyl group. This reaction is commonly known as **Dakins reaction**[1]. For example, salicylaldehyde or o-hydroxyacetophenone are converted into catechol in about 70% yield.

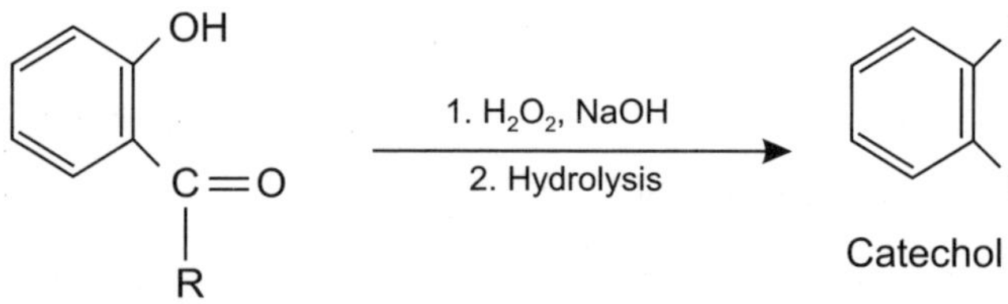

OH C=O R —— 1. H_2O_2, NaOH / 2. Hydrolysis ——> OH OH Catechol

Salicylaldehyde R = H
o-Hydroxyacetophenone, R = CH_3

The mechanism of Dakins reaction is not certain. However, a mechanism analogous to **Baeyer-villiger oxidation** has been suggested[2].

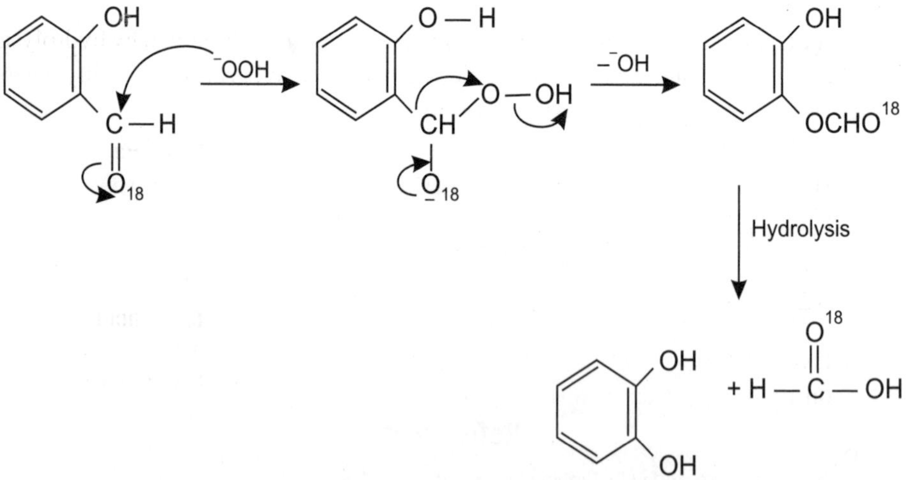

OH C—H ‖ O_{18} —— ^-OOH ——> O—H CH—O—OH O_{18} —— $-^-OH$ ——> OH $OCHO^{18}$

Hydrolysis

OH OH + H—C(=O^{18})—OH

The above mechanism has been confirmed[3] by taking o-hydroxybenzaldehyde in which the oxygen of the CHO group is labelled by ^{18}O as shown above.

(*ii*) Per acids can be obtained from carboxylic acids[3a]

$$\underset{\text{Carboxylic acid}}{RCOOH} + H_2O_2 \longrightarrow \underset{\text{Peracids}}{RCOOOH} + H_2O$$

(*iii*) Hydrogen peroxide oxidises alkyl boranes to alkyl borates, which on hydrolysis yield alcohols[4].

$$R_3B \xrightarrow[^{-}OH]{H_2O_2} (RO)_3B \xrightarrow{H_2O} 3ROH + B(OH)_3$$

(*iv*) Hydrogen peroxide oxidises nitriles to amides

$$RC{\equiv}N \xrightarrow[^{-}OH]{H_2O_2} RCONH_2$$

(*v*) Hydrogen peroxide oxidises primary amines to nitroso compounds[5], secondary amines to hydroxylamines[6], and tertiary amines to amine oxides[7].

$$R-NH_2 \xrightarrow{H_2O_2} [R-NH\,OH] \longrightarrow \underset{\text{Nitroso compound}}{R-N{=}O}$$

$$\underset{\text{2° Amine}}{(C_2H_5)_2NH} \xrightarrow{H_2O_2} \left[(C_2H_5)_2\overset{+}{N}(H)-\bar{O}\right] \longrightarrow \underset{\text{N, N-Diethyl hydroxylamine}}{(C_2H_5)_2\,NOH}$$

$$\underset{\text{3° Amine}}{R_3N} \xrightarrow{H_2O_2} \underset{\text{Amine oxide}}{R_3\overset{+}{N}-\bar{O}}$$

Aromatic amines are oxidised to nitro compounds via the intermediate formation of N-phenylhydroxylamine.

$$C_6H_5NH_2 \xrightarrow{H_2O_2} [C_6H_5\,NH\,OH] \xrightarrow{H_2O_2} C_6H_5\,NO_2$$

(*vi*) Hydrogen peroxide oxidises thiols to disulphides and disulphides to sulfonic acids[9]

$$\underset{\text{Thiol}}{2RSH} \xrightarrow{H_2O_2} \underset{\text{Disulphides}}{R-S-S-R} \xrightarrow{H_2O_2} \underset{\text{Sulfonic acid}}{2RSO_3H}$$

(*vii*) Hydrogen peroxide oxidises sulphides to sulphoxide and sulfones depending on the concentration of hydrogen peroxide.

$$\underset{\text{Dialkyl sulphide}}{R_2S} \xrightarrow[25°C]{H_2O_2,\ CH_3COOH} \underset{\text{Dialkyl sulphoxide}}{R_2S{=}O} \xrightarrow[CH_3COOH]{H_2O_2} \underset{\text{Dialkyl sulphone}}{R_2S({=}O)_2}$$

(*viii*) Sodium peroxide (Na_2O_2) is used for the conversion of aldoximes to carboxylic acids[10].

(*ix*) **Reaction with alkenes:** Hydrogen peroxide alone in acetonitrile oxidises alkenes to epoxides[11]. However, in presence of osmic acid syn hydroxylation takes place[12] and tungstic acid catalyses antihydroxylation.[13]

Hydrogen peroxide has also been used in the formation of urea-hydrogen peroxide complex (UHP) for the oxidation of alcohols to carbonyl compounds. For details *see* section 1.154a.

References

1. H.D. Dakin, OS, 1941, **7**, 149; J.E. Lettler, *Chem. Rev.*, 1949, **45**, 385.
2. V.K. Ahluwalia and R.K. Parashar, *Organic Reaction Mechanisms*, 2007, Narosa Publishing House, Page 433.
3. C.A. Bunton in Peroxide Reaction Mechanism, edited by J.O. Edwards, Interscience.

3a. G.B. Payne, *Org. Synth.*, Collective Volume, 1973, **5**, 805; L.S. Silbert, E. Siegel and D. Swern *Org. Synth.*, Collective Volume, 1975, **5**, 904.

4. M.S. Newman, A.Arkell and T. Fukunaga, *J.Am. Chem. Soc.*, 1960, **82**, 2498; R.L. Kidwell, M. Murphy and S.D. Davling, *Org. Synth, Collective* Volume, 1973, **5**, 918.
5. H.R. Holmes and R.P. Bayer, *J. Am. Chem. Soc.*, 1960, **82**, 3454.
6. W.R. Dunstan and E. Goulding, *J. Chem. Soc.*, 1899, **75**, 1004.
7. A.C. Cope and E. Ciganek, *Org. Synth.*, Collective Volume, 1963, **4**, 612; E.C. Tayler and A.J. Crovetic, *Org. Synth.*, Collective Volume, 1963, **4**, 704.
8. C.C. Price and G.W. Stacy, *Org. Synth.*, Collective Volume, 1955, **3**, 86.
9. T.H. Chao and W.B. Hardy, *Chem. Ind.* (London), 1965, 81.
10. T.-L. HO and G.A. Olah, *Synthesis,* 1976, 807.
11. G.B. Payne, P.H. Deming and P.H Williams, *J. Org. Chem.*, 1961, **26**, 659.
12. N.A. Milas and S. Sussman, *J. Am. Chem. Soc.*, 1936, **58**, 1302.
13. M. Murgdan and D.P. Young, *J. Chem. Soc.*, 1949, 2988.

1.57 IODINE, I_2

Iodine along with benzoyl peroxide is used in benzoxylation[1]. Thus, stilbene on heating with the benzoyl peroxide and iodine in carbon tetrachloride at 80°C for 48hr. gives[1] an 83% yield of the dibenzoyl ester of hydrobenzoin (1, 2-dibenzoxy-1, 2-diphenylethane).

$$\underset{\text{Stilbene}}{C_6H_5CH{=}CHC_6H_5} \xrightarrow[CCl_4,\ 80°C,\ 48\ hr.]{I_2,\ \text{benzoyl peroxide}} \underset{\text{1,2-Dibenzoxy–1, 2––diphenyl ethane}}{C_6H_5\underset{\underset{C_6H_5CO\,O}{|}}{CH}-\underset{\underset{OCOC_6H_5}{|}}{CH}-C_6H_5}$$

Iodine along with silver oxide is used for epoxidation[2] of alkenes and cyclo alkenes.

Cyclohexene $\xrightarrow[\text{RT, 7 hr.}]{I_2,\ Ag_2O,\ \text{dioxane } H_2O\ (12:1)}$ Cyclohexane epoxide 80%

Stereospecific hydroxylation can be accomplished via silver acetate[3, 4] or silver benzoate[5] and iodine followed by hydrolysis of the intermediate esters.

B_2OAg, I_2, C_6H_5, 3 days

1. AcOAg, I_2, AcOH, 40 min
2. H_2O, 20°C, 12 hr
3. LAH, Et_2O, reflux 1 hr.

at 20°C	4%	30%	81%
at 80°C	15%	29%	

α-Iodoketones are obtained[6] by the reaction of alkenes and cycloalkenes with silver chromate and iodine.

Cyclohexene $\xrightarrow[CH_2Cl_2,\ 0\text{–}5°C,\ 25\ \text{min.}]{Ag_2CrO_4,\ I_2,\ C_5H_5N}$ 2-Iodocyclohexanone 60%

$Ag_2CrO_4 + I_2 \rightarrow AgI + AgO-CrO_2-OI$ → $C_6H_5CH{=}CH_2$ + $AgO-CrO_2-O-I$ → $C_6H_5C(-O-CrO_2OAg)(H)-CH_2I$ → $C_6H_5COCH_2I + AgHCrO_3$

Iodine is also used for the oxidation of thiols and thiol-group-containing compounds to the corresponding disulphides[7].

An important application of iodine is in the conversion of methyl ketones into iodoform by the so called **Haloform reaction** (*see* sodium hypoiodite).

References

1. A. Perret and R. Perrot, *Helv. Chem. Acta,* 1945, **28**, 558.
2. M. Parrilli, G Barone, M. Adinolti and L. Mangoni, *Tetrahedron Lett.,* 1976,207.
3. R.B. Woodward and F.V. Brutcher, Jr., *J. Am. Chem. Soc.,* 1958, **80**, 209.
4. F.D. Gunstone and L.J. Morris, *J. Chem. Soc.,* 1957, 487.
5. E. Späth, F. Kuffner and F. Keszler, *Chem. Ber.,* 1936, **69**, 378.
6. G. Cardillo and M. Shimitzu, *J. Org. Chem.,* 1977, **42**, 4268.

1.58 IODINE-PYRIDINE

A solution of iodine in pyridine oxidises methyl ketones of the type Ar $COCH_3$ and $R_2CHCOCH_3$ to give[1,2] the pyridinium iodides in 95% yield. The salts are cleaved with sodium hydroxide to give carboxylic acids.

$COCH_3$ (benzene ring) $\xrightarrow{I_2,\ Py}$ C_6H_5–C(=O)–CH_2–N^+(pyridinium) I^- $\xrightarrow[2.\ H^+]{1.\ NaOH\ Reflux}$ C_6H_5COOH

Pyridinium iodide salt

Benzoic acid

References

1. I.C. King, *J. Am. Chem. Soc.,* 1944, **66**, 894, 1612.
2. F. Kronnke, Ber., 1933, **66**, 1386.

1.59 IODINE AND SILVER CARBOXYLATES

A solution of iodine in carbon tetrachloride and equimolar amount of silver acetate or silver benzoate, under anhydrous conditions convert the alkenes into the diacetyl derivatives of the *trans* gycol, which an saponification gives *trans* diods. This reaction is known as **prevost reaction**[1]. The reaction proceeds via the formation of an iodonium salt, followed by the reaction with silver acetate to give *trans* diacetate. The cyclic oxonium salt is opened by silver acetate (route a) to give the trans diacetate, which on saponification yielded *trans* diol. The prevost reaction gives *trans* diols in contrast to permanganate and osmium tetroxide hydroxylation which give *cis* diols. As already stated in prevost reaction, the *trans* diacetates are formed under anhydrous conditions. However, in the presence of water, *cis* diacetate is obtained, which on saponification gives *cis* diols. This is called **Woodward modification[2] of the Prevost reaction**. In aqueous solution, the oxonium salt reacts with water (route *b*) to give *cis*-1, 2-diol in 65% overall yield.

Cyclohexene + I—I + $\overset{+}{Ag}\ \bar{O}COCH_3$ (Silver acetate) $\xrightarrow{-AgI}$ Iodonium salt ($\bar{O}COCH_3$) → Iodoacetate → oxonium salt

oxonium salt $\xrightarrow[\text{route a}]{CH_3CO\bar{O}}$ Diacetate of *trans*-1,2-dihydroxy cyclohexane $\xrightarrow{\text{saponification}}$ *trans*-1,2-Dihydroxy cyclohexane

oxonium salt $\xrightarrow[\text{route b}]{H_2\ddot{O}}$ (cyclic orthoacetate, CH_3, OH) $\xrightarrow{\overset{+}{Ag}\ \bar{O}COCH_3}$ Diacetate of *cis*-1,2-dihydroxy cyclohexane $\xrightarrow{\text{saponification}}$ *cis*-1,2-Dihydroxy cyclohexane

In a similar way, oleic acid on oxidation under prevost conditions gives threo-9, 10-dihydroxy stearic acid, while by Woodward procedure, erythro isomer results.

Oleic acid ($CH_3(CH_2)_7CH=CH(CH_2)_7COOH$) $\xrightarrow[\text{Prevost conditions (Anti hydroxylation)}]{[O]}$ Threo-9,10-dihydroxystearic acid

Oleic acid $\xrightarrow{\text{Woodward modification of prevost reaction (syn hydroxylation)}}$ Erythro-9,10-dihydroxystearic acid

References

1. C. Provost, Compt. Rend., 1933, **196**, 1129.
2. R.B. Woodward and F.V. Bruscher, *J. Am. Chem. Soc.*, 1958, **80**, 209.

1.60 IODINE AND SILVER OXIDE

It oxidises alkenes and cycloalkenes to epoxides. For details *see* Iodine.

1.61 IODINE TRIACETATE, $I(OCOCH_3)_3$

It is prepared[1] from iodine trichloride and silver acetate in acetic acid at room temperature.

It is used for the epoxidation[2] of alkenes and for the cleavage of vicinal diols to dialdehydes[1].

References

1. R.C. Cambie, D. Chambers, P.S. Rutledge and P.D. Woodgate, *J. Chem. Soc.*, Perkin Trans I, 1978, 1483.
2. M. Linkeseder and E. Zbiral, *Justus Liebigs Ann. Chem.*, 1978, 1076.

1.62 IODINE TRIS(TRIFLUOROACETATE), $I(OCOCF_3)$

It is used for the epoxidation of certain alkenes[1].

References

1. M. Linskeseder and E. Zbiral, *Justus Liebigs Ann. Chem.*, 1978, 1076.

1.63 IODOBENZENE DIACETATE, $C_6H_5I(OCOCH_3)_2$

It is obtained by the treatment of iodobenzene with peroxyacetic acid[1] or with 30% hydrogen peroxide and acetic anhydride at 40°C. It is used for cleaving vicinal diols[2].

$$C_6H_5\underset{OH}{CH}-\underset{OH}{CH}C_6H_5 \xrightarrow[RT]{PhI(OAc)_2,\ 50\%\ AcOH} \underset{62\%}{C_6H_5CHO}$$

Iodobenzene diacetate hydroxylate[3], the esters in the α-position with respect to ester group.

$$C_6H_5CH_2CO_2Me \xrightarrow{PhI(OAc)_2,\ 3KOH} C_6H_5\underset{OH}{CH}CO_2Me \quad 50\%$$

Thiaxanthone is oxidised 50% to thiaxanthone oxide in 78% yield[4] by refluxing for 4hr. with iodobenzene diacetate in acetic acid.

$$\text{Thiaxanthone} \xrightarrow[\text{reflux 4 hr.}]{PhI(OAc)_2,\ AcOH} \text{Thiaxanthone oxide}$$

Iodobenzene diacetate (IBD) supported on neutral alumina has been used[5] for the oxidation of alcohols to carbonyl compounds under solvent-free conditions and MW irradiation in quantitative yield.

$$RR_1CH{-}OH \xrightarrow[\text{IBD/Neutral Alumina}]{\text{MW, 1-3 min}} RR_1C{=}O$$

References

1. K.H. Pausacker, *J. Chem. Soc.*, 1953, 107.
2. R. Criegee and H. Beucker, *Justus Liebigs Ann. Chem.*, 1939, **541**, 218.
3. R.M. Moriarty and H. Hu, *Tetrahedron Lett.*, 1981, **22**, 2747.
4. J.P.A. Castrillón and H.H. Szmant, *J. Org. Chem.*, 1967, **32**, 976.
5. R.S. Varma, R. Dahiya and R.K. Saini, *Tetrahedron Lett.*, 1997,**38**,7029.

1.64 IODOBENZENE DICHLORIDE, $C_6H_5ICl_2$

It oxidises selenides to selenoxides[1].

$$\underset{\text{Benzylphenyl selenide}}{C_6H_5CH_2SeC_6H_5} \xrightarrow[-40°]{C_6H_5ICl_2,\ \text{aq pyridine}} \underset{\text{85\% Benzylphenyl selenoxide}}{C_6H_5CH_2Se(=O)C_6H_5}$$

Reference

1. M. Cinquini, S. Colonna and R. Giovini, Chem. Ind. (London), 1969, 1737.

1.65 IODO COMPOUNDS

Same of the iodo compounds like iodosobenzene diacetate, 2-iodo-4, 5-dinitro benzoic acid and iodoxybenzene are good oxidising agents and are all prepared from iodobenzene as shown ahead:

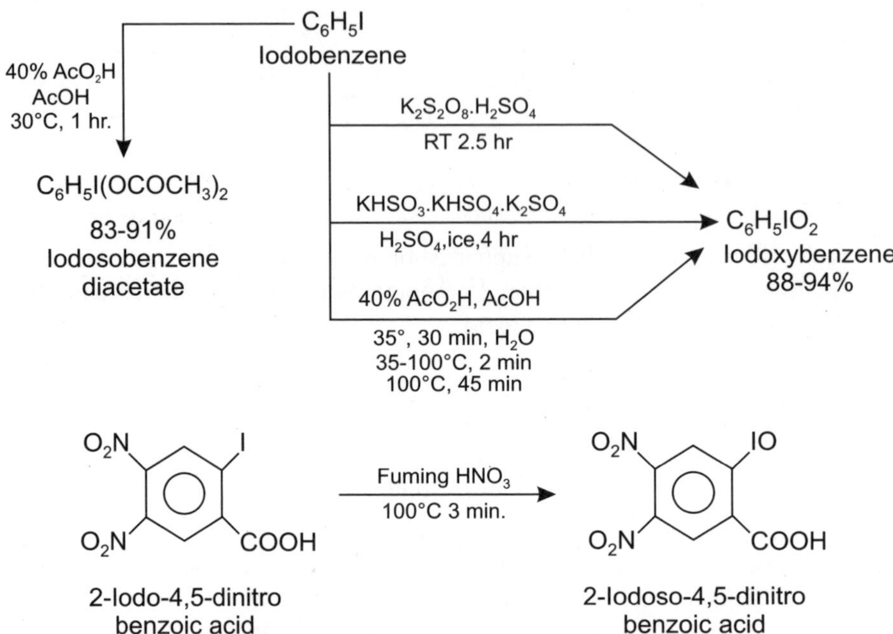

The best iodine containing reagent is **Dess-Martin Periodinane**. Its preparation from o-iodobenzoic acid has been described (*See* Dess-Martin Periodinone).

The oxidising properties of all these reagents along with some other related oxidising agents are described in separate sections.

1.66 IODOSOBENZENE, C_6H_5IO

It is prepared[1] by way of iodobenzene diacetate. Alkyl acetylenyl ethers on treatment with iodosobenzene in dichloromethane in presence of 1% tris (triphenylphosphine) ruthenium dichloride give ketoesters[2].

$$C_6H_5C{\equiv}C{-}OC_2H_5 \xrightarrow[\text{RT, 15 min}]{\text{PhIO/(Ph}_3\text{P)}_3\text{ RuCl}_2,\ \text{CH}_2\text{Cl}_2} C_6H_5COCO_2C_2H_5$$

Iodosobenzene oxidises aldehydes to carboxylic[3] acids in presence of 1% *tris* (triphenyl phophine) ruthenium chloride.

$$\text{PhCHO} \xrightarrow[\text{RT, 24 hr.}]{\text{2PhIO/(Ph}_3\text{P)}_3\text{ RuCl}_2,\ \text{CH}_2\text{Cl}_2} \underset{75\%}{C_6H_5COOH}$$

It oxidises vicinal diols to carbonyl compounds[4].

$$C_6H_5\underset{\text{OH}}{\underset{|}{C}}H{-}\underset{\text{OH}}{\underset{|}{C}}H{-}CH_3 \xrightarrow[\text{3 hr.}]{\text{PhIO, C}_6\text{H}_6\text{, RT}} \underset{78\%}{C_6H_5CHO}$$

Iodsobenzone oxidises alcohols to aldehydes or ketones[3].

Similar reactions can be peformed by p-iodosotoluene or o-iodosobenzoic acid[5].

References

1. J.G. Sharefkin and H. Saltzman, *Org. Synth.* Collective Volume, 1973, **5**, 658, 660.
2. P. Müller and G. Godoy, *Tetrahedron Lett.*, 1982, **23**, 3661.
3. Püller and J. Godoy, *Tetrahedron Lett.*, 1981, **22**, 2361.
4. R. Crigee and H. Beucker, *Justus Liebigs. Ann. Chem.*, 1939, **541**, 218.
5. P. Askenasy and V. Meyer, *Chem. Ber.*, 1983, **26**, 1354.

1.67 IODOXYBENZENE, $C_6H_5IO_2$

It is prepared[1] by the oxidation of iodobenzene with peroxyacetic acid. It cleaves vicinal diols[2].

$$C_6H_5CH(OH)-CH(OH)C_6H_5 \xrightarrow[\text{RT, 45 min.}]{PhIO_2,\ CCl_3CO_2H} \underset{90\%}{C_6H_5CHO}$$

Cholestan-3-ol is converted into 1, 4-cholestadien-3-one on treatment with iodoxybenzene and a catalytic amount of benzeneselenic anhydride, which is generated *in situ* from diphenyldiselenide[3].

$PhIO_2$, $(PhSe)_2$, PhMe, reflux, 2 min → $(PhSeO)_2O$; reflux, 24 hr

Cholestan-3-ol → 81% 1,4-Cholestadien-3-one

References

1. J.G. Shanefkin and H. Saltzman, *Org. Synth.* Collective Volume, 1973, **5**, 665.
2. D.H.R. Barton, C.R.A. Godfrey, J.W. Morzycki, W.B. Motherwell and A. Stobil, *Tetrahedron Lett.*, 1982, **23**, 957.
3. D.H.R. Bartan, J.W. Morzycki and W.B. Motherwell *Chem. Commun.*, 1981, 1044.

1.68 *m*-IODOXYBENZOIC ACID, *m*-$O_2IC_6H_4CO_2H$

As in the case of iodoxybenzene, *m*-iodoxybenzoic acid in presence of catalytic amount of phenylselenic anhydride is used for dehydrogenation of ketones to α, β-unsaturated ketones. Thus, chlolestan-3-ol is converted into

1,4-cholestadien-3-one by *m*-iodoxybenzoic acid in presence of catalytic amount of phenylselenic anhydride, which is generated *in situ* from diphenyldiselenide (*see* iodoxybenzene).

Reference

1. D.H.R. Bartan, J.W. Morzycki and W.B. Motherwell, *Chem. Commun,* 1981, 1044.

1.69 LEAD DIOXIDE (LEAD SUPEROXIDE), PbO_2

It is mainly used for the oxidation of phenols to quinones. Thus, 3, 4-dimethyl phenol is oxidised[1] to orthoquinone in about 50% yield.

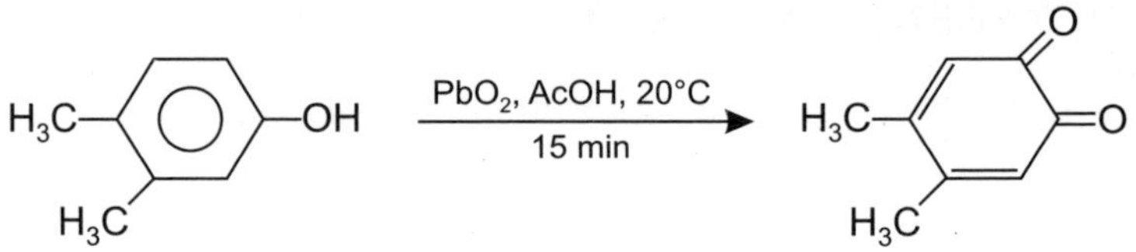
H3C
OH
PbO2, AcOH, 20°C
15 min
H3C
O
O
H3C
H3C

Sometimes, the oxidation of quinones is accompanied by doubling of the molecule[1].

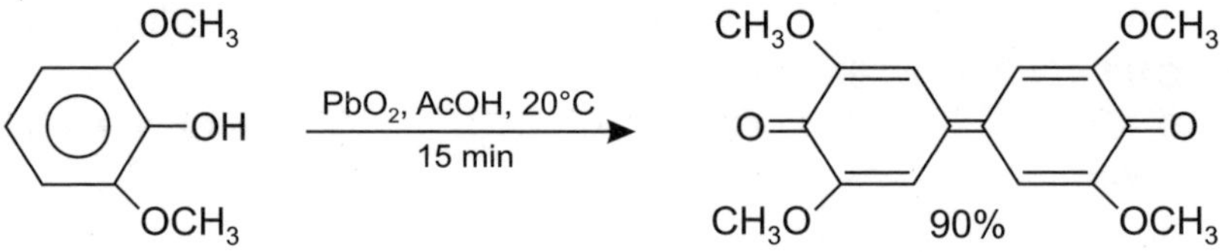
OCH3
OH
OCH3
PbO2, AcOH, 20°C
15 min
CH3O
OCH3
O
O
CH3O
90%
OCH3

Lead dioxide is used[2] for the preparation of 2, 3-dichloro-5, 6-dicyano-p-benzoquinone (DDQ) by oxidation of 2, 3-dichloro-5, 6-dicyano hydroquinone.

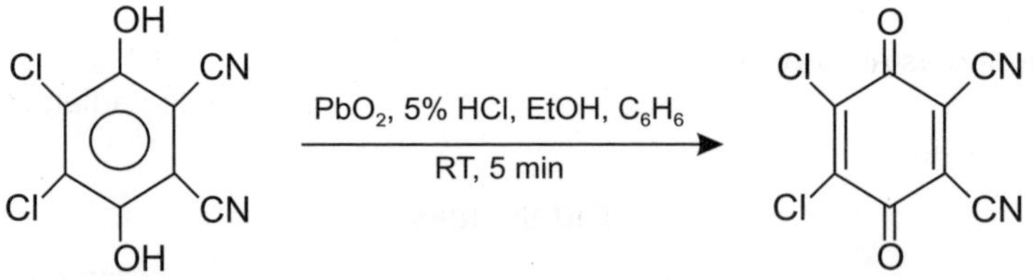
OH
Cl
CN
Cl
CN
OH
PbO2, 5% HCl, EtOH, C6H6
RT, 5 min
O
Cl
CN
Cl
CN
O

2,3-Dichloro-
-5,6-dicyano
hydroquinone

83%
DDQ

p, p′-Dihydroxyazobenzene (azophenol) is oxidised into quinone azine with lead dioxide[3]; in this case anhydrous Na_2SO_4 is added to remove the water of the reaction.

$HO-C_6H_4-N=N-C_6H_4-OH \xrightarrow[Et_2O,\ RT]{PbO_2,\ Na_2SO_4} O=C_6H_4=N-N=C_6H_4=O$

Azophenol

70%

Oxidation of N, N-diphenyl-N′-(2, 4, 6-trinitrophenyl) hydrazine with lead dioxide gives[4] diphenylpicrylhydrazyl, which is used for the dehydrogenation of 1, 4-dihydronaphthalene to naphthalene and the hydrazobenzene to azobenzene[5].

$(C_6H_5)_2NHN-C_6H_2(NO_2)_3 \xrightarrow{PbO_2} (C_6H_5)_2N\dot{N}-C_6H_2(NO_2)_3$

N, N-Diphenyl-N′-(2,4,6-trinitrophenyl hydrazine)

Diphenyl picryl hydrazyl

References

1. C.R.H.I. de Jone, H.M. Van Dort and L. Volbracht, *Tetrahedron Lett.,* 1970, 1881.
2. D. Todd and A.E. Martell, *Org. Synth.* Collective Volume, 1973, **5**, 617.
3. R. Willstätter and M. Benz, *Chem. Ber.,* 1906, **39**, 3482.
4. S. Goldschmidt and K. Renn, *Chem. Ber.,* 1922, **55**, 628.
5. E.P. Brande, A.G. Brook and R.P. Linstead, *J. Chem. Soc.,* 1954, 3574.

1.70 LEAD NITRATE, $Pb(NO_3)_2$

It is used to convert benzylic halides into the corresponding carbonyl compounds[1]. Thus, o-fluorobenzyl bromide is converted into o-fluorobenzaldehyde.

$o\text{-}F-C_6H_4-CH_2Br \xrightarrow[\text{reflux 5-6 hr.}]{Pb(NO_3)_2,\ H_2O} o\text{-}F-C_6H_4-CHO$

Reference

1. M. Hudlicky, Collect. Czech. *Chem. Commun.,* 1960, **25**, 1199.

1.71 LEAD TETRAACETATE, $Pb(OCOCH_3)_4$

It is a white hydroscopic crystalline substance and is readily hydrolysed with water to lead dioxide. In view of this, the oxidations with lead tetracetate are carried out in anhydrous solvents like benzene, dioxane and glacial acetic acid.

In lead tetracetate oxidations, the reactive species are the acetoxyl radicals. The addition of acetoxy radicals across double bonds produces vicinal diacetates[1].

Thus, lead tetraacetate in acetic acid at 70°C converts 1, 2-dihydronaphthalene to trans-1, 2-diacetoxy-1, 2, 3, 4-tetrahydronaphthalene in 72% yield.[1]

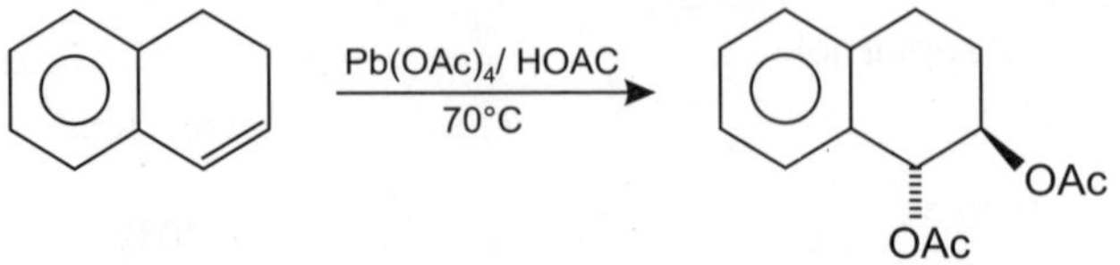

$Pb(OAc)_4$/ HOAC, 70°C

1,2-Dihydro naphthalene → 1,2-Diacetoxy 1,2,3,4-tetrahydro naphthalene

Lead tetraacetate replaces the benzylic hydrogen by an acetoxy group[2, 3]. One example is given below.

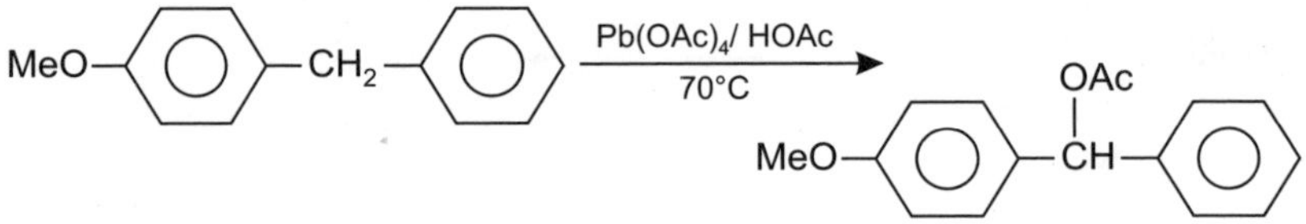

MeO–C_6H_4–CH_2–C_6H_5 → ($Pb(OAc)_4$/ HOAc, 70°C) → MeO–C_6H_4–CH(OAc)–C_6H_5

80%

Esters of hydroxy ketones are prepared by acetyoxylation[4] with lead tetracetate in benzene. Thus, cyclohexanone furnishes a 75% yield of 2-acetoxycy clohexanone and 3% yield of *cis*-2, 6-diacetoxy cyclohexanone.

$Pb(OAc)_4$/ HOAc, 80°C, 8 hr.

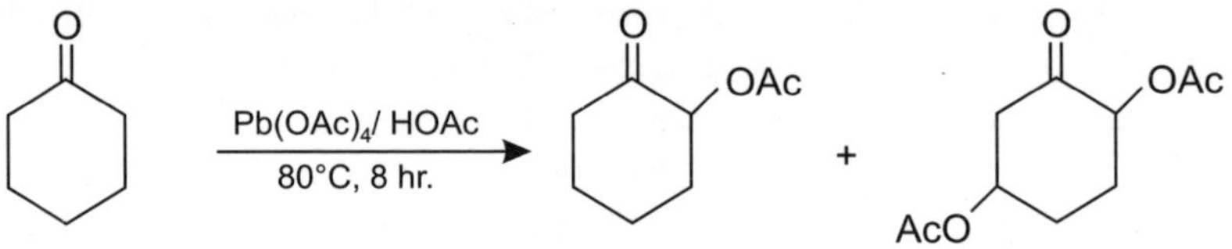

Cyclohexanone → 2-Acetoxy cyclohexanone 75% + 2,6-Diacetoxy cyclohexanone 3%

Lead tetraacetate also causes one-electron oxidation resulting in the formation of quinones from phenols[5, 6]. Thus, p, p′-dihydroxybiphenyls are converted into diphenoquinones.

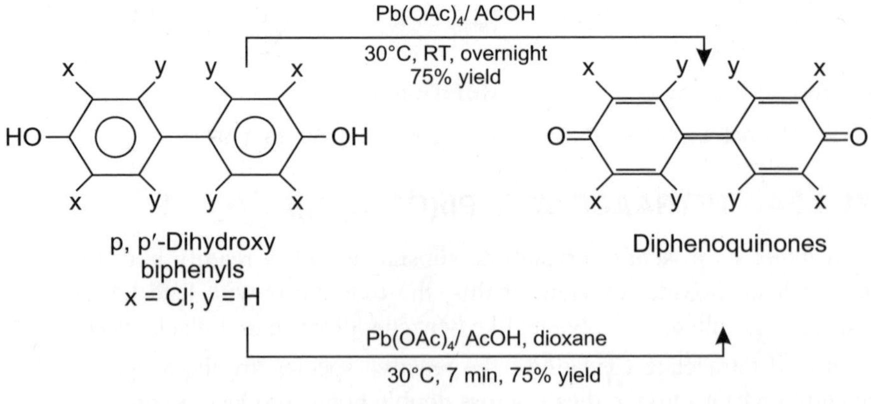

$Pb(OAc)_4$/ ACOH, 30°C, RT, overnight, 75% yield

p, p′-Dihydroxy biphenyls x = Cl; y = H → Diphenoquinones

$Pb(OAc)_4$/ AcOH, dioxane, 30°C, 7 min, 75% yield

Primary allylic alcohols are oxidised into the corresponding aldehydes without affecting the double bond.

$$CH_2{=}CHCH_2OH \xrightarrow[\text{RT, 10 hr.}]{Pb(OAc)_4,\ C_5H_5N} \underset{35\%}{CH_2{=}CHCHO}$$

$$C_6H_5CH_2OH \xrightarrow[\text{RT, few hours}]{Pb(OAc)_4,\ C_5H_5N} \underset{85\%}{C_6H_5\,CHO}$$

Lead tetracetate affects dehydrogenation of primary amines having the hydrogen atoms on the adjacent carbons to nitriles[8].

$$C_6H_{13}CH_2NH_2 \xrightarrow[60°C,\ 30\ min]{Pb(OAc)_4, C_6H_6,\ C_5H_5N} \underset{65\%}{C_6H_{13}\,CN}$$

Ketone hydrazones are transformed into diazo compounds[8, 9].

$$\underset{\text{Ketone hydrazones}}{R_2C{=}NNH_2} \xrightarrow{Pb(OAc)_4, PhCN,\ 0–25°C} \underset{76–79\%}{R_2C{=}N{=}N}$$

The oxidation of primary amides with lead tetraacetate in DMF in presence Et_3N gives isocyanates[10].

$$(CH_3)_3CCONH_2 \xrightarrow[50–60°C]{Pb(OAc)_4, DMF,\ Et_3N} \underset{44–96\%}{(CH_3)_3CNCO}$$

Lead tetraoxide oxidises secondary alcohols to ketones.

$$RR'C{-}OH \xrightarrow[C_5H_5N,\ rt]{Pb(OAc)_4} RR'C{=}O$$

The oxidative cleavage of the carbon-tin bond is possible with lead tetraacetate[11]. Thus, phenoxymethyltributylstannate on refluxing in dichloro methane with lead tetracetate gives phenoxymethyl acetate in 58% yield.

$$\underset{\text{Phenoxymethyltributyl stannate}}{C_6H_5OCH_2Sn(C_4H_9)_3} \xrightarrow[\text{reflux 32 hr.}]{Pb(OAc)_4,\ CH_2Cl_2} \underset{\text{58\% Phenoxymethylacetate}}{C_6H_5OCH_2OCOCH_3}$$

An important reaction of lead tetraacetate is the cleavage[12, 13] of vicinal diols to carbonyl compounds.

$$RO_2C\,CH(OH){-}CH(OH)\,CO_2R \xrightarrow[30°C,\ 1\ hr.]{Pb(OAc)_4,\ C_6H_6} \underset{77–87\%}{2RO_2CCHO}$$

$$R{=}C_4H_9$$

$$\text{1,1'-bicyclohexyl-1,1'-diol (OH OH)} \xrightarrow[25°C]{Pb(OAc)_4,\ 75\%\ HOAc} \underset{94\%}{\text{cyclohexanone}\ (C{=}O)}$$

Lead tetraacetate oxidises monocarboxylic acid in presence of copper(II) salts to give alkenes[14].

$$-\underset{H}{\overset{|}{C}}-\underset{COOH}{\overset{|}{C}}- \xrightarrow[Cu(OAc)_2]{Pb(OAc)_4} -\overset{|}{C}=\overset{|}{C}-$$

Cyclobutane carboxylic acid $\xrightarrow[C_6H_6,\ reflux]{Pb(OAc)_4/Cu(OAc)_2}$ Cyclobutene 77%

In a similar way ortho dicarboxylic acids on reaction with lead tetraacetate in presence of oxygen and pyridine give the corresponding unsaturated product.[15]

Cyclohexene 3,4-dicarboxylic acid $\xrightarrow[Py/67°C]{Pb(OAc)_4/O_2}$ Cyclohexadiene

References

1. R. Criegee, *Justus Liebigs Ann. Chem.*, 1930, **481**, 263
2. J. Cason, *Org. Synth* Collective Volume, 1955, **3**, 3.
3. G.W.K. Cavill and D.H. Solomon, *J. Chem. Soc.,* 1954, 3943.
4. G.W.K. Cavill and D.H. Solomon, *J. Chem. Soc.,* 1955, 4426.
5. K.-H. König, W. Schulze and G. Müller, *Chem. Ber.*, 1960, **93**, 554.
6. B.R. Brown and A.R. Todd, *J. Chem. Soc.,* 1954, 1280.
7. R.E. Partch, *Tetrahedron Lett.*, 1964, 3071.
8. A. Stojiljkovic', V. Andrejevic' and M.L. Mihailovic', *Tetrahedron*, 1967, **23**, 721.
9. W.J. Middleton and D.M. Gale, *Org. Synth.* Collective Volume, 1988, **6**, 161.
10. H.E. Baumgarten and A. Staklis, *J. Am. Chem. Soc.*, 1965, **87**, 1141.
11. Y. Yamamoto, H. Izukawa, M. Saiki and Y. Yamada, *Chem. Commun,* 1988, 560.
12. F.J. Wolf and J. Weijlard, *Org. Synth.* Collective Volume, 1963, **4**, 124.
13. W.S. Trahanovsky, L.H. Young and M.H. Bierman, *J. Org. Chem.,* 1969, **34**, 869.
14. R.J. Anderson and C.H. Henrick, *J. Am. Chem. Soc.,* 1975, **97**, 4327.
15. Kochi, *J. Am. Chem. Soc.*, 1965, **87**, 1811.

1.72 LEAD TETRAKIS (TRIFLUOROACETATE), $Pb(OCOCF_3)_4$

It is prepared[1] from red lead oxide (Pb_3O_4), trifluoroacetic acid and trifluoro acetic anhydride.

It is useful for replacement of benzylic hydrogen with an acyloxy group. Thus, 10-methylanthrone is converted to 10-acyloxy-10-methyl anthrane in 67% yield.

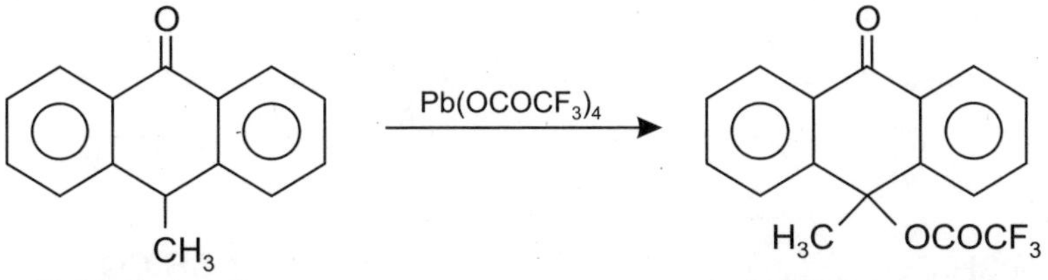

O O

Pb(OCOCF$_3$)$_4$

CH_3 H_3C $OCOCF_3$

10-Methyl anthrone 10-Trifluoroacetoxy-10-methyl anthrone

It converts alkylbenzenes into benzylic trifluoroacetates[1] and benzene and its derivatives into phenyltrifluoroacetates[1, 2].

References

1. R.E. Partch, *J. Am. Chem. Soc.,* 1967, **89**, 3662.
2. J.R. Campbell, J.R. Kalman, J.T. Pinhey and S. Sternhell, *Tetrahedron Lett.*, 1972, 1763.

1.73 MAGNESIUM PERMANGANATE $Mg(MnO_4)_2$ and ZINC PERMANGANATE, $Zn(MnO_4)_2$

Both these oxidising agents are much more reactive than potassium permanganate. Both these cause ignition on addition to alcohol, tetrahydrofuran, acetone and acetic acid. However, these can be used safely[1] when supported on silica gel in dichloromethane or chloroform at room or refluxing temperatures.

These are specially used for the conversion of acetylenes into α-diketones, of cyclic ethers to lactones and amines into to amides[1].

Reference

1. S. Wolfe and C.F. Ingold, *J. Am. Chem. Soc.,* 1983, **105**, 7755.

1.74 MESITYLGLYOXAL

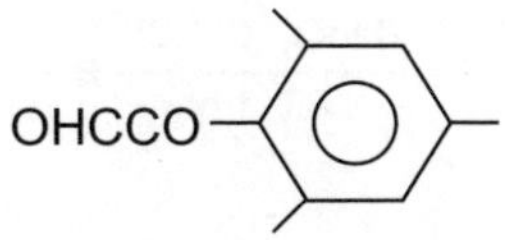

OHCCO

It is obtained[1] from 2-acetylmesitylene by oxidation with selenium dioxide.

Mesitylglyoxal is used for the conversion of primary amines (having amino group on secondary carbon) into ketones[2] via schiff bases at 0–23°C in high yields. The process is called **transamination**.

$>CHNH_2 + OHCCO$–(mesityl) ⟶ $>CHN{=}CH\ CO$–(mesityl) ⟶

$>C{=}N\ CH_2CO$–(mesityl) $\xrightarrow{H_2O}$ $>C{=}O + H_2NCH_2CO$–(mesityl)

In place of mesitylglyoxal, 3-nitromesitylglyoxal, 3, 5-dinitromesitylglyoxal and 3, 5-di-tert-butyl-o-benzoquinone can also be used.

References

1. A.R. Gray and R.C. Fusan, *J. Am. Chem. Soc.,* 1934, **56**, 739.
2. E.J. Corey and K. Achiwa, *J. Am. Chem. Soc.,* 1969, **91**, 1429.

1.75 MANGANESE DIOXIDE, MnO_2

It is commercially available and is prepared[1] by simultaneous addition of 42.5% aqueous solution of manganese sulphate tetrahydrate and a 40% solution of sodium hydroxide to a hot stirred 14% aqueous solution of potassium permanganate. The separated manganese dioxide is washed with water, dried (100 – 120°C) and ground to a fine powder. The activity of manganese dioxide can be modified by changing the particle size by using different drying temperatures (220 – 280°C). It is a by product of oxidation with potassium permanganate. It cleaves viccinal diols to carbonyl compounds. Following are given some of the oxidation reactions conducted by manganese dioxide:

1. *cis*-Decalin-9, 10-diol on treatment with manganese dioxide produces[2] cyclodecane-1, 6-dione in 90% yield.

$\xrightarrow[\text{RT, 1 hr.}]{MnO_2,\ CH_2Cl_2}$

cis-Decalin-9,10-diol Cyclodecane 1, 6-dione 90%

2. Manganese dioxide is used for dehydrogenation in six-membered nitrogen-containing compounds[3].

$\xrightarrow[80°C]{MnO_2,\ C_6H_6}$

79%

3. Secondary amines having hydrogen on the α-carbon atom are dehydrogenated to imines[3]

$$C_6H_5CH_2NH\ C_6H_5 \xrightarrow[80°C]{MnO_2,\ C_6H_6} C_6H_5CH{=}NC_6H_5$$

84%

4. Coupling takes place easily at benzylic positions. Thus, diphenyl methanes on oxidation with manganese dioxide give tetraphenyl ethane.

$$(C_6H_5)_2CH_2 \xrightarrow[211°C]{MnO_2,\ C_6H_6} (C_6H_5)_2CHCH\ (C_6H_5)_2$$

Diphenyl methane → 81% 1,1,2,2-Tetraphenyl ethane

5. Tertiary amines containing a methyl group bound to nitrogen are demethylated. Thus, codeine is converted into norcodeine in 75% yield[4].

Codeine $\xrightarrow[80°C]{MnO_2.\ C_6H_6}$ Norcodeine (75%)

6. In some cases, tertiary amines are cleaved to carbonyl compounds[5].

N,N-Dimethyl cyclohexylamine $\xrightarrow[20°C,\ 8\ hr.\ yield\ 85\%]{MnO_2,\ C_6H_6}$ Cyclohexanone $\xleftarrow[20°C,\ 8\ hr.\ yield\ 51\%]{MnO_2,\ C_6H_6}$ N,N,-Diethyl cyclohexyl amine

7. Normally manganese dioxide converts toluene into benzaldehyde.[5a] However, toluene on refluxing with manganese dioxide in petroleum ether yields benzoic acid in 75% yield[6].
8. Cinnamaldehyde on reaction with sodium cyanide gives the cyanohydrin, which on oxidation with MnO_2 in methanol gives cinnamic acid or methyl cinnamate depending on the reaction conditions[6a].

$$C_6H_5CH{=}CH-CHO \xrightarrow[\text{MeOH}]{\text{NaCN}} \left[C_6H_5CH{=}CH\,\underset{}{CH}(OH)-CN\right]$$

Cinnamaldehyde

$$\xrightarrow[\text{MeOH}]{-H_2O,\ \text{MnO}_2} [C_6H_5CH{=}CHCOCN]$$

20–25°C, 12 hr. → $C_6H_5CH{=}CHCO_2CH_3$ > 95% Methyl cinnamate

25°C, 2 hr. → $C_6H_5CH{=}CHCO_2H$ > 90% Cinnamic acid

9. Manganese dioxide oxidises cinnamyl alcohol to the corresponding aldehyde[1, 6b].

$$C_6H_5CH{=}CHCH_2OH \xrightarrow[\text{RT, 30 min}]{\text{MnO}_2,\ \text{CCl}_4} C_6H_5CH{=}CHCHO$$

Cinmamyl alcohol → Cinmamaldehyde

$$PhCH_2OH \xrightarrow[\text{24 hr, chloroform}]{\text{MnO}_2,\ \text{RT}} PhCHO\ \ 89\%$$

$$CH_2{=}CHCH_2OH \xrightarrow[\text{RT, 19 min}]{\text{MnO}_2,\ \text{Petroleum elter}} CH_2{=}CHCHO$$

$$(CH_3)_2C{=}CHCH_2CH_2C(CH_3){=}CHCH_2OH \xrightarrow[\text{RT, 96 hr}]{\text{MnO}_2,\ C_6H_6} (CH_3)_2C{=}CHCH_2CH_2C(CH_3){=}CHCHO\ \ 63\%$$

Under certain conditions, the oxidation of primary alcohols gives carboxylic acids[7].

3-hydroxy-4,5-bis(hydroxymethyl)-2-methylpyridine (CH_2OH, CH_2OH, HO, H_3C, N H^+Cl^-) $\xrightarrow[\text{RT, 3 hr.}]{\text{MnO}_2,\ \text{KOH, EtOH}}$ 4-carboxy product (CO_2H, CH_2OH, HO, H_3C, N) 82%

Under appropriate conditions[8] primary alcohols can also be converted into esters.

$$\underset{\text{Cinnamyl alcohol}}{C_6H_5CH{=}CHCH_2OH} \xrightarrow[0°C,\ 30\ min]{MnO_2/C_6H_{14}} \underset{\text{Cinnamaldehyde}}{C_6H_5CH{=}CHCHO}$$

$$\downarrow \text{NaCN, AcOH, 20-25°C; } MnO_2\text{, MeOH, 12 hr}$$

$$\left[C_6H_5CH{=}CHCH(OH)CN \longrightarrow C_6H_5CO{=}CHCOCN \right]$$

$$\downarrow \text{MeOH}$$

$$C_6H_5CH{=}CHCO_2Me$$

95%
Methyl cinnamate

10. Oxidation of appropriately substitute secondary alcohols yield acetylenic ketones[9].

$$CH_3CH{=}CHCH(OH)C{\equiv}COEt \xrightarrow[CH_2Cl_2,\ RT,\ 4.25\ hr]{16\ mol\ MnO_2} CH_3CH{=}CHCOC{\equiv}COEt \quad 40\%$$

Allylic alcohols in steroids are oxidised by manganese dioxide[10, 11].

$$\text{(3-hydroxy-}\Delta^4\text{-steroid, 17-OH)} \xrightarrow[RT,\ 4\text{-}5\ hr.]{MnO_2,\ CH_2Cl_2} \text{(3-oxo-}\Delta^4\text{-steroid, 17-OH)} \quad 70\%$$

Another example[12] of oxidation of secondary alcohol to ketone is given below.

$$CH_3CH(OH)C_6H_{13} \xrightarrow[distillation]{MnO_2,\ C_6H_6} CH_3COC_6H_{13} \quad 42\text{-}65\%$$

11. Hydrazo compounds are oxidised to azo compounds[13].

$$C_6H_5NHNHC_6H_5 \xrightarrow[80°C,\ 24\ hr.]{MnO_2,\ C_6H_6} C_6H_5N{=}NC_6H_5$$

12. Hydroxylamines are oxidised to nitroso compounds[14].

$$C_6H_5NHOH \xrightarrow[0°C,\ 3\ hr.]{MnO_2,\ H_2O} C_6H_5NO \quad 40\%$$

13. Thiols (mercaptans) on reaction with manganese dioxide leads to coupled products (disulphides)[14].

$$CH_2{=}CHCH_2SH \xrightarrow[\text{reflux, 5-6 hr}]{MnO_2,\ CHCl_3} CH_2{=}CHCH_2SSCH_2CH{=}CH_2$$

66%

14. Methylene groups adjacent to aromatic rings are oxidised to keto compounds. Thus, diphenyl methane on heating with manganese dioxide gives benzophenone in 74% yield[15].

$$C_6H_5{-}CH_2{-}C_6H_5 \xrightarrow[\text{6 hr.}]{MnO_2,\ 125°C} C_6H_5{-}CO{-}C_6H_5$$

Diphenylmethane

74%
Benzophenone

15. Oximes are oxidised to nitro compounds[16].

$$(EtO_2C)_2C{=}NOH \rightleftharpoons (EtO_2C)_2CHNO \xrightarrow[\text{30°C, 24 hr.}]{MnO_2,\ AcOH} (EtO_2C)_2CHNO_2$$

88%

16. Tertiary phosphines are readily converted into the corresponding oxides[17].

$$(C_6H_5)_3P \xrightarrow[\text{reflux, 5 hr.}]{MnO_2,\ \text{Petroleum ether}} (C_6H_5)_3P{=}O$$

100%

17. Aromatic primary amines are converted into azo compounds[17].

$$C_6H_5NH_2 \xrightarrow[\text{reflux, 6 hr.}]{MnO_2,\ \text{Petroleum ether}} C_6H_5N{=}NC_6H_5$$

87%

18. Manganese dioxide oxidises dibutyl sulphide exclusively to dibutyl sulfoxide[18].

$$C_4H_9SC_4H_9 \xrightarrow[\text{RT, 85 hr.}]{MnO_2,\ \text{Petroleum ether}} C_4H_9S(=O)C_4H_9$$

Dibutyl sulphide → Dibutyl sulfoxide
83%

19. Using manganese dioxide-silica, an expedicus and high yield route to carbonyl compound has been developed[19]. Thus, benzyl alcohols are selectively oxidised to carbonyl compounds using 35% MnO_2 'doped' silica under MW irradiation conditions.

$$R_2R_1CH{-}OH \xrightarrow[\text{MW, 20–60 sec.}]{MnO_2\text{-Silica}} R_2R_1C{=}O$$

67–96%

R_1 = Ph, p-MeC_6H_4, p-MeOC_6H_5, Ph CH = CH; R_2 = H
R_1 = Ph; R_2 = Et, Ph, PhCO.
R_1 = p-MeOC_6H_4; R_2 = p-MeO C_6H_4CO

References

1. J. Attenburrow, A.F.B. Carmeron, J.H. Chapman, R.M. Evans, B.A. Hems, A.B.A. Janson and T. Walker, *J. Chem. Soc.*, 1952, 1094.
2. G. Ohloff and W. Giersch, *Angew. Chem.*, 1973, **85**, 401.
3. E.F. Pratt and Mc Govern, *J. Org. Chem.,* 1964, **29**, 1540.
4. H.B. Henbest and M.J.W. Stratford, *J. Chem. Soc.,* 1966, 995.
5. E.F. Curragh, M.B. Henbest and A. Thomas, *J. Chem. Soc.,* 1960, 3559.

5a. X. Gilliard, P. Monnet and X. Cartier, Ger. Patent, 101 221; *Chem. Zentr.,* 1899, **1**, 960; Ger. Patent, 107 722; *Chem. Zentr.,* 1900, **1**, 1113.

6. M.Z. Barakat, M.F. Abdel-Wahab and M.M. Elsadr, *J.Chem. Soc.,* 1956, 4685.

6a. E.J. Corey, N. W. Gilman and B.E. Ganem, *J. Am. Chem. Soc.,* 1968, **90**, 5616.

6b. R.J. Highet and W.C. Wildman, *J. Am. Chem. Soc.,* 1955, **77**, 4399; J.R. Holum, J. Org. Chem., 1961, **26**, 4814; R.J. Gritter and T.J. Wallace, *J. Org. Chem.,* 1959, **24**, 1051.

7. M.P. Dogle, W.J. Patrie and S.B. Williams, *J. Org. Chem.,* 1979, **44**, 2955.
8. L.K. Sydnes, I.C. Burkow and S.H. Hansen, *Tetrahedron*, 1985, **41**, 5703.
9. G.. Stork and M. Tomasz, *J. Am. Chem. Soc.*, 1964, **86**, 475.
10. S. Cacchi, *Synthesis*, 1978, 848.
11. D. Burn, V. Petrow and G.O. Weston, *Tetrahedron Lett.*, 1960, 14.
12. E.F. Pratt and J.F. Van de Castle, *J. Org. Chem.,* 1961, **26**, 2973.
13. E.F. Pratt and T.P. McGovern, *J. Org. Chem.,* 1964, **29**, 1540.
14. E.P. Papadopoulos, A. Jarror and C.H. Issidorides, *J. Org. Chem.,* 1966, **31**, 615.
15. E.F. Pvatt and S.P. Suskind, *J. Org. Chem.,* 1963, **28**, 638.
16. L. Canonica, Gyazz, *Chem. Ital.,* 1947, **77**, 92.
17. H. Firouzabadi and Z. Mostafavipoor, Bull *Chem. Soc.,* Jpan., 1983, **56**, 914.
18. D. Edwards and J.B. Slenlake, *J. Chem. Soc.,* 1954, 3272.
19. R.S. Varma and R. Dahiya, *Tetrahedron Lett.,* 1997, **38**, 7823.

1.76 MANGANESE TRIACETATE, $Mn(OCOCH_3)_3$

It is prepared by refluxing a solution of manganese acetate tetrahydrate in acetic acid with potassium permanganate.

The reagent forms lactones from terminal alkenes[1, 2].

$$C_6H_5C(CH_3)=CH_2 \xrightarrow[\text{reflux, 45 min}]{Mn(OAc)_3.2H_2O,\ AcOH,\ Ac_2O} \underset{74\%}{C_6H_5C(CH_3)-CH_2-CH_2-CO-O \text{ (ring)}}$$

This reagent also hydroxylates benzylic methylene groups[3].

References

1. J.B. Bush, Jr., and H. Finkbeiner, *J. Am. Chem. Soc.*, 1968, **90**, 5903.
2. E.I. Heiba R.M. Dessau, A.L. Williams and P.G. Rodewald, *Org. Synth,* 1983, **61,** 22.
3. E. Baciocchi L. Mandolini and C. Rol, *J. Org.Chem.,* 1980, **45**, 3906.

1.77 MANGANIC SULPHATE $Mn_2(SO_4)_3$

It is obtained[1] by the oxidation of manganese sulphate (tetrahydrate) in 6M sulphuric acid by aqueous potassium permanganate at 0 – 25°C. The reagent is used for the oxidation of aromatic hydrocarbons to quinones.

Reference

1. M. Periasamy and M.V. Bhatt, *Tetrahedron Lett.*, 1978, 4561.

1.78 MERCURIC ACETATE, $Hg(OCOCH_3)_2$

It is used for dehydrogenations, resulting in the introduction of double bonds into the α-position with respect to conjugated systems[1, 2]. Thus, ergosterol or ergosterol acetate gives dehydroergosterol or its acetate in about 40% yield by refluxing in acetic acid with mercuric acetate.

$Hg(OAc)_2$, AcOH / EtOH, reflux / 40 min

Ergosterol R = H
Ergosterol acetate R = CH_3CO

40%
Dehydro ergosterol
or its acetate

The amines are dehydrogenated to imines[3] via quaternary salts. Double bonds between carbon and nitrogen in secondary and tertiary cyclic amines are formed on treatment with mercuric acetate[3].

Secondary amines having hydrogens on the α-carbon, one dehydrogenated to imines by mercuric oxide[4].

$Hg(OAc)_2$, 10% AcOH / reflux, 20 hr.

R = Me, Pr, Bu,
t-Bu

30-75% yield

Mercuric acetate can also demethylate tertiary amines to secondary amines[5] and introduce acetoxyl groups into the α-position with respect to double bonds[6].

$$C_6H_5CH_2CH{=}CH_2 \xrightarrow[\text{reflux, 50 hr.}]{Hg(OAc)_2,\ AcOH} C_6H_5CH{=}CHCH_2OCOCH_3$$

Allyl benzene 68%

References

1. A. Windaus and O. Linsert, *Justus Liebigs Ann. Chem.,* 1928, **465**, 148.
2. A. Züvcher, H. Heusser, O. Jeger and P. Geistlich, *Helv Chem. Acta,* 1954, **37**, 1562.
3. N.J. Leonard, A.S. Hay, R.W. Fulmer and V.W. Gash, *J. Am. Chem. Soc.,* 1955, **77**, 439; N.J. Leonard, and F.P. Hauck, Jr., *J. Am. Chem. Soc.,* 1957, **79**, 5279.
4. M.F. Grundon and B.E. Reynolds, *J. Chem. Soc.,* 1964, 2445.
5. N.J. Leonard and D.F. Morrow, *J. Am. Chem. Soc.,* 1958, **80**, 371.
6. Z. Rappoport, S. Winstein and W.G. Young, *J. Am. Chem. Soc.,* 1972, **94**, 2320.

1.79 MERCURIC BROMIDE, $HgBr_2$

It oxidises methylene groups that are adjacent to double bonds or aromatic rings to carbonyls[1].

$$C_6H_5CH_2CH_3 \xrightarrow{HgBr_2} C_6H_5COCH_3$$

Ethylbenzene → Acetophenone

Reference

1. N. Friedman, M. Gorodetsky and Y. Mazur, *Chem. Commun,* 1971, 874.

1.80 MERCURIC OXIDE, HgO

The yellow mercuric oxide is more reactive than the red mercuric oxide. It resembles silver oxide in its oxidising properties. Some of its applications are given below:

(*i*) Oxidation of phenols to quinones.

The oxidation of 2, 5-di-tert-butyl-4-methoxyphenol to 2, 5-di-tert-butyl-p-benzoquinone is achieved by mercuric oxide[1].

$$\text{2,5-Di-tert-butyl-4-methoxyphenol} \xrightarrow{HgO,\ MeOH,\ RT} \text{2,5-Di-tert-butyl-p-benzoquinone (86\%)}$$

2,5-Di-tert-butyl-4-methoxyphenol — 86% 2,5-Di-tert-butyl-p-benzoquinone

(*ii*) Oxidation of hydrazones into diazo compounds[2]

$$C_6H_5C(=NNH_2)COC_6H_5 \xrightarrow[RT.,\ 10\text{–}15\ min]{HgO,\ Et_2O,\ Na_2SO_4} C_6H_5C(=N_2)COC_6H_5$$

87–94%

(*iii*) Oxidation of dihydrazones of α-diketones to acetylenes[3, 4, 5].

$C_6H_5C(=NNH_2)-C(=NNH_2)C_6H_5 \xrightarrow[\text{reflux 1 hr. RT overnight}]{HgO,\ C_6H_6} C_6H_5C\equiv C-C_6H_5$ 67–73%

$(CH_2)_8(C=NNH_2)_2 \xrightarrow[\text{PhMe, reflux 2.5 hr}]{HgO,\ KOH (S),\ Na_2SO_4} (CH_2)_8(C\equiv C)$

Mercuric oxide dehydrogenates N-aminopiperidines to tetrazenes[6] or converted into hydrocarbons[7].

References

1. A. McKillop and D.W. Young, *Synth. Commun.,* 1977, **7**, 467.
2. C.D. Neritzescu and E. Solomonica, *Org. Synth.* Collective Volume, 1943, **2**, 496.
3. A.C. Cope, D.S. Smith and R.J. Cotter, *Org. Synth.* Collective Volume, 1963, **4**, 377.
4. A. Blomquist, R.E. Burge, Jr., and A.C. Sucsy, *J. Am. Chem. Soc.,* 1952, **74**, 3636.
5. V. Prelog, K. Schenker and H.H. Günthard, *Helv. Chem, Acta,* 1952, **35**, 1598.
6. C.G. Overberger, L.C. Palmer, B.S. Marks and N.R. Byrd, *J. Am. Chem. Soc.,* 1955, **77**, 4100.
7. C.G. Overberger, J.G. Lombardino and R.G. Hiskey, *J. Am. Chem. Soc.,* 1957, **79**, 6430.

1.81 MERCURIC TRIFLUOROACETATE, $Hg(OCOCF_3)_2$

It resembles mercuric oxide in its action. It oxidises hydroquinones to quinones.[1] Phenols are converted into ortho[2] and para[3] quinones by mercuric trifluoroacetate.

3,4-Dimethyl phenol $\xrightarrow{\text{Mercuric trifluoroacetate}}$ Orthoquinone (49–50%)

2,3,6-Trimethyl phenol $\xrightarrow{\text{Mercuric trifluoroacetate}}$ p-Quinone (77-79%)

It also oxidises hydrazones to diazo compounds.[4]

$C_6H_5C(=NNH_2)CH_2C_6H_5 \xrightarrow{\text{Mercuric trifluoroacetate}} C_6H_5C(=N_2)CH_2C_6H_5$

References

1. R. Baltzly, N.B. Mehta, P.B. Russel, R.E. Brooks, E.M. Grivsky and A.M. Sleinberg *J. Org. Chem.,* 1961, **26**, 3669.
2. H. J. Teuber, *Org. Synth.* Collective Volume, 1988, **6** , 480.
3. P.A. Wehrli and F. Pigott, *Org. Synth.* Collective Volume 1988, **6**, 1010.
4. M.S. Newman and A. Arkell, *J. Org. Chem.,* 1959, **24**, 385.

1.82 MERCUROUS TRIFLUOROACETATE, $HgOCOCF_3$

Benzil monohydrazone on refluxing with mercurous trifluoroacetate in ether for 2 hr. yields diphenyl acetylene[1].

$$C_6H_5-\underset{\underset{NNH_2}{\|}}{C}-\overset{\overset{O}{\|}}{C}-C_6H_5 \xrightarrow[\text{ether, reflux, 2 hr.}]{HgO\ COCF_3} C_6H_5C\equiv CC_6H_5$$

Benzil Monohydrazone → Diphenyl acetylene 43%

Reference

1. R.J. Theis and R.E. Dessy, *J. Org. Chem.,* 1966, **31**, 624.

1.83 MOLYBDENUM HEXAFLUORIDE, MoF_6

It is suitable for the recovery of ketones from their hydrazones and tosylhydrazones under very mild conditions[1].

Cyclohexanone hydrazone ($=NNH_2$) $\xrightarrow[0°C\ 78\%\ yield]{MoF_6,\ C_2Cl_3F_3}$ Cyclohexanone ($=O$) $\xleftarrow[0°C\ 74\%\ yield]{MoF_6,\ C_2Cl_3F_3}$ Cyclohexanone tosylhydrazone ($=NNHToS$)

In place of molybdenum hexafluoride, molybdenum oxychloride ($MoOCl_3$) can also be used. It is prepared by the partial hydrolysis of molybdenum pentachloride[1].

Reference

1. G.A. Olah, J. Welch, G.K.S. Prakash and T.-L. HO, *Synthesis,* 1976, 808.

1.84 NICKEL PEROXIDE

It is prepared from nickel sulphate hexahydrate by oxidation in alkaline medium with an ozone-oxygen mixture[1] or with sodium hypochlorite[2].

The unusual oxidant nickel peroxide converts aromatic aldehydes into carboxylic acids[3].

$$C_6H_5CHO \xrightarrow[30°C\ 1.5\ hr.]{NiO_2,\ NaOH,\ H_2O} \underset{94\%}{C_6H_5CO_2H}$$

Aromatic side chains are oxidised to carboxylic acids. Thus, 2-methylquinoline-3-carboxylic acid is oxidised quantitatively by nickel peroxide in sodium hydroxide to quinoline 2, 3-dicarboxylic acid[4].

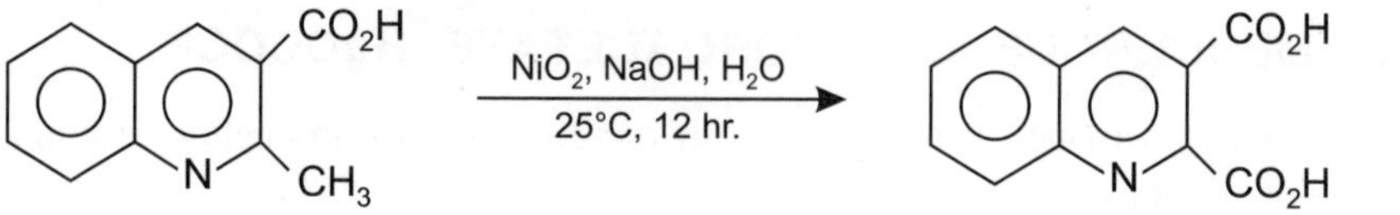

$$\text{2-Methyl quinoline-3-carboxylic acid} \xrightarrow[25°C,\ 12\ hr.]{NiO_2,\ NaOH,\ H_2O} \text{Quinoline-2,3-dicarboxylic acid (100\%)}$$

2-Methyl quinoline
-3-carboxylic acid

100%
Quinoline-2,3-dicarboxylic
acid

Hydrazones are oxidised to diazo compounds[5].

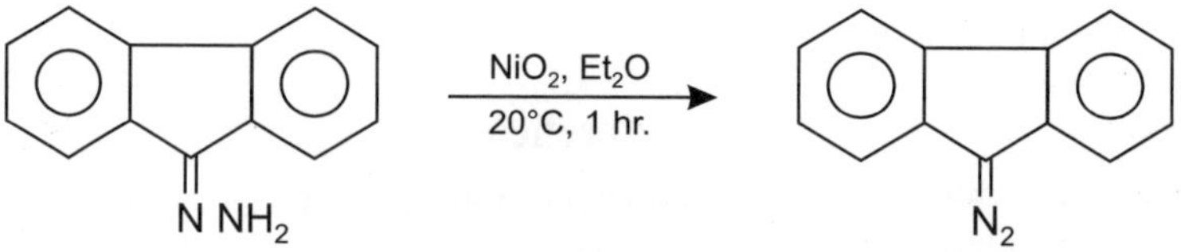

$$\text{(fluorenone hydrazone, } C{=}N\,NH_2) \xrightarrow[20°C,\ 1\ hr.]{NiO_2,\ Et_2O} \text{(9-diazofluorene, } C{=}N_2) \quad 92\%$$

Primary alcohols are oxidised to aldehydes[1] amides[6] and carboxylic acids[7].

$$C_6H_5CH_2OH \xrightarrow[reflux,\ 1\ hr.]{NiO_2,\ C_6H_6} \underset{100\%}{C_6H_5CHO}$$

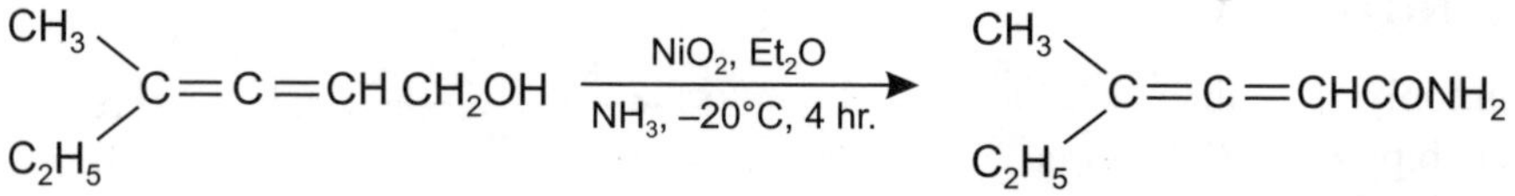

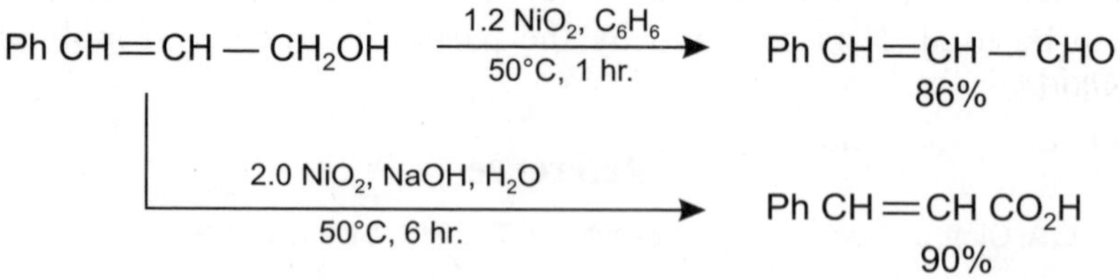

$$(CH_3)(C_2H_5)C{=}C{=}CH\,CH_2OH \xrightarrow[NH_3,\ -20°C,\ 4\ hr.]{NiO_2,\ Et_2O} (CH_3)(C_2H_5)C{=}C{=}CHCONH_2$$

$$Ph\,CH{=}CH-CH_2OH \xrightarrow[50°C,\ 1\ hr.]{1.2\ NiO_2,\ C_6H_6} \underset{86\%}{Ph\,CH{=}CH-CHO}$$

$$Ph\,CH{=}CH-CH_2OH \xrightarrow[50°C,\ 6\ hr.]{2.0\ NiO_2,\ NaOH,\ H_2O} \underset{90\%}{Ph\,CH{=}CH\,CO_2H}$$

$$CH{\equiv}CCH_2OH \xrightarrow[5°C,\ 0.5\ hr.]{NiO_2,\ NaOH,\ H_2O} \underset{50\%}{CH{\equiv}CCO_2H}$$

Primary amines are oxidised to azo compounds[8] or to nitriles[8].

NH$_2$ / NO$_2$ (p-Nitroaniline) $\xrightarrow[\text{reflux, 6 hr.}]{NiO_2,\ C_6H_6}$ O$_2$N–C$_6$H$_4$–N=N–C$_6$H$_4$–NO$_2$

p-Nitroaniline

44%
p,p′-Dinitro azobenzene

$$CH_3(CH_2)_5NH_2 \xrightarrow[\text{reflux, 6 hr.}]{NiO_2,\ C_6H_6} CH_3(CH_2)_4CN$$

Hexylamine → 73% Capronitrile

References

1. J.S. Belew and C. Tek-Liny, *Chem. Commun.,* 1967, 1100.
2. R.N. Warrener and E.N. Coin, *Aust. J. Chem.,* 1971, **24**, 785.
3. K. Nakagawa, S. Mineo and S. Kawamura, *Chem. Pharm. Bull,* 1978, **26**, 229.
4. D.W. Ladner, *Synth. Commun.,* 1986, **16**, 157.
5. K. Nakagawa, H. Onoue and K. Minami, *Chem. Commun.,* 1966, 730.
6. M. Bertrand, G. Gil and J. Viala, *Tetrahedron Lett.,* 1979, 1595.
7. K. Nakagawa, R. Konaka and T. Nakata, *J. Org. Chem.,* 1962, **27**, 1597.
8. K.Nakagawa and T. Tussji, *Chem. Pharm. Bull.*, 1963, **11**, 296.

1.85 NITRIC ACID, HNO_3

Commonly available nitric acid is about 68% (d 1.41, azeotropic mixture with water, b.p. 120.5°C). Some of the reactions in which nitric acid is used as an oxidising agent are given below.

(*i*) Oxidation of cyclic ketones.

Oxidation of cyclopentanone with nitric acid gives glutaric acid[1].

$$\text{Cyclopentanone} \xrightarrow{13\%\ HNO_3,\ 100°C} HOOC{-}CH_2{-}CH_2{-}CH_2{-}COOH$$

Cyclopentanone

82%
Glutaric acid

Similary, cyclohexanone on oxidation with nitric acid gives adipic acid[2].

$$\text{Cyclohexanone} \xrightarrow[\text{Warm, gradual addn.}]{33\%\ HNO_3} \begin{array}{l} CH_2CH_2CO_2H \\ | \\ CH_2CH_2CO_2H \end{array}$$

Cyclohexanone

100%
Adipic acid

(*ii*) Dehydrogenation (aromatization) of dihydropyridines[3].

Dihydropyridine derivatives resulting from the Hantzch synthesis are dehydrogenated to pyridines by heating with dilute mixture of nitric acid and sulphuric acid.

$$\text{4-R-3,5-}(CO_2Et)_2\text{-2,6-}(CH_3)_2\text{-1,4-dihydropyridine (N–H)} \xrightarrow[H_2SO_4,\ \Delta]{HNO_3} \text{4-R-3,5-}(CO_2Et)_2\text{-2,6-}(CH_3)_2\text{-pyridine}$$

(*iii*) Oxidation of aldehydes to carboxylic acids.

Some aldehydes are oxidised to acids by strong oxidising agent. Thus, β-chloropropionaldehyde gives β-chloropropionic acid[4] with fuming nitric acid (100% nitric acid).

$$ClCH_2CH_2CHO \xrightarrow[30\text{–}35°C,\ 20\ min]{\text{Fuming } HNO_3} ClCH_2CH_2CO_2H$$

β-Chloropropionaldehyde β-Chloropropionic acid

The oxidation of both the aldehyde group and the terminal primary alcoholic group in sugars is generally carried out with dilute nitric acid at 100°C and gives dicarboxylic acid (aldaric acid).

$$\begin{array}{c} CHO \\ \vdash OH \\ HO \dashv \\ \vdash OH \\ \vdash OH \\ CH_2OH \end{array} \xrightarrow{\text{dil } HNO_3} \begin{array}{c} COOH \\ \vdash OH \\ HO \dashv \\ \vdash OH \\ \vdash OH \\ COOH \end{array}$$

Glucose Aldaric acid

(*iv*) Oxidation of aromatic amines to quinones.

Aminonaphthols are converted into naphthoquinones by nitric acid[5].

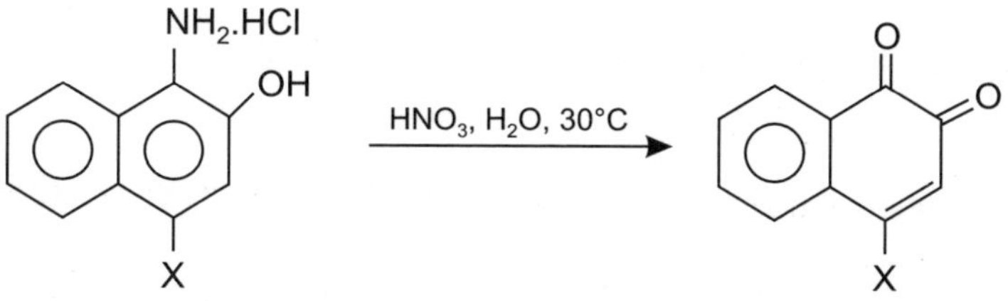

$NH_2.HCl$, OH, X — HNO_3, H_2O, 30°C → O, O, X

X = SO_3H 94–98%

In the above reaction, the product is isolated as the ammonium salt.

(*v*) Oxidation of aromatic side chains.

Not only methyl group, but practically any other alkyl group attached to aromatic ring are oxidised to carboxylic acids by nitric acid.

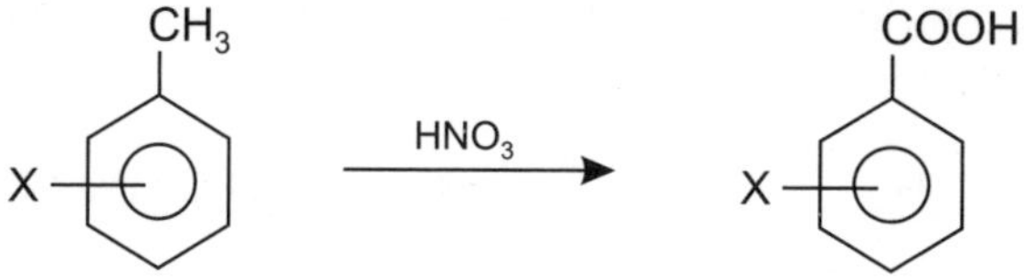

CH_3, X — HNO_3 → COOH, X

X=H — 15% HNO_3, 170–200°C, 30–50 atm. → Benzoic acid[6] (85–90%)

X=o—CH_3 — dil. HNO_3 (1 : 2), 145–550°C, reflux, 5 hr. → Phthalic acid[7] (53–55%)

X=p—CH_3, o—F — 22% HNO_3, 200°C, 52 atm. 15 min → 2–Fluoro benzene 1, 4-dicarboxylic acid[8] (79)%

X=p—F — 20% HNO_3, 190–95°C, 70 atm. 6 hr. → p-Fluorobenzoic acid[9] (96%)

In a similar way, methyl substituted pyridines are oxidised to the corresponding carboxylic acids.

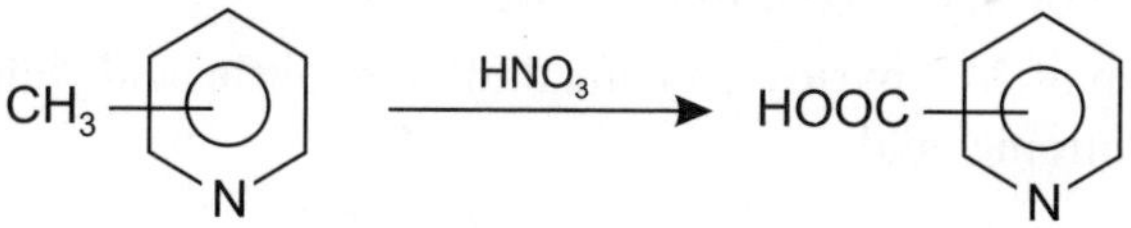

CH_3, N — HNO_3 → HOOC, N

β-Methyl pyridine — 15% HNO_3, 200°C, 35–40 atm. → Pyridine–3–carboxylic acid[6]

γ-Methyl pyridine — 10% HNO_3, H_3PO_4, 230°C, 40 atm. → Pyridine–4–carboxylic acid[6]

(*vi*) Oxidsation of 3, 4-dihydro-2H pyran with dilute nitric acid at 100°C gives glutaric acid[10] in 70 – 75% yield.

(*vii*) Oxidation of esters to carboxylic acids. 2-Fluoroheptylacetate on heating with nitric acid at 48°C for 25 hr. gives 2-fluoroheptanoic acid[11] along with acetic acid.

$$\underset{\text{2-Fluoroheptyl acetate}}{CH_3CH_2CH_2CH_2CH_2CH_2\underset{F}{\underset{|}{CH}}\ COOCH_3} \xrightarrow[48°C,\ 25\ hr.]{HNO_3} \underset{\substack{\text{2-Fluoroheptanoic acid} \\ 78\text{–}84\%}}{CH_3CH_2CH_2CH_2CH_2CH_2\underset{F}{\underset{|}{CH}}\ COOH} + \underset{\text{Acetic acid}}{CH_3COOH}$$

(*viii*) Oxidation of hydrazo compounds to azo compounds[12].

Diethyl hydrazodicarboxylate is dehydrogenated to azodicarboxylate with fuming nitric acid at 5°C for 2hr.

$$\underset{\text{Diethyl hydrazodicarboxylate}}{C_2H_5O_2CNH\ NH\ CO_2C_2H_5} \xrightarrow[50,\ 2\ hr.]{\text{Fuming } HNO_3} \underset{\substack{\text{Diethyl azo-dicarboxylate} \\ 70\text{–}80\%}}{C_2H_5O_2C\ CN{=}N\ CO_2C_2H_5}$$

(*ix*) Oxidation of iodo compounds to idoso compounds[13].

2-Iodo-4, 5-dinitrobenzoic acid an heating with fuming nitric acid at 100°C for 3 min gives 2-iodoso-4, 5-dinitrobenzoic acid in 92% yield.

2-Iodo-4,5-dinitro benzoic acid (COOH, I, O_2N, NO_2) $\xrightarrow[100°C,\ 3\ min]{\text{Fuming } HNO_3}$ 2-Iodoso-4,5-dinitro-benzoic acid (COOH, IO_2, O_2N, NO_2)

(*x*) Oxidation of thiols to sulfonic acids[14].

Oxidation of 2-(2-pyridyl) ethane thiol with nitric acid yields 2-(2-pyridyl) ethane sulfonic acid.

2-(2-Pyridyl) ethane thiol (HCl salt) (N, CH_2CH_2SH, HCl salt) $\xrightarrow[\text{exothermic, } 100°C]{HNO_3,\ (d.\ 1.4)}$ 2-(2-Pyridyl) ethane sulfonic acid (N, $CH_2CH_2SO_3H$), 90%

(*xi*) Oxidation of perfluoro aromatic compounds to quinones[15, 16].

Perfluorinated aromatic compounds are oxidised to quinones with nitric acid. Two examples are given below.

$$C_6F_6 \xrightarrow[\text{reflux 80°C, 6 hr.}]{HNO_3} \text{tetrafluoro-}p\text{-benzoquinone (30\%)}$$

$$\text{Heptafluoroisoquinoline} \xrightarrow[\text{100°C, 20 hr.}]{\text{100\% } HNO_3} \text{tetrafluoro homophthalimide dione (92\%)}$$

(*xii*) Oxidation of primary alcohols to carboxylic acids[17, 18].

Oxidation of 3-chloropropanol to 3-chloropropanoic acid in 78-79% is accomplished by conc. nitric acid (d 1.42) at 25 – 30°C.

$$\underset{\text{3-Chloropropanol}}{ClCH_2CH_2CH_2OH} \xrightarrow[\text{5–10 hrs.}]{\text{Conc. } HNO_3 \text{ (d. 1.42)}} \underset{\substack{\text{3-Chloropropanoic acid} \\ 78\text{–}79\%}}{ClCH_2CH_2CH_2COOH}$$

$$\underset{\text{6-Bromohexanol}}{BrCH_2CH_2CH_2CH_2CH_2CH_2OH} \xrightarrow[\text{100°C, 45 min}]{\text{Conc. } HNO_3 \text{ (d. 1.42)}} \underset{\substack{\text{6-Brmohexanoic acid} \\ 91\%}}{BrCH_2CH_2CH_2CH_2CH_2COOH}$$

(*xiii*) Oxidation of secondary alcohols to carboxylic acids[19, 20].

Oxidation of secondary alcohols like cyclohexanol gives initially cyclohexanone, which gets oxidised into adipic acid.

$$\underset{\text{Cyclohexanol}}{C_6H_{11}\text{–}OH} \xrightarrow[\text{55–60°C, > 1 hr.}]{\text{50\% } HNO_3\text{/ } NH_4VO_3} \underset{\substack{\text{Adipic acid} \\ 58\text{–}60\%}}{HOOC(CH_2)_4COOH}$$

(*xiv*) Oxidation of sulphides to sulfoxides[21].

Thiacyclohexanes substituted at position 4 on oxidation with nitric acid give a mixture of two possible stereoisomers of sulfoxides (*cis* and *trans*).

$(CH_3)_3C$ — 4-tert. butyl thiacyclohexane $\xrightarrow[0°C]{HNO_3,\ Ac_2O}$ cis 67% + trans 33%

(*xv*) Nitric acid is also used for the synthesis of a large number of intermediates, which are required for organic synthesis. Two such intermediates are given below.

Pyrogallol trimethyl ether $\xrightarrow[CH_3COOH]{[O]\ HNO_3\ (d\ 1.2)}$ (quinone, CH_3O, OCH_3) $\xrightarrow[(2)\ Methylation]{(1)\ Redn.}$ 1,2,3,5-Tetramethoxy benzene

Pyrogallol tribenzoate $\xrightarrow[CH_3COOH]{HNO_3\ (d\ 1.19)}$ amnine (BzO, OBz) $\xrightarrow[(2)\ Methylation\ (3)\ Debenzylation]{(1)\ Redn.}$ 1,3-Dihydroxy-2,5-dimethoxy benzene

References

1. E. Boedker, *J. Pharm. Chem.,* 1932, **15**, 225.
2. E. Looft, *Chem. Ber.,* 1894, **27**, 1542.
3. A. Singer and S.M. McElvain, *Org. Synth.* Collective Volume, 1943, **2**, 214.
4. C. Moureu and R. Chaux, *Org. Synth.* Collective Volume, 1932, **1**, 166.
5. E.L. Martin and L.F. Fieser, *Org. Synth.* Collective Volume, 1955, **3**, 633.
6. E.B. Bengtsson, *Acta Chem.* Scand., 1953, **7**, 774; 1954, **8**, 842; 1955, **9**, 832.
7. H.E. Zaugg and R.T. Rapala, *Org. Synth.* Collective Volume, 1955, **3**, 820.
8. G. Volkanas and H. Hopff, *J. Chem. Soc.,* 1963, 3475.
9. H. Hopff and G. Valkanas, *J. Org. Chem.,* 1962, **27**, 2923.
10. J. English, Jr. and J.E. Dayan, *Org. Synth.* Collective Volume 1963, **4**, 499.
11. F. H. Dean, J.H. Amin and F.L.M. Pattison, *Org. Synth.* Collective Volume, 1973, **5**, 580.
12. J.C. Kauer, *Org. Synth.* Collective Volume, 1963, **4**, 411.
13. H. Goldstein and R. Jaunin, *Hel. Chem. Acta,* 1951, **34**, 2222.
14. L. Bauer and L.A. Gardella, Jr. *J. Org. Chem.,* 1961, **26**, 82.
15. P. Sartori, K. Anlers and H.-J. Frohn, *J. Fluorine Chem.* 1976, **8**, 457.
16. G.G. Yakobson, V.D. Shteingarts and N.N. Vorozhtov, Zh. Vses. Khim. Ova. im. D.I. Mendeleeva, 1964, **9**, 701; *Chem. Abstr,* 1965, **62**, 9052C.
17. S.G. Powell, E.H. Huntress and E.B. Hersberg, *Org. Synth.* Collective Volume, 1932, **1**, 168.
18. E.F. Degering and L.G. Boatright, *J. Am. Chem. Soc.,* 1950, **72**, 5137.
19. R.R. Coates and J.W. Cook, *J. Chem. Soc.,* 1942, 559.
20. B.A. Ellis, *Org. Synth.* Collective Volume, 1932, **1**, 18.
21. C.R. Johnson and D. McCants, Jr., *J. Am. Chem. Soc.,* 1965, **87**, 1109.

1.86 NITROSOBENZENE, C_6H_5NO

It is obtained by the oxidation of phenylhydroxylamine and is a specific oxidising agent for the preparation[1,2] of aromatic aldehydes from benzyl halides or tosylates and of α-dicarbonyl compounds from α-haloketones.

$$ArCH_2X \xrightarrow{C_5H_5N} ArCH_2-\overset{+}{N}C_5H_5\ X^- \xrightarrow{^-OH}$$

X=Cl, Br, I or TosO

$$\longrightarrow Ar\overset{-}{C}H-\overset{+}{N}C_5H_5 \ (+\ R-\ddot{N}=O) \xrightarrow[-H_2O]{-C_5H_5N} \begin{matrix} ArCH \\ \| \\ R-N-O \end{matrix} \xrightarrow{H_2O} ArCHO + RNHOH$$

$$RCOCH_2X \xrightarrow{C_5H_5N} RCOCH_2-\overset{+}{N}C_5H_5\ X^- \xrightarrow{^-OH}$$

X=Cl, Br, I or TosO

$$\longrightarrow RCO\overset{-}{C}H-\overset{+}{N}C_5H_5 \ (+\ R'-\ddot{N}=O) \xrightarrow[-H_2O]{-C_5H_5N} RCOCH=\overset{O}{\overset{|}{N}}R' \xrightarrow{H_2O} RCOCHO + R'NHOH$$

Ketoaldehyde

References

1. F. Kröhnke, Angew, *Chem.* Int. Ed. Engl. 1963, **2**, 380.
2. F. Kröhuke and E. Börner, *Chem.* Ber., 1936, **69**, 2006.

1.87 P-NITROSODIMETHYLANILINE, P-$(CH_3)_2$ NC_6H_4NO

It is prepared by the nitrosation of dimethylaniline. Like nitrosobenzene, the reagent is also specific for the oxidation of aldehydes from benzyl halides or tosylates and α-dicarbonyl compounds from α-haloketones (*see* nitrosobenzene).

1.88 OXYGEN

The ordinary oxygen is called the ground-state oxygen (triplet oxygen). It has two electrons with parallel spins and behaves like a diradical and is paramagnetic [↑· O : O · ↑]. However, in excited state oxygen (called singlet oxygen), the two odd electrons possess antiparallel spins [↑ · O : O · ↓]. Such a molecule is unstable with a half life of 10^{-6} sec. and is diamagnetic (for more details see singlet oxygen). Each form reacts differently with organic molecules.

Ordinary oxygen (ground state oxygen) is not used as an oxidising agent. A classcial example is the autooxidation of benzaldehyde to benzoic acid. However, oxygen is used for various oxidations in presence of certain catalysts. This process is known as catalytic oxidation and is used for industrial preparation of a number of important products. Some of the processes are described below:

(*i*) Catalytic oxidation of methane into methyl alcohol.

$$\underset{\text{Methane}}{CH_4} + \frac{1}{2}O_2 \xrightarrow[\text{100 atm. 260°C}]{\text{Cu catalyst}} \underset{\text{Methyl alcohol}}{CH_3OH}$$

(*ii*) Controlled oxidation of methane into formaldehyde.

$$\underset{\text{Methane}}{CH_4} + O_2 \xrightarrow[\text{MoO, 400°C, 200 atm.}]{\text{Controlled oxidation}} \underset{\text{Formaldehyde}}{HCHO} + H_2O$$

(*iii*) Oxidation of ethyl alcohol into acetaldehyde

$$\underset{\text{Ethyl alcohol}}{CH_3CH_2OH} \xrightarrow[\text{or Cu/300°C}]{\text{Ag/Air/250°C}} \underset{\text{Acetaldehyde}}{CH_3CHO} + H_2O$$

(*iv*) Oxidation of ethylene into acetaldehyde

$$\underset{\text{Ethylene}}{H_2C{=}CH_2} + \frac{1}{2}O_2 \xrightarrow[CuCl_2]{PdCl_2/H_2O} \underset{\text{Acetaldehyde}}{CH_3CHO}$$

(*v*) Oxidation of toluene in vapour phase to benzaldehyde.

$$\underset{\text{Toluene}}{C_6H_5CH_3} + O_2 \xrightarrow[\text{500°C}]{MnO_2/Mo_2O_3/ZrO_2} \underset{\text{Benzaldehyde}}{C_6H_5CHO}$$

(*vi*) Oxidation of naphthalene into phthalic anhydride

$$\text{Naphthalene} + 4\tfrac{1}{2}O_2(\text{air}) \xrightarrow[500°C]{V_2O_5} \text{Phthalic anhydride}$$

Naphthalene   Phthalic anhydride

Alternatively o-xylene could be oxidised into phthalic anhydride

$$\text{o-Xylene} \xrightarrow[V_2O_5,\ 500°C]{[O]} \text{Phthalic anhydride}$$

o-Xylene   Phthalic anhydride

(*vii*) Oxidation of furfural into the corresponding carboxylic acid.

$$\text{Furfural (–CHO)} + O_2 \xrightarrow{Ag_2O} \text{Furoic acid (–COOH)}$$

Furfural   Furoic acid

(*viii*) Conversion or benzene into maleic anhydride.

$$\text{Benzene} + 4\tfrac{1}{2}O_2 \xrightarrow[400–500°C]{V_2O_5} \text{Maleic anhydride} \xrightarrow[\Delta]{H_2O} \text{Maleic acid (CHCOOH=CHCOOH)}$$

Benzene   Maleic anhydride   Maleic acid

Maleic anhydride is used for the manufacture of maleic acid.

(*ix*) Manufacture of chlorine by oxidation of hydrogen chloride (Deacon process).

$$4HCl + O_2 \underset{440°C}{\overset{CuCl_2,\ catalyst}{\rightleftharpoons}} 2Cl_2\uparrow + 2H_2O + heat$$

In place of $CuCl_2$ catalyst, didymium promoted catalyst, Dm_2O_3 and $CuCl_2$ are used at lower temperature (400°C).

(*x*) Manufacture of nitric acid by the reaction of nitrogen and oxygen (Birkeland-Eyde process).

$$N_2 + O_2 \xrightarrow{spark} NO \xrightarrow{O_2} NO_2 \xrightarrow{H_2O} HNO_3$$

This process is no longer used due to huge cost of electricity.

Alternatively, nitric acid is obtained from ammonia (Ostwald's process).

$$4NH_3 + 5O_2 \xrightarrow[\text{5 atms. 850°C}]{\text{Pt-Rh catalyst}} 4NO + 6H_2O$$

$$2NO + O_2 \longrightarrow 2NO_2$$

$$2NO_2 + H_2O \longrightarrow HNO_3 + HNO_2$$

(*xi*) Manufacture of sulphuric acid from sulphur via the formation of SO_2 and SO_3.

$$S + O_2 \longrightarrow SO_2 \xrightarrow[\text{Pt catalyst}]{[O]} SO_3 \xrightarrow{H_2O} H_2SO_4$$

Oxidations by oxygen and catalysts are also used for the conversion of alkanes into alcohols, ketones or acids[1]; for the epoxidation[2] of alkene; for the formation of hydroperoxides[3]; conversion of terminal alkynes into methyl ketones[4]; for the oxidation of aromatic compounds to quinones[5] or carboxylic acids[6]; for dehydrogenation of alcohols to aldehydes[7], and ketones into carboxylic acids[8]; for the oxidation of primary amines to nitriles[9], of thiols to disulphides[10] or sulfonic acids[11] *etc. see* also singlet oxygen.

References

1. F.F. Rust and W.E. Vaugham, *Ind. Eng. Chem.,* 1949, **41**, 2595.
2. N. Shimizu and P.D. Bartett, *J. Am. Chem. Soc.,* 1976, **98**, 4193.
3. E.H. Farmer and A. Sundralingam, *I Chem. Soc.,* 1942, 121.
4. W.H. Clement and C.M. Selwitz, *J. Org. Chem.,* 1964, **29**, 241.
5. C.E. Senseman and O.A. Nelsan, *Ind. Eng. Chem.,* 1923, **15**, 521.
6. A.S. Hay, J.W. Eustance and H.S. Blanchad, *J. Org. Chem.,* 1960, **25**, 616.
7. G. de Vries and A. Schors, *Tetrahedron Lett.,* 1968, 5689.
8. S. Ito and M. Matsumoto, *J. Org. Chem.*, 1983, **48**, 1133.
9. J. Tusji and H. Takayanagi, *Org. Synth*. Collective Volume, 1988, **6**, 662.
10. K.-T. Liu and Y.-C., Tong, *Synthesis*, 1978, 669.
11. T.J. Wallace and A. Schriesheim, *Tetrahedron Lett.*, 1963, 1131.

1.89 OSMIUM TETROXIDE (OSMIC ACID), OSO_4

It is used only in catalytic amount in the presence of oxidants that regenerate osmium tetroxide, such as hydrogen peroxide, sodium chlorate, potassium chlorate, barium chlorate and amine oxides. It is expensive and very toxic.

An important application of osmic acid is the cleavage of a double bond. The reaction involves[1] hydroxylation of the double bond followed by fission of the formed diol.

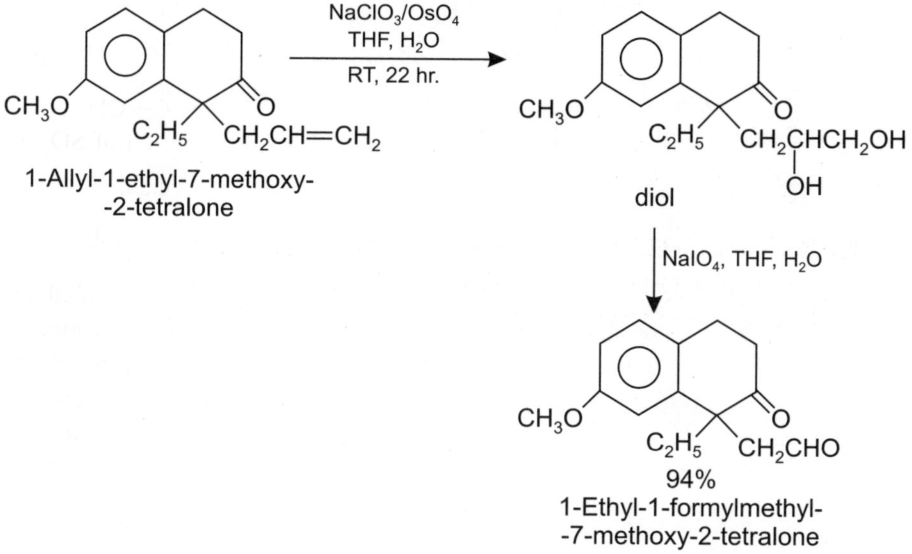

$NaClO_3/OsO_4$, THF, H_2O, RT, 22 hr.

1-Allyl-1-ethyl-7-methoxy--2-tetralone → diol

$NaIO_4$, THF, H_2O

94%

1-Ethyl-1-formylmethyl--7-methoxy-2-tetralone

The hydroxylation of alkenes with osmium tetroxide proceeds via the formation of five-membered cyclic ester[2] of vicinal diols (osmate ester, 1) as an intermediate, which is then hydrolysed to yield *cis*-diols. In most of the cases, the osmate ester is not isolated and the reaction is carried out in the presence of tert-butyl alcohol-water mixture, which hydrolyses the osmate ester to the diol.

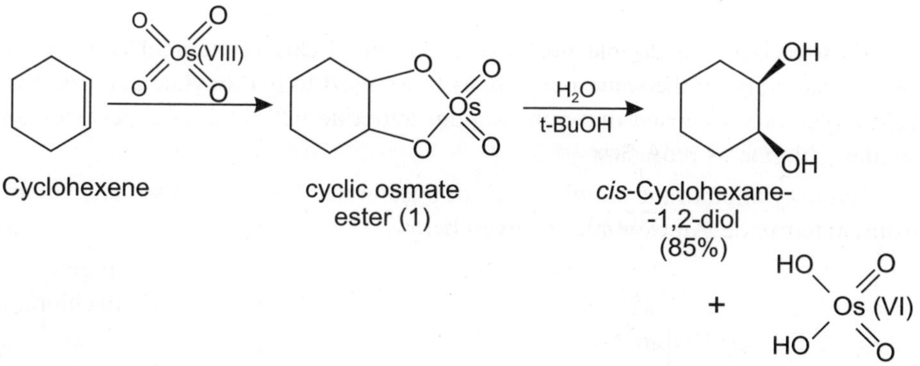

Cyclohexene → cyclic osmate ester (1) → (H_2O, t-BuOH) *cis*-Cyclohexane--1,2-diol (85%) + $(HO)_2OsO_2$ (Os (VI))

When the reaction is carried out in the presence of pyridine, an ester complex (2) is formed by coordination with two equivalents of the ligand (pyridine), and it can sometimes be isolated. The ester then is hydrolysed to give a *cis*-diol.

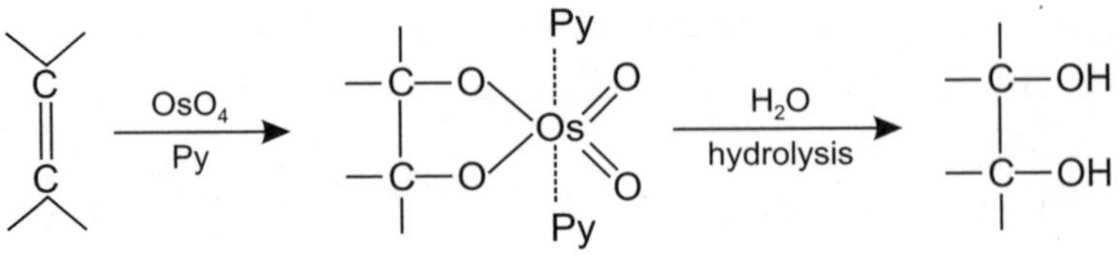
OsO4
Py
Py
Os
Py
H2O
hydrolysis
OH
OH

(2) cis diol

Tertiary amine-oxide, such as N-methylmorpholine (NMQ) (3) are very effective catalyst and oxidise Os(VI) back to Os(VIII). In these reactions[3], where NMO is used, only 1 mol % of OsO_4 is required.

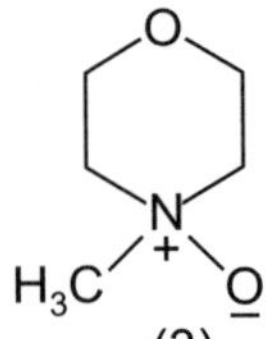

(NMO)

(3)

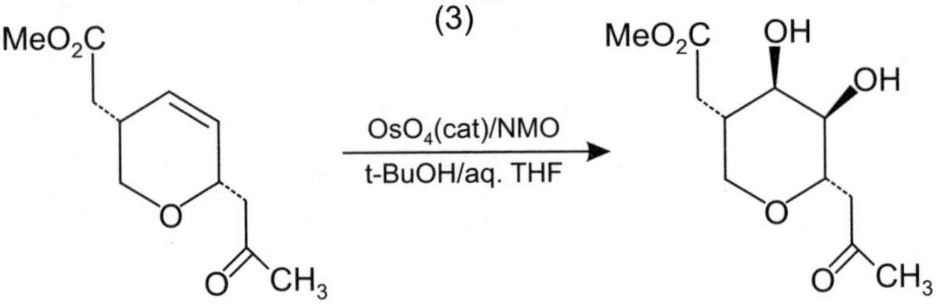
MeO2C
CH3
OsO4(cat)/NMO
t-BuOH/aq. THF
MeO2C
OH
OH
CH3

Hydroxylation at double bonds of unsaturated carboxylic acid is achieved stereoselectively by the same reagents as those used to hydroxylate alkenes. Syn Hydroxylation[3] is carried out with osmium tetroxide with hydrogen peroxide, or sodium chlorate as reoxidant.

Hydroxylation of the double bonds of unsaturated ketones is also achieved by osmium tetroxide. An example[4] is given below.

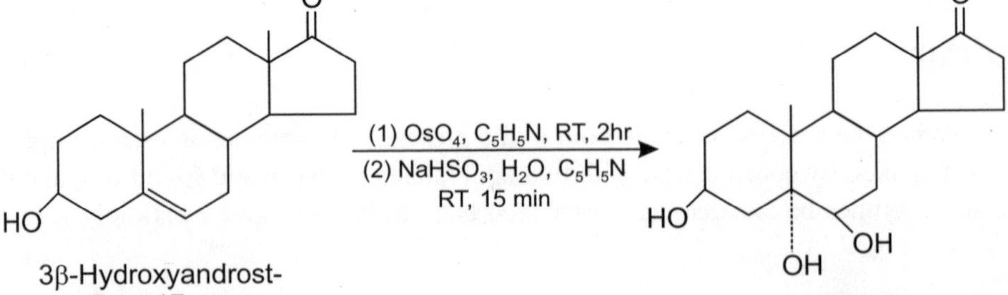
HO
(1) OsO4, C5H5N, RT, 2hr
(2) NaHSO3, H2O, C5H5N
RT, 15 min
HO
OH
OH

3β-Hydroxyandrost- -5-en-17-one

86%

3β,5,6-Trihydroxy-androst -17-one

When Osmium tetroxide is used in combination with sodium periodate[5], it is known as **Lemieux-Johnson reagent** and it gives the same products as is obtained in the ozonlysis procedure. Some examples are given below:

$$RCH=CR_2 \xrightarrow{OsO_4} \text{Osmate ester} \xrightarrow{2NaIO_4} RCHO + R_2CO + OsO_4 + 2NaIO_3$$

Osmate ester

$$CH_3(CH_2)_6CH{=}CH_2 \xrightarrow[H_2O,\ \text{dioxane}]{OsO_4/NaIO_4} CH_3(CH_2)_6CHO + H{-}\overset{O}{\overset{\|}{C}}{-}H$$

1-Dodecene → Undecanal + Formaldehyde

$$\text{Cyclohexene} \xrightarrow[NaIO_4/H_2O/\text{ether}]{OsO_4} H{-}\overset{O}{\overset{\|}{C}}{-}(CH_2)_4{-}\overset{O}{\overset{\|}{C}}{-}H$$

77%

Cyclohexene → Hexanedial

Alkynes react with osmium tetroxide in presence of tertiary amines, such as pyridine or isoquinoline to give osmium (VI) ester complex, which on hydrolysis with sodium sulphite gives the corresponding 1, 2-diketones[6].

$$R{-}C{\equiv}C{-}R' \longrightarrow 2OsO_4 + 4L \longrightarrow \text{osmium(VI) ester complex} \xrightarrow{Na_2SO_3} RCOCOR'$$

L = tert-amine

Asymmetric Hydroxylation[6]

Normally, osmylation of alkenes with osmium tetroxide gives racemic mixture of 1, 2-diols. Sharpless obtained enantiomeric excess (% ee) of the diol product by adding chiral substrate to the osmylation reagent. Thus, trans stilbene with osmium tetroxide and N-methylmorpholine-N-oxide (NMO) gave the diol with 99% ee in the presence of cincona alkaloids, especially esters of dihydroquinidines (4). High diastereoselectivity is obtained only with regiochemically pure alkenes.

Ph Ph Trans stilbene; 0.4% OsO_4/NMO (4); OH Ph Ph OH

MeO N N H O O OH (4)

References

1. K. Wiesner, K.K. Chan and C.Demerson, *Tetrahedron Lett.,* 1965, 2893.
2. R. Griegee, B. Marchand and H. Wannowius, *Ann.* 1942, **99**, 550.
3. N.A. Milas and S. Sussman, *J. Am. Chem. Soc.,* 1936, **58**, 1302.
4. J.S. Baran, *J. Org. Chem.,* 1960, **25 ,** 257.
5. R. Pappo, D.S. Allen, Jr., R.U. Lemieux and W.S. Johnson, *J. Org. Chem.,* 1956, **21**, 478.
6. V.K. Anluwalia and R.K. Parashar, *Organic Reaction Mechanism*, Narosa Publishing House, 2007, Page 583.

1.90 OZONE

Ozone [O = O – O :] is a blue gas or dark blue liquid and is used in a mixture with oxygen. It is prepared by passing oxygen through a silent elecrical discharge. An instrument, called as ozonizer is used for this purpose. Normally, the commonly reached concentration of ozone in oxygen in 2–3%.

The most common reaction of ozone is with alkenes, which first add ozone to form unstable molozonides, which break down into two separate species; these species recombine to form true ozonides[1].

Alkene $\xrightarrow{O_3}$ Molozonide ⟶ Ozonide

Ozonolysis is generally carried out by bubbling a stream of oxygen containing 2 – 3% ozone (this mixture is obtained directly from an ozoniser) into a suspension or solution of alkene into methyl alcohol or any other solvent, which does not react with ozone. The reaction is carried out at room temperature or slightly below room temperature. The formed product (ozonide) is not isolated and is treated with appropriate reagent. The product formed depends on the structure of the alkene and the reagent used for reacting with the ozonide. Following products can be obtained:

***(i)* Formation of Aldehydes:** Catalytic hydrogenation of the ozonide over palladium on carbon[2], platinum oxide[3] or Raney nickel[4] or with Zn in acetic acid[5] result in the formation of aldehydes.

$$RCH{=}CHR + O_3 \longrightarrow \text{(ozonide: R(H)C}\langle\text{O, O–O}\rangle\text{C(H)R)} \xrightarrow{Zn/HOAc} 2RCHO$$

***(ii)* Formation of Alcohols:** Reduction of ozonides with lithium aluminium hydride[6] or sodium borohydride[7] gives alcohols. In this case, the formed aldehyde is reduced to the corresponding alcohol.

***(iii)* Formation of Carboxylic Acids:** The oxidative cleavage of ozonides to carboxylic acids is achieved with hydrogen peroxide in sodium hydroxide solution[6].

In case only one molecule of a carboxylic acid is expected from the ozonide, boiling the ozonide with acetic acid is a good procedue.

In addition to the ozonolysis of alkenes, ozone oxidises some other groups. Thus, saturated hydrocarbons containing tertiary hydrogen atoms are converted into tertiary alcohols[7a].

- Some alkenes are transformed into epoxides[8] or α, β-unsaturated ketones[9].
- Benzene rings are oxidised to carboxyl groups[10].
- Ethers[11] and aldehyde acetates[12] are oxidised to esters.
- Aldehydes are converted into peroxy acids[13].
- Sulphides are converted[14] to sulfoxides and sulfones.
- Organomercury compounds are converted into ketones or carboxylic acids[15].

References

1. R. Criegee, G. Blust and H.Zinke, *Chem. Ber.,* 1954, 87, 766; R.W. Murray, R. D. Youssefyeh and P.R. Story, *J. Am. Chem. Soc.,* 1967, **89**, 2429.
2. F.G. Fischer, H. Düll and L. Ertel, *Chem. Ber.,* 1932, **65**, 1467.
3. R. Criegee and M. Lederer, *Justus Liebigs Ann. Chem.,* 1953, **583**, 1, 29.
4. N.C. Cook and F.C. Whitmore, *J. Am. Chem. Soc.,* 1941, **63**, 3540.
5. W.G. Overend, M.Stacey and L.F. Wiggins, *J. Chem. Soc.,* 1949, 1358.

6. B. Witkop andJ.B. Patrick, *J. Am. Chem. Soc.,* 1952, **74**, 3855.
7. D.G.M. Diaper and D.L. Mitchell, *Can. J. Chem.,* 1960, **38**, 1976.
7a. Z. Cohen, E. Keinan, Y. Mazur and T.H. Varkony, *J. Org. Chem.,* 1975, **40**, 2141.
8. P.S. Bailey and A.G. Lane, *J. Am. Chem. Soc.,* 1967, **89**, 4473.
9. J.R. Durland and H. Adkins, *J. Am. Chem. Soc.,* 1939, **61**, 429.
10. R. Criegee and H. Höver, *Chem. Ber.,* 1960, **93**, 2521.
11. F.G. Fischer, *Justus Liebigs Ann. Chem.,* 1929, **476 ,** 233.
12. P. Deslongchamps and C. Moreau, *Can. J. Chem.,* 1971, **49**, 2465.
13. C.R. Dick and R.F. Hanna, *J. Org. Chem.,* 1964, **29**, 1218.
14. Q.E. Thompsan, *J. An. Chem. Soc.,* 1961, **83**, 845.
15. P.E. Pike, P.G. Marsh, R.E. Erikson and W.L. Walers, *Tetrahedron Lett.,* 1970, 2679.

1.91 OXODIPEROXYMOLYBDENUM-PYRIDINE-HEXAMETHYLPHOSPHORIC TRIAMIDE $MoO_5 \cdot C_5H_5N \cdot$HMPA(MoOPH)

This reagent is prepared[1, 2] from molybdenum trioxide, MoO_3. It hydoxylates the enolates of ketones (aromatic aliphatic ketones and cyclic ketones) and esters in the α-position with respect to carbonyl groups[1, 2, 3] giving α-hydroxy ketones in 62% yield.

$$C_5H_5COCH_2C_3H_8 \xrightarrow[\text{(2) } MoO_5.C_5H_5N.\text{HMPA, } -44°C]{\text{(1) LDA, THF, } -44°C} C_6H_5CO\underset{\substack{| \\ OH}}{C}HC_3H_8$$

62%

In presence of mercuric acetate the reagent, MoOPH converts acetylenes into α-dicarbonyl compounds.[4]

References

1. E. Vedejs, D.A. Engler and J.E. Telschow, *J. Org. Chem.,* 1978, **43**, 189.
2. E. Vedejs and S. Larsen, *Org. Synth.,* 1985, **64**, 127.
3. E. Vedej's, *J. Am. Chem. Soc.,* 1974, **96**, 5944.
4. F.P. Ballisteri, S. Failla, G.A. Tomaselli and R. Curci, *Tetrahedron Lett.,* 1986, **27**, 5139.

1.92 OXONE

Oxone (a registered trade mark of E. 1. du pont de Nemours company) is a mixture of $2KHSO_5 \cdot KHSO_4 \cdot K_2SO_4$. It is a versalite oxidising agent[1]. It is used in the following oxidations:

(*i*) Epoxidation of olefins[2].

$$\underset{\text{Olefin}}{(C_6H_5)HC{=}CH(C_6H_5)} \xrightarrow[\text{22°C, 2 hr.}]{\text{Oxone, } Me_2CO} \underset{\text{epoxide}}{(C_6H_5)HC\overset{O}{—}CH(C_6H_5)} + \underset{\text{Dimethyl dioxirane}}{Me_2C\langle{}^{O}_{O}|}$$

(*ii*) Epoxidation of carbon-nitrogen bond[3].

Oxone

(*iii*) Oxidation of side chain to caboxyl group[1].

Not only methyl groups, but practically any other alkyl group attached to aromatic rings are oxidised to carboxyl group. Nitric acid can also be used for such oxidations, but the disadvantage is the formation of nitrated byproducts.

The partial oxidation[4] of o-xylene to o-toluic acid is affected by heating with oxone and oxygen at 115–120°C in presence of cobalt acetate in acetic acid (yield 77%).

(*iv*) Oxidation of iodo compounds to iodoxy compounds can be effected with oxone.

$$C_6H_5I \xrightarrow{\text{Oxone}} C_6H_5IO_2$$

(*v*) Oxidation of primary alcohol to esters

$$\underset{\text{Ethanol}}{CH_3—CH_2—OH} \xrightarrow[\text{conc. } H_2SO_4\text{, 70°C}]{\text{Oxone}} \underset{\text{Ethyl acetate}}{CH_3CH_2OC_2H_5}$$

(*vi*) Oxidation of primary amines to nitroso compounds.

p-Phenylenediamine (1, 4-diamino benzene) is oxidised by oxone in an aqueous suspension at room temperature to give *p*-dinitrosobenzene in quantitative yield[1].

$H_2N—C_6H_4—NH_2$ (*p*-Phenylene diamine) $\xrightarrow[\text{aq. suspension}]{\text{Oxone}}$ $ON—C_6H_4—NO$ (*p*-Dinitroso benzene)

(*vii*) Oxidation of sulphides to sulfoxides.

Diphenyl sulphide on oxidation with oxone in presence of sulphuric acid and acetic acid give diphenyl sulfoxide in 975 yield[1].

$$\underset{\text{Diphenyl sulphide}}{C_6H_5SC_6H_5} \xrightarrow[\text{conc. } H_2SO_4\text{, AcOH, RT, 5 hr.}]{\text{Oxone}} \underset{\text{Diphenyl sulfoxide}}{C_6H_5S(=O)_2—C_6H_5}\ (97\%)$$

(*viii*) Preparation of dioxiranes.

Dioxiranes are prepared from acetone and other aliphatic ketones by treatment with oxone[2, 5].

$$R_2CO \xrightarrow[NaHCO_3,\ H_2O]{Oxone} R_2C(O_2)$$

Ketone Dioxirane

(*ix*) Oxidation of aldehydes to carboxylic acid.

Oxone oxidises benzaldehyde in chloroform and dilute sulphuric acid at room temperature within 2 days to benzoic acid in 70% yield[6].

References

1. R.J. Kennedy and a.M. Stock, *J. Org. Chem.,* 1960, **25**,1901.
2. R.W. Murray and R. Jeyaraman, *J. Org. Chem.,* 1985, **50**, 2847.
3. F.A. Davis, M.S. Haque, T.G. Ulatowski and J.C. Towsan, *J. Org.Chem.,*1986, **51**, 2402.
4. A.S. Hay, J.W. Eustance and H.S. Blanchard, *J. Org. Chem.,* 1960, **25**, 616.
5. R.W. Murray, R. Jeyaraman and L. Mohan, *Tetrahedron Lett.,* 1986, **27**, 2335.
6. R. J. Kennedy and A.M. Stock, *J. Org. Chem.,* 1960, **25**, 1901.

1.93 1-OXO-2, 2, 6, 6-TETRAMETHYLPIPERIDINEUM CHLORIDE

It is prepared[1] from chlorine and 2, 2, 6, 6-tetramethyl piperidinyloxyl (a free radial tempo).

$$\text{TEMPO (N}^\bullet) \xrightarrow{Cl_2} \overset{+}{N}{=}O\ \bar{Cl}$$

It is an exotic reagent and converts diols into acyloins or lactones[1]. In these oxidants 4-methoxy-2, 2, 6, 6-tetramethyl-1-oxopiperidinium chloride (which is prepared by the treatment of 4-methoxy-2,2,6,6-tetramethylpiperidin-1-oxyl (a stable for redical) with chlorine. Thus, 1, 4-butanediol and 1, 5-pentane diol are oxidised[1] to α-butyrolactone and α-valeolactone, respectively.

$$HOCH_2(CH_2)_nCH_2OH \xrightarrow[CH_2Cl_2]{\text{MeO-TEMPO}^+{=}O\ \bar{Cl}} \text{lactone: } O,\ CH_2,\ CO,\ (CH_2)_n$$

$n = 2$, 10 min 81% yield

$n = 3$, 30 min 40% yield

Reference

1. T. Miyazawa and T. Endo, *J. Org. Chem.* 1985, **50**, 3930.

1.94 PALLADIUM, Pd, PLATINUM, Pt

Both palladium and platinum are used on activated carbon or asbestos for dehydrogenative aromatisation. Such catalytic dehydrogenations require high temperatures (250–350°C) and are accompanied by side reactions, such as rearrangements *etc*. These dehydrogenations are used for the structure elucidation of terpenes and alkaloids. Some examples are given below:

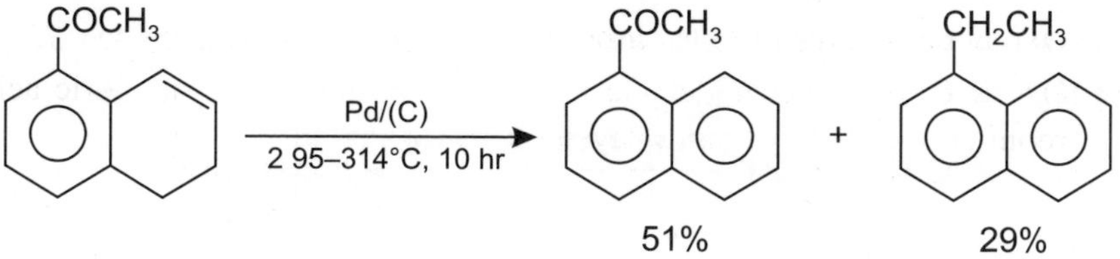

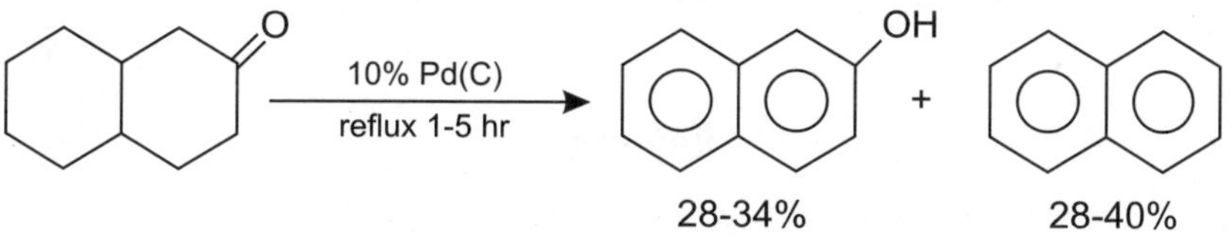

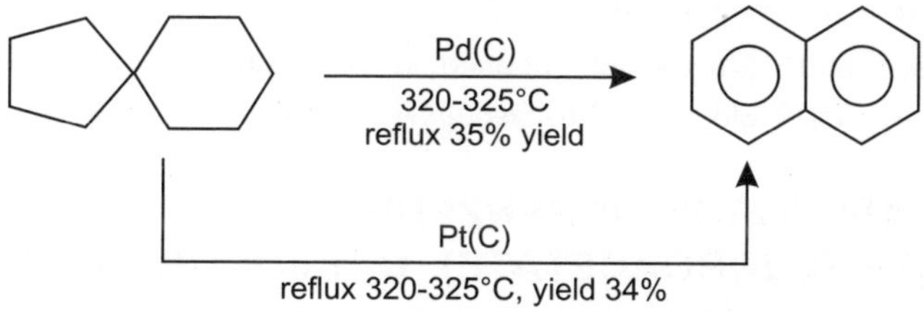

$COCH_3$ (dihydronaphthalene) —Pd/(C), 2 95–314°C, 10 hr→ $COCH_3$ (naphthalene) 51% + CH_2CH_3 (naphthalene) 29%

(decalone) —10% Pd(C), reflux 1-5 hr→ OH (naphthol) 28-34% + (naphthalene) 28-40%

(spiro compound) —Pd(C), 320-325°C, reflux 35% yield→ (naphthalene)

Pt(C), reflux 320-325°C, yield 34%

Platinum or better platinum oxide is used[4] for the oxidation of primary alcohols to aldehydes.

$$CH_3CH{=}C(CH_3)CH_2OH \xrightarrow[60°C,\ 2\ hr.]{O_2/PtO_2,\ C_7H_{16}} CH_3CH{=}C(CH_3)CHO$$

77%

Platinum catalyses the oxidation of sugars to aldonic acids[5].

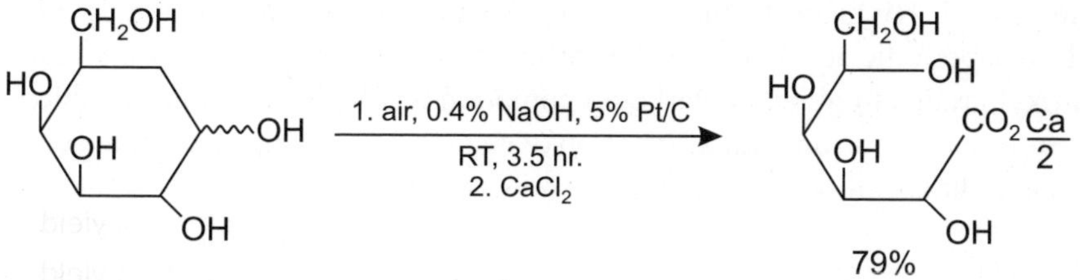

CH_2OH, HO, OH, OH, OH (sugar) —1. air, 0.4% NaOH, 5% Pt/C, RT, 3.5 hr. 2. $CaCl_2$→ CH_2OH, HO, OH, OH, $CO_2\frac{Ca}{2}$, OH

79%

Platinum oxide along with oxygen oxidises[4] primary alcohols to aldehydes or carboxylic acids depending on the reaction conditions.

$$C_{11}H_{23}CH_2OH \xrightarrow[60°C,\ 15\ min]{O_2/PtO_2,\ C_7H_{16}} C_{11}H_{23}CHO \quad 70\%$$

$$C_{11}H_{23}CH_2OH \xrightarrow[60°C,\ 15\ min]{O_2/PtO_2,\ C_7H_{16}} C_{11}H_{23}CO_2H \quad 96\%$$

Some other examples of oxidation[5] of primary alcohols to aldehydes are given below

$$C_7H_{15}CH_2OH \xrightarrow{O_2/Pt,\ AcOEt,\ RT,\ 20\ hr} C_7H_{17}CHO \quad 21\%$$

$$C_6H_5CH_2OH \xrightarrow{O_2/Pt,\ AcOEt,\ RT,\ 20\ hr} C_6H_5CHO \quad 72\%$$

References

1. M.S. Newman and T.S. Bye, *J. Am. Chem. Soc.,* 1952, **74**, 905.
2. R.P. Linstead and K.O.A. Michaelis, *J. Chem. Soc.,* 1940, 1134.
3. C.S. Marvel and L.A. Brooks, *J. Am. Chem. Soc.,* 1941, **63**, 2630.
4. K. Heyns and L. Blazejewicz, *Tetrahedron,* 1960, **9**, 67.
5. K. Heyns and O Stöckel, *Justus Liebegs Ann. Chem.,* 1947, **558**, 192.

1.95 PENTAFLUOROIODOBENZENE BIS (TRIFLUOROACETATE), $C_6H_5I(OCOCF_3)_2$

It is obtained by the treatment of iodopentafluorobenzene with trifluoroacetic anhydride and fuming nitric acid. It converts acetylenes into carboxylic acids.

References

1. M. Schmeisser, K. Dahmen and P. Sartori, *Chem. Ber.,* 1967, **100**, 1633.

1.96 PERIODIC ACID, HIO_4

Periodic acid (HIO_4) or its dihydrate (H_5IO_6) are commercially available. Periodic acid (metaperiodic acid) cleaves the carbon-carbon bond between two vicinal free hydroxyl groups to give two carbonyl compounds. This reaction is valuable in the structure elucidation and transformation of sugars. The mechanism[1] of the cleavage of vicinal diols can be represented as shown below.

$$\begin{matrix} -C-OH \\ -C-OH \end{matrix} \xrightarrow[-H_2O]{HIO_4} \text{cyclic periodate ester} \longrightarrow \begin{matrix} -C=O \\ + \\ -C=O \end{matrix} + HNO_3$$

The cleavage of secondary straight chain diols is a good method of preparation of aldehydes.

H_3C–$C_6H_2(CH_3)_2$–CH(C$_6$H$_5$)–CH(OH)CH$_2$OH $\xrightarrow[\text{H}_2\text{O, EtOH, 18°C, 30 min; 33°C, 1 hr.}]{\text{KIO}_4\text{, H}_2\text{SO}_4}$ H_3C–$C_6H_2(CH_3)_2$–CH(C$_6$H$_5$)–CHO

100%

Periodic acid is capable of dehydrogenative coupling of such compounds in which initial abstraction of a hydrogen from one carbon, usually by a free radical process is possible. Such abstraction of hydrogen occurs easily when the remaining free radical is stabilised by resonance. Thus, pyrene furnishes[3] 1, 1′-bipyrene in 70 – 75% yield when heated with a solution of periodic acid in acetic acid to 48 – 51°C.

Pyrene $\xrightarrow[\text{48–51°C, }\Delta]{\text{HIO}_4\text{/HOAc}}$ 1,1′-Bipyrene (70-75%)

Methyl homologues of aromatic compounds are oxidised to aldehydes by periodic acid[4]. In fact, selectivity is noticed in the oxidation of 2, 3-dimethylindole. However, in this case the methyl group in position 2 is oxidised to formyl group in low yield.

2,3-Dimethylindole $\xrightarrow[\text{ice-salt bath, 0.5 hr.}]{\text{HIO}_4\text{, H}_2\text{O, MeOH}}$ 2-Formyl-3-methylindole (10%)

Epoxides (oxiranes) are oxidised by periodic acid into dicarbonyl compounds. Thus, cyclohexane epoxide give hexane 1, 6-dial in 82% yield[5].

$$\text{Cyclohexane epoxide} \xrightarrow{HIO_4,\ H_2O,\ 45°C} \text{OHC} \quad \text{CHO (Hexane 1,6-diol, 82\%)}$$

Cyclohexane epoxide

Hexane 1,6-diol
82%

Trimethylsilyl ethers of tertiary alcohols are degraded[6] by periodic acid to carboxylic acids with shorter chain.

$$(CH_3)_2CHCH(OH)\text{—}CH(CH)\text{—}COC(CH_3)_2OSi(CH_3)_3 \xrightarrow[\text{0-5°C, 15 min; 25°C, 1.5 hr}]{HIO_4.2H_2O} (CH_3)_2CHCH(OH)\text{—}CH(CH_3).COOH$$

77-83%

Periodic acid is used as a catalyst (as a reoxidant) in the oxidation of steroids by DDQ.

References

1. J. Roček and F.H. Westheimer, *J. Am. Chem. Soc.,* 1962, **84**, 224.
2. L. Vagha and M. Remenyi *J. Chem. Soc.,* 1951, 1068.
3. A.J. Fatiadi, *J.Org. Chem.,* 1967, **32**, 3903.
4. L.J. Dolby and D.L. Booth, *J. Am. Chem. Soc.,* 1966, **88**, 1049.
5. J.P. Nagarkatti and K.R. Ashley, *Tetrahedron Lett.,* 1973, 4599.
6. B.Bal, C.T. Buse, K. Smith and C.H. Heathcock, *Org. Synth.*, 1984, **63**, 89.

1.97 PEROXY ACIDS

A number of per acids (RCO_3H) have been used to effect oxidation. Some of the common per acids are performic acid (HCO_3H), peracetic acid (CH_3CO_3H), trifluoroacetic acid (CF_3CO_3H), perbenzoic acid ($C_6H_5CO_3H$), m-chloroperbenzoic acid(m-$ClC_6H_5CO_3H$) p-nitroperbenzoic acid (p-$NO_2C_6H_4CO_3H$), peroxyphthalic acid ([O–C_6H_4(COOH) CO_3H], peroxymaleic acid (*cis*-HO_2C CH = $CHCO_3H$), peroxypropionic acid ($CH_3CH_2CO_3H$) and benzeneperoxy selenic acid ($C_6H_5Se(O)OOH$). A brief account of different peroxy acids is given below.

(*i*) Performic Acid (HCO_3H): It is prepared[1] *in situ* from 25 – 30% hydrogen peroxide and 98 – 100% formic acid.

$$HCOOH + H_2O_2 \rightleftharpoons HCO_3H + H_2O$$

Performic acid is prepared (*in situ*) in presence of organic compounds (like alkenes) to be oxidised. The temperature is maintained below 40°C. The reaction products are the formyl esters of diols (not isolated), which on heating with alkali give vicinal diols with antistereochemistry[2].

$$>C{=}C< \xrightarrow{HCO_3H} >C(OCHO){-}C(OH)< \xrightarrow{NaOH} >C(OH){-}C(OH)<$$

alkene — Formyl ester — *trans* diol

The formation of *trans* diols from alkene is the most important reaction of performic acid.

***(ii)* Peroxyacetic Acid (CH_3CO_3H) :** It is commercialy available as 40 – 45% solution in acetic acid. It can be prepared in situ from 30% hdrogen peroxide and acetic acid in presence of catalytic amount of sulphuric or perchloric acid. The earlier addition of substrate to the reaction mxiture makes the oxidation to go to completion at 25 – 70°C.

The most important application of peroxyacetic acid are the epoxidations[3] and antrhydroxylation of double bonds[4] (as in the case of performic acid).

The epoxidation is believed to take place by transfer of oxygen atom from per acid to the alkene. The reaction is stereospecific leading to the syn-addition of the oxygen atom.

$$>C{=}C< + H{-}O{-}O{-}C(=O){-}CH_3 \longrightarrow \text{epoxide} + CH_3COOH$$

Alkene — Peracetic acid — Epoxide (oxirane)

***(iii)* Trifluoroacetic Acid (CF_3CO_3H):** It is prepared from 90% hydrogen peroxide, trifluoroacetic acid and usually trifluoroacetic anhydride. It reacts much faster than other peroxyacids. It is used for epoxidation and anti hydroxylation alkenes.

***(iv)* Peroxybenzoic Acid ($C_6H_5CO_3H$):** It is commerically available (M.P. 41 – 43°C). However, it can be prepared from dibenzoyl peoxide[5] as given below.

$$\underset{\text{Benzoyl peroxide}}{(C_6H_5COO)_2} \xrightarrow{CH_3ONa} C_6H_5CO_2CH_3 + C_6H_5CO_3Na$$

$$C_6H_5CO_3Na \xrightarrow{H_2SO_4} C_6H_5CO_3H$$

It can also be obtained by the treatment of benzoic acid with 70% hydrogen peroxide in presence of methanesulfonic acid[6].

Oxidations with peroxybenzoic acid are carried out is solutions in dichloromethane, chloroform, benzene, ethyl acetate at or below room temperature. Its reactions include epoxidation of double bonds, oxidation of benzaldehyde to cabroxylic acids or phenols and the Baeyer Villiger reaction of ketones. It is also used for anti hydroxylation of double bonds, the oxidation of pyrrolidines to pyrrolidones and of pyrroles to succinimides.

***(v) m*-Chloroperbenzoic Acid:** It is commercially available (m.p.92 – 94°C). However, it is prepared from *m*-chlorobenzoyl chloride and sodium peroxide[7] or sodium hydrogen peroxide[8]. The oxidations are usually carried out in solution in hexane, dichloromethane, chloroform at temperature. ranging from –78 to 40°C.

m-Chloroperbenzoic acid is used for epoxidations and the Baeyer-villiger reaction. Secondary alcohols are oxidised to ketones in the presence of hydrogen chloride[9].

***(vi) p*-Nitroperbenzoic Acid:** It is prepared from *p*-nitrobenzaldehyde and ozone[10] or form *p*-nitrobenzoylchloride and sodium peroxide[11]. It is used for the stereoselective epoxidations of allyl alcohols[12] and also for highly regio-and stereoselective hydroxylation of hydrocarbons at tertiary carbon[13].

***(vii)* Peroxyphthalic Acid:** It is obtained from phthalic anhydride and hydrogen peroxide[14] or sodium hydrogen peoxide[15]. It is used in ethereal solutions for epoxidations and Baeyer villiger reaction.

Peroxymaleic Acid

It is prepared from 90% hydrogen peroxide and maleic anhydride[16]. It oxidises alkenes to epoxides and ketones to ester.

Benzeneperoxyselenic Acid

It is prepared *in situ* from benzene selenic acid and 30% hydrogen peoxide in tetrahydofuran or methylene chloride. It is useful for the oxidation of cyclic ketones to lactones[17].

As already stated, most of the peracids are used for the formation of epoxides of olefins and their anti hydroxylation.

In epoxidation, the terminal alkenes react slowly with peracids as compared to alkyl substituted alkenes like butene[2].

$$\underset{\text{(Terminal alkene)}}{R{-}CH{=}CH_2} \xrightarrow[CH_2Cl_2,\ \text{reflux}]{m\text{-}ClC_6H_4CO_3H} R{-}\overset{\;\;O}{CH{-}CH_2}$$

$$\underset{\text{2-Butene}}{H_3C{-}CH{=}CH{-}CH_3} \xrightarrow[\text{dioxane, 0°C}]{m\text{-}ClC_6H_4CO_3H} \underset{75\%}{H_3C(H)C{-}C(H)CH_3\ (\text{epoxide, }O)}$$

The epoxidations with peroxy acids are stereoselective, taking place by syn-addition to the double bond. This is established by the X-ray analysis of the formed products. This, *cis* alkene gives *cis*-epoxide and *trans*-alkene gives *trans*-epoxide; the reaction is concerted.

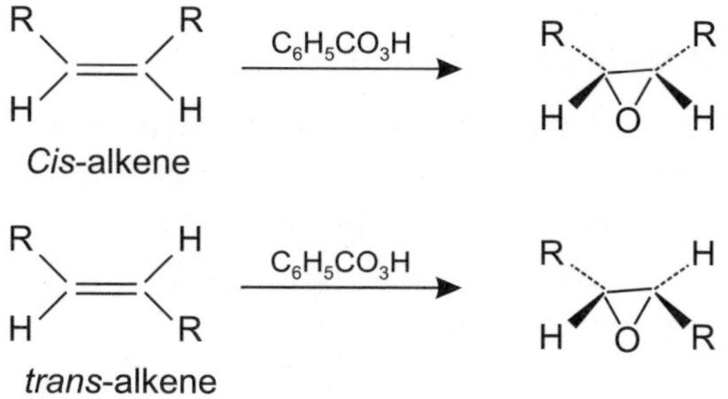
R
R
$C_6H_5CO_3H$
R
R
H
H
H
O
H
Cis-alkene
R
H
$C_6H_5CO_3H$
R
H
H
R
H
O
R
trans-alkene

The epoxidation of alkenes can also be effected with *t-butyl* hydroperoxide (*see t*-butyl hydroperoxide).

In case of cyclic alkenes, the reagent approaches from the less hindered side of the double bond. Thus, the epoxidation of norbornene takes place in the following way:

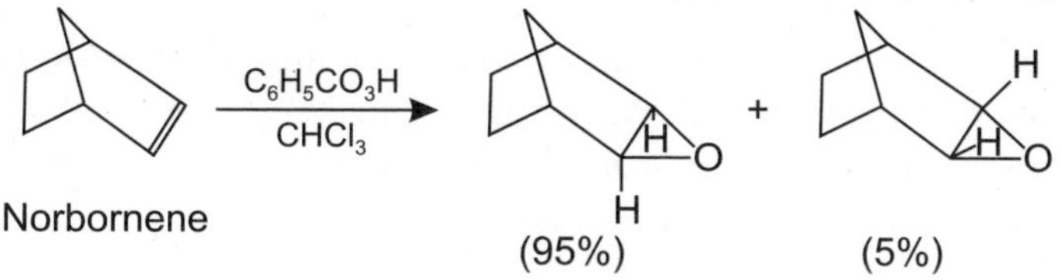
$C_6H_5CO_3H$
$CHCl_3$
H
O
+
H
H
O
Norbornene
H
(95%)
(5%)

In case of unhindered olefins, a mixture of epoxides is obtained. This is because the two potential modes are not markedly different.

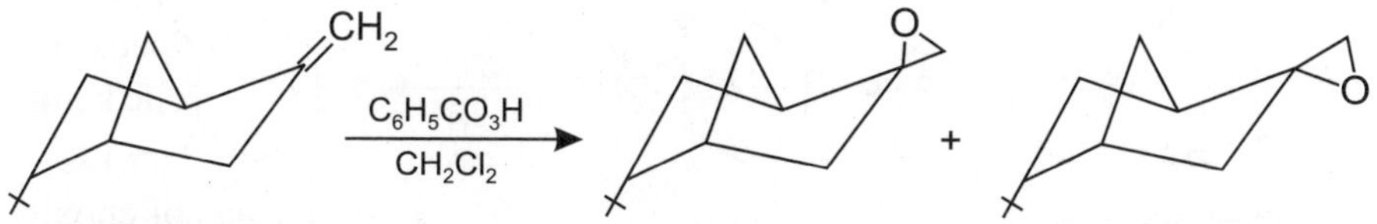
CH_2
$C_6H_5CO_3H$
CH_2Cl_2
O
+
O

It has already been stated that alkylsubstituted alkenes react faster than the terminal alkenes. Thus, in case of *p*-mentha-1, 4(8) diene there is preferential epoxidation.

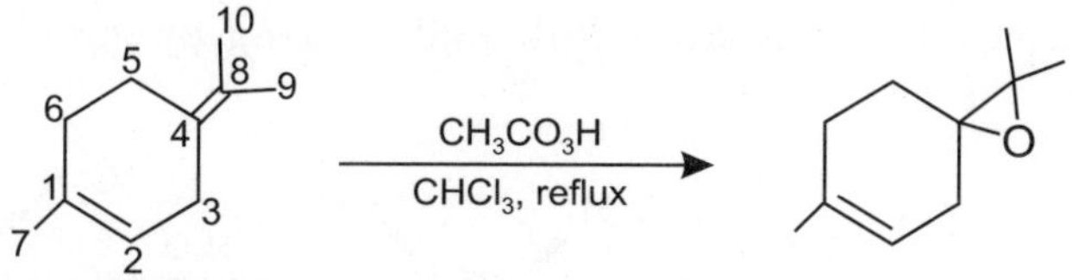
10
5
8
9
6
4
CH_3CO_3H
$CHCl_3$, reflux
O
1
3
7
2

p-Mentha-1,4(8) diene

The isolated alkenes (unconjugated) are epoxidised in preference to conjugated alkenes. Thus, epoxidation of β-myrecene takes place in the following way:

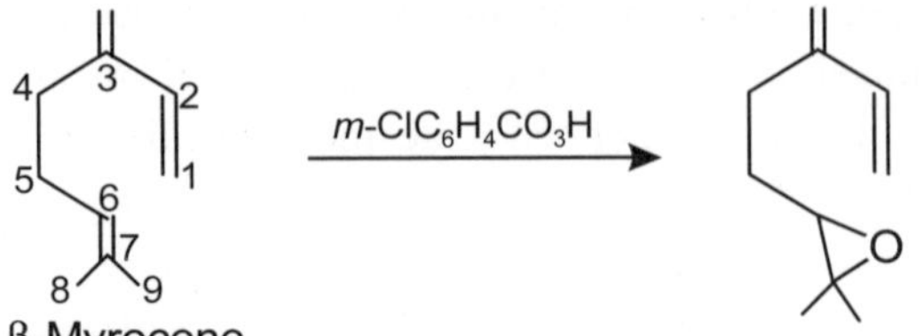

$\xrightarrow{m\text{-}ClC_6H_4CO_3H}$

β-Myrecene

In the case of cumulated dienes (allenes) of the type RHC = C =CHR, the two double bonds are in different planes at right angle to each other. The epoxidation of allene takes place from the less hindered face of the molecule[18].

Epoxidation of allenes can also be affected by dimethyldioxirane[19] (*see* dimethyldioxerane) to give spirodioxides[20].

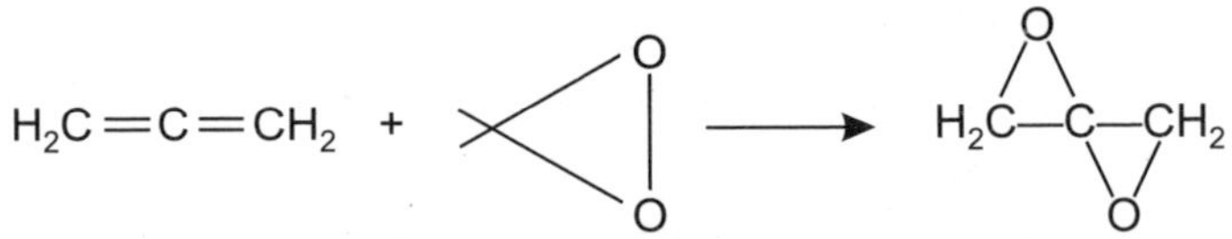

$$H_2C{=}C{=}CH_2 + \text{Dimethyl dioxirane} \longrightarrow \text{Spirodioxide}$$

Allene Dimethyl dioxirane Spirodioxide

Ring Opening of Epoxides

The epoxides are useful in organic synthesis, as they react with a number of nucleophiles resulting in ring opening[21]. Thus, the epoxides are essential intermediates for the synthesis of *trans*-1, 2-dihydroxycyclohexane by ring opening of cyclohexane epoxide with dilute acids.

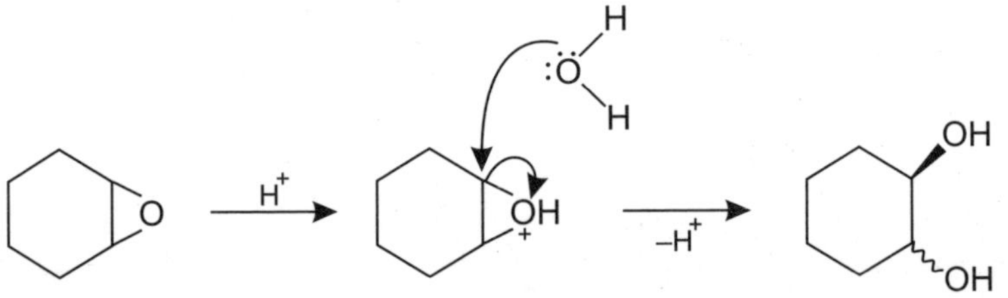

$\xrightarrow{H^+}$ $\xrightarrow[-H^+]{}$

Cyclohexene epoxide

trans-1,2-dihydroxy cyclohexane

Thus, the epoxidation followed by hydrolysis of the epoxide is a routine procedure for the preparation of *trans*–1,2–diols from alkenes. For example, *cis*–2–hexene on epoxidation followed by ring opening gives racemic 2,3–dihydroxyhexane

$\xrightarrow{RCO_3H}$ $\xrightarrow{H_3O^+}$

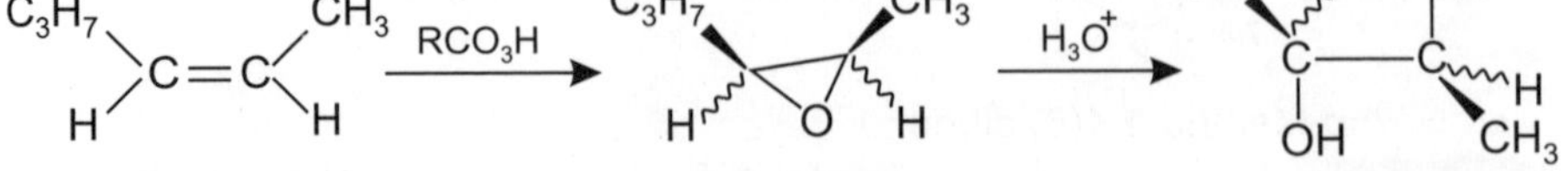

cis-2-Hexene Epoxide Racemic 2,3-dihydroxy hexane

Ally alcohols can also be obtained from epoxides by treatment with potassium t-butoxide or lithium alkylamides.

$$\text{2,3-Epoxyheptane} \xrightarrow[\text{ether, reflux}]{LiN(C_2H_5)_2} \text{allylic alcohol}$$

The epoxide ring opening can also be affected by reaction with grigrard reagent. In this case a new carbon-carbon bond is formed. Thus, cyclohexane epoxide on reaction with phenylmagnesium bromide gives *trans* 2-phenylcyclohexanol in presence of cataylic amount of CuI[22].

$$\text{Cyclohexane epoxide} \xrightarrow[\text{CuI, THF, 0°C}]{C_6H_5MgBr} \textit{trans}\text{-2-phenyl cyclohexanol}$$

The oxide ring opening can also be affected with BF_3-etherate. In this case, there is rearrangement to give carbonyl compounds. A typical example is given below.

$$\underset{\text{1,2-Diphenyl ethane epoxide}}{C_6H_5\text{—}CH\text{—}CHC_6H_5 \ (\text{O bridged})} \xrightarrow[\text{etherate}]{BF_3} C_6H_5\overset{+}{C}H\text{—}\underset{O\bar{B}F_3}{CH}\text{—}C_6H_5 \longrightarrow \underset{\text{Diphenyl acetaldehyde (80\%)}}{(C_6H_5)_2CHCHO}$$

A very important application of peracids is the oxidation of ketones to esters. The reaction is known as **Baeyer-Villiger oxidation**[23].

$$\underset{\text{Ketone}}{R_2C{=}O} \xrightarrow{C_6H_5COOOH} \underset{\text{Ester}}{R\text{—}\overset{O}{\overset{\|}{C}}\text{—}OR} + C_6H_5COOH$$

The mechanism of Baeyer-Villiger oxidation is not clear. However, it is believed that the reaction proceeds as follows:

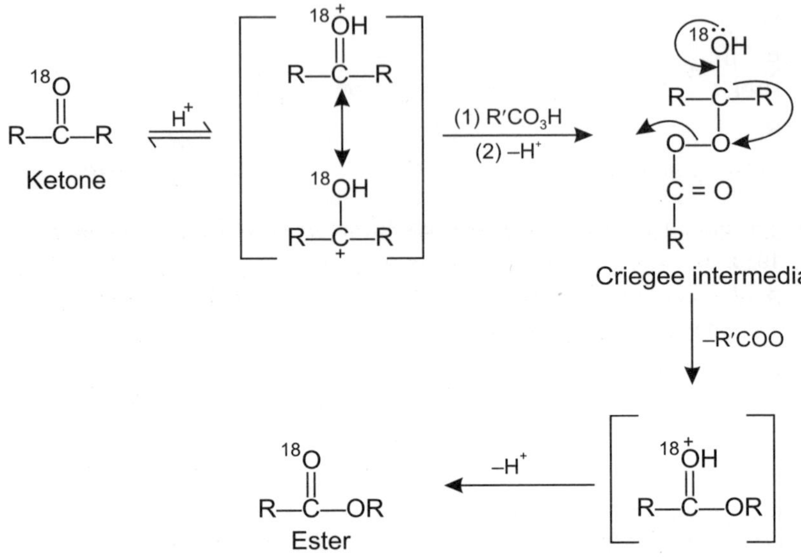

Besides the oxidation reactions by peracid mentioned above, the peracids are also used in:

- The oxidation of amines to nitroso[24] or nitrocompounds[25].
- The oxidation of tertiary amines to amine oxides[26].
- The oxidation of Sulphides to sulfoxides and sulfones[27].
- The oxidation of iodo compounds to iodoso or iodoxy compounds[28].
- The oxidation of alcohols to ketones[29].

References

1. F. P. Greenspan and D. G. Mackellar. *Anal. Chem.,* 1948, 20, 1061.
2. D. Swern, G. N. Billen and J. T. Scanlan, *J. Am. Chem.* Soc., 1946, 68, 1504; A. Roebuck and H. Adkins, *Org. Synth.* Collective Volume, 1955, **3**, 217.
3. W. Dittmann and F. Stürzenhofecker, *Justius Liebigs Ann. Chem.,* 1965, **688**, 57. E.; Wenkert and M. Rubin, *Nature* (London), 1952, 170, 708.
4. D. Swern, G. N. Billen, T. W. Findley and T. J. Scanlan, *J. Am. Chem. soc.,* 1945,67,178; M. Korach, D. R. Nielsen and W. H. Rideout, *Org. Synth.,* Collective Volume, 1973, **5**, 414.
5. G. Braun, *Org. Synth.* Collective Volume, 1932, **I**, 431.
6. L. S. Silbert, E. Siegel and D. Swern, *Org. Synth.,* Collective Volume, 1973, **5**, 904.
7. J. R. Moyer and N. C. Manley, *J. Org. Chem.,* 1964, **29**, 2099.
8. R. N. McDonald, R. N. Steppel and J. E. Dorsey, *Org. Synth.* Collective Volume 1988, **6**, 276.
9. J. A. Cella, J. P. McGrath and S. L. Regen, *Tetrahedron Lett.,* 1975, 4115.
10. C. R. Dick and R. F. Hanna, *J. Org. Chem.,* 1964, **29**, 1218.
11. C. C. Price and E. Krebs, *Org. Synth.* Collective Volume, 1955, **3**, 649.

12. S. Tanaka, H. Yamamolo, H. Nozaki, K. B. Sharpless, R. C. Michaelson and J. D. Cutting, *J. Am. Chem. Soc.,* 1974, **96**, 5254.
13. W. Müller and H. –J. Schneider., *Angew. Chem.* Int. Ed. Engl., 1979, **18**, 407.
14. E. E. Royals and L. H. Harrer. Jr., *J. Am. Chem. Soc.,* 1955, **77**, 3405.
15. H. Böhme, *Org. Synth.* Collective Volume, 1955, **3**, 619.
16. R. W. White and W. D. Emmons, *Tetrahedron*, 1961, **17**, 31.
17. U. M. Dzhemilev, N. S. Vostrikov, A. M. Moiseenkov and G. A. Tolstikov, Izv. Akad. Nauk SSSR, 1981,1320; Chem. Abtr., 1981, **95**, 167877.
18. J. K. Grandall, W. W. Conover, J. B. Kowin and W. H. Mach, *J. Am. Chem. Soc.,* 1968, **90**, 7347.
19. R. W. Murray, *Chem. Rev.,* 1989, **89**, 1187.
20. J. K. Crandall and E. Rambo, *J. Org. Chem.,* 1990, **55**, 5229.
21. A. S. Rao, S. K. Patnikar and J. G. Klirtane, *Tetrahedron*, 1983, **39**, 2323.
22. C. Huynth, F. Derguini-Boumechal and G. Linstrumelle, *Tet. Lett.,* 1979, 1503.
23. S. L. Friess, J. Am. Chem., Soc., 1949,71,2571; T. Yokoyama and F. Nohara, *Bull. Chem., Soc. Jpn.,* 1965, **38**, 1498.
24. R. R. Holmes and R. P. Bayer, *J. Am. Chem. Soc.,* 1960, **82**, 34545.
25. W. D. Emmos *J. Am. Chem. Soc.* 1957, **79**, 5528.
26. H. S. Mosher, L. Turner and A. Carlsmith, *Org. Synth.* Collective Volume 1963, **4**, 828.
27. M. J. Böeseken and E. Arrias, *Rec. Trav. Chim.* Pays-Bas, 1935, **54**, 711.
28. J. G. Sharefkin and H. Saltzman. *Org. Synth.* Collective Volume, 1973, **5**, 658,660.
29. E. J. Corey, E. –P. Barrette and P. A. Magriotis, *Tetrahedron Lett.,* 1985, **26**, 5885.

1.98 PERSULPHURIC ACID, H_2SO_5

Also known as sulphomonoperacid, peroxymonosulphuric or **Caro's acid**. It is prepard *in situ* either from hydrogen peroxide[1], potassium persulphate ($K_2S_2O_8$) or ammonium persulphate [$(NH_4)_2S_2O_8$][1] (*see* also potassium persulphate)

$$(NH_4)_2S_2O_8 + H_2SO_4 \longrightarrow (NH_4)_2SO_4 + H_2SO_5 + SO_3$$

It can also be made from chlorosulphonic acid

$$\underset{\text{Chloro sulphonic acid}}{(HO)ClSO_2} + H_2O_2 \longrightarrow \underset{\text{Caro's acid}}{H_2SO_5} + HCl$$

It oxidizes α, β-unsaturated aldehydes into esters[1,2] (in the presence of alcohol)

$$CH_2{=}C(CH_3)CHO \xrightarrow{H_2SO_5\ CH_3OH,\ <10°\text{-}15°C} CH_2{=}C(CH_3)COOCH_3$$

Methyl methacrylate (71-97%)

It oxidises[3] disulphides to sulphonic acids

$$\text{2,2'-Diaminodiphenyl disulphide} \xrightarrow[25°C,\ 1\ hr]{H_2SO_5} 2\ \text{o-Aniline sulphonic acid (80\%)}$$

2,2′-Diaminodiphenyl disulphide

o-Aniline sulphonic acid 80%

It oxidises[4] primary aromatic amines to nitroso compounds.

$$\xrightarrow[aq.\ alcohol]{H_2SO_5}$$

3-Nitro-o-toludine, $x = NO_2$; $y = H$ — 3-Nitro-2-nitrosotoluene, $x = NO_2$; $y = H$

5-Nitro-o-toludine, $x = H$; $y = NO_2$ — 5-Nitro-2-nitrosotoluene, $x = H$; $y = NO_2$

50-60% yield

Besides the above oxidations, persulphunic acid also oxidises ketones into lactones or esters[5], primary aromatic amines to azoxy,[6] nitroso[4,6] or nitro compounds

Addition of some secondary and tertiary alcohols to concentrated presulphuric acid causes explosion[6].

The use of persulphuric acid is limited since organic peroxy acids are now used (*see* peroxyacids).

References

1. A. Nishihara and I. Kabota, *J. Org. Chem,* 1968, 33, 2525.
2. E. J. Corey and Y.W. J. Fleet, *Tetrahedron Lett.,* 1973, 4499.
3. T.H. Chao and W.B. Hardy, *Chem.* Ind (London), 1965, 81.
4. J. Meenheimer and E. Hesse, *Chem, Ber.,* 1919, **52**, 1161.
5. G. Büchi and O. Jeger, *Helv. Chim. Acta,* 1949, **32**, 538.
6. C.M. Atkinson and J.C.E. Simpson, *J. Chem. Soc.,* 1947, 1649.

1.99 PHENYLSELENYL CHLORIDE, C_6H_5SeCl

Phenylselenyl chloride along with hydrogen peroxide is used for the conversion of aldehydes, ketones and esters into their α, β, unsaturated analogues[1].

(1) PhSeCl
AcOEt, 1 hr
(2) 30% H_2O_2, THF,
35°, 1 hr

Cholestan-3-one

96%
1-Cholesten-3-one

Reference

1. K.B. Sharpless, R.F. Lauer and A.Y. Teranishi, *J. Am. Chem Soc.* 1973, 95, 6137.

1.100 POLY (VINYL PYRIDINIUM) CHLOROCHROMATE (PVPCC)

It is obtained[1] by adding chromium trioxide and concentrated hydrochloric acid to an aqueous sespension of cross-linked poly (vinyl pyridine) (PVP) resin (50–100 mesh). It oxidised primary and secondary alcohols to carbonyl compounds by heating with slight excess of the reagent at 75–80°C in cyclohexane

H
OH

Poly (vinylpyridine)
CrO_3, cyclohexane
75°C, 24 hr.

O

Cyclohexanol

Cyclohexanone

Reference

1. J.M.J. Fréchet, J. Warnock and M.J. Farrall, *J.Org. Chem.*, 1978, **43**, 2618.

1.101 POTASSIUM BROMATE, $KBrO_3$

Potassium bromate along with hydrobromic acid are used in the *in situ* generation of bromine for the oxidation of primary alcohols to esters[1].

It is used for the oxidation at o-iodobenzoic acid to **Dess-Martin periodinane**[2]

O C OH I

$KBrO_3$
H_2SO_4

O C O I O OH

Ac_2O, AcOH
100°C,
40 min

O C O I CH_3COO $OCOCH_3$ $OCOCH_3$

93%

93%
Dess-Martin
reagent

Potassium bromate is used for the preparation of p-benzoquinone from hydroquinone.

OH—⟨O⟩—OH $\xrightarrow[60°–80°C]{KBrO_3,\ 1N\ H_2SO_4}$ O=⟨ ⟩=O

Hydroquinone 71.4% *p*-Benzoquinone

References

1. L. Farkas and O.Schächter., *J. Am. Chem. Soc.,* 1949, **71**, 2827.
2. V.K. Ahluwalia and Renu Aggarwal, Comprehensive practical organic chemistry: Preparations and quantitative analysis, University Press, 2000, Page 169.

1.102 POTASSIUM CHROMATE, K_2CrO_4

It is rarely used. It converts hydrazomethane to azomethane[1]. In presence of hexamethylphosphoramide and crown ether, it transforms allylic and benzylic bromides into α, β-unsaturated aldehydes[2]. It oxidises chlorohydroquinone to chloro-p-benzoquinone[3] in presence of dilute sulphuric acid.

Cl
HO—⟨O⟩—OH $\xrightarrow{K_2CrO_4,\ dil\ H_2SO_4}$ O=⟨ ⟩=O (Cl)

Chlorohydroquinone Chlorohydroquinone

References

1. K. Langheld, *Chem. Ber.,* 1909, **42**, 392.
2. G. Cardillo, M. Orena and S.Sandri, *Chem. Commun,* 1976, 190.
3. A J. den Hollander, *Rec. Trav. Chim.* Pays-Bas, 1920, **39**, 481.

1.103 POTASSIUM DICHROMATE, $K_2Cr_2O_7$

It possesses almost identical oxidative properties as sodium dichromate. Thus, it oxidises benzene rings to quinones[1–3], methylene groups adjacent to aromatic rings to carbonyls[4], primary alcohols to aldehydes[5–7], secondary alcohols to ketones[6,8] and aldehydes to acids[9]. Also phenyl hydroxylamine is transformed into nitrosobenzene[10].

Potassium dichromate converts aromatic nitroso compounds into nitro compounds[11].

CH_3
O_2N—⟨O⟩—NO $\xrightarrow[65°C,\ 1.75\ hr.]{K_2Cr_2O_7,\ H_2SO_4,\ H_2O}$ O_2N—⟨O⟩—NO_2 (CH_3)

55–66%

Potassium dichromate oxidises aromatic amines to quinones. Thus, treatment of anilines with potassium dichromate and sulphuric acid gives p-benzoquninone[1] in 86% yield. This is the classical industrial method for the preparation of p-benzoquinone. Aminonaphthols are converted into naphthoquinones by potassium dichromate-sulphuric acid.

NH_2HCl (4-amino-1-naphthol) $\xrightarrow[100°C]{K_2Cr_2O_7,\ H_2SO_4,\ H_2O}$ Naphthoquinone 78–81%

In presence of dilute sulphuric acid, both potassium and sodium dichromate liberate chromic acid, H_2CrO_4, the most common oxidant. Standard 10% solution of CrO_3 are known as a **Kiliani mixture** or a **Beckmann mixture**. The Kiliani mixture is prepared by dissolving 60g. of $Na_2Cr_2O_7.2H_2O$ in dilute sulphuric acid (obtained by adding 80 g of concentrated sulphuric acid to 270 ml water) not the other way around). The Beckmann mixture is obtained by dissolving 60g. of $K_2Cr_2O_7$ in dilute sulphuric acid (80g. of concentrated sulphuric acid and 270 ml. of water)

References

1. K. Schniter, *Chem. Ber.,* 1887, **20**, 2282.
2. F. Kehrmann and T.E. Stiller, *Chem. Ber.,* 1912, **45**, 3346.
3. R.P. Linstead and P. Levine, *J. Am Chem. Soc.,* 1942, **64**, 2022.
4. R. Glenn and J.R. Bailey, *J. Am. Chem- Soc.,* 1941, **63**, 639.
5. R.O. Hutchins, N.R. Natale and W J. Cook, *Tetrahedron Lett.,* 1977, 4167.
6. D.Pletcher and S.J.D. Tail, *Tetrahedron Lett.,* 1978, 1601.
7. C.D. Hurd and R.N. Meinert, *Org. Synth.* Collective Volume, 1943, **2**, 541.
8. L. I. Sandborn, *Org. Synth.* Collective Volume, 1932, **1**, 340.
9. C.D. Hurd, J. W. Garrett and E.N. Osborne, *J. Am Chem. Soc.,* 1933, **55**, 1082.
10. E. Bamberger, *Chem. Ber.,* 1894, **27**, 1548.
11. W.D. Langley, *Org. Synth.* Collective Volume., 1955, **3**, 334.
12. L.F. Fieser, *Org. Synth.* Collective Volume, 1932, **1**, 383.

1.104 POTASSIUM FERRATE, K_2FeO_4

It is prepared[1] from ferric nitrate and alkaline hypochlorites. It is used for selective oxidation of primary alcohols to adehydes[2,3] (not acids) and of secondary alcohols to Ketones[2]

$$C_6H_5CH_2OH \xrightarrow[4\ min]{K_2FeO_4,\ H_2,\ RT} C_6H_5CHO$$

80%

$$PhCH{=}CHCH_2OH \xrightarrow[\text{OR } K_2FeO_4,\ 10\%\ KOH,\ t\text{-}BuOH,\ RT,\ 1\ hr.\ \text{yield } 96\%]{K_2FeO_4,\ H_2O,\ RT,\ 7\ min\ \text{yield } 75\%} PhCH{=}CH{-}CHO$$

$$C_6H_5CH(OH)CH_3 \xrightarrow[RT,\ 18\ min]{K_2FeO_4,\ H_2O} C_6H_5COCH_3\ (90\%)$$

Potassium ferrate dehydrogenates primary amines to aldehydes[2]

$$C_6H_5CH_2NH_2 \xrightarrow[RT,\ 1\ min]{K_2FeO_4,\ H_2O} C_6H_5CHO\ (70\%)$$

References

1. M. Szwarc and A. Shaw, *J. Am. Chem. Soc.*, 1951, **73**, 1379.
2. R.J. Audette, J.W. Quail and P.J. Swith, *Tetrahedron Lett.*, 1971, 279.
3. Y. Tsuda and S.Nakajima, *Chem. Lett.*, 1978, 1397.

1.105 POTASSIUM FERRICYANIDE, $K_3Fe(CN)_6$

It is used in dehydrogenation of heterocyclic compounds[1] and primary amines[2]

[1,2,3,4-Tetrahydroquinoline (N–H) $\xrightarrow{K_3Fe(CN)_6}$ quinoline (N)]

$$C_6H_{13}CH_2NH_2 \xrightarrow{K_3Fe(CN)_6} C_6H_{13}CN\ (55\%)$$

Dehydrogenative cyclisations are accomplished[3] with potassium ferricyanide

[Structure: OH, OH, $N(CH_3)_2I$, OH, OH] $\xrightarrow[HCl]{K_3Fe(CN)_6}$ [Structure: OH, OH, $N(CH_3)_2Cl$, OH, OH]

Laudonosoline methiodide → Apomorphine hydrochloride

Hydrazides are converted into aldehydes by potassium ferricyanide in aqueous ammonia

$$C_6H_5CONHNH_2 \xrightarrow[RT]{K_3Fe(CN)_6/NH_4OH} \underset{64\%}{C_6H_5CHO}$$

Tertiary amines are demethylated to secondary amines. Thus, dicyclohexylmethylamine on treatment with potassium ferricyanide is potassium hydroxide at room temperature gives 87% dicyclohexylamine[5]

$$\underset{\text{Dicyclohexyl methylamine}}{C_6H_{11}N(CH_3)C_6H_{11}} \xrightarrow[\text{RT, over night}]{K_3Fe(CN)_6/KOH} \underset{\text{87\% Dicyclohexyl amine}}{C_6H_{11}NHC_6H_{11}}$$

Potassium ferricyanide affects one electron oxidation of sterically hindered phenols to phenoxy radicals[6,9]

$$\text{2,4,6-}R_3C_6H_2OH \xrightarrow[\text{RT, 20 min}]{K_3Fe(CN)_6,\ NaOH,\ H_2O,\ Et_2O} \text{2,4,6-}R_3C_6H_2O^{\bullet}$$

$R = CH_3$ or C_6H_5

91–100% Phenoxy radical

References

1. V. Merz and C. Ris, *Chem Bev.*, 1887, **20**, 1190.
2. G.I. Nikashin, E.I. Troyanskii and V.A. Joffe, Izv. Akad. Nauk. SSSR, 1982, 2758; *Chem. Abstr.*, 1983, **98**, 142911.
3. J.D. Bower and Y.R. Ramage, *J. Chem. Soc.*, 1957, 4506.
4. L. Kalb and O. Gross, *Chem. Ber.*, 1926, **59**, 727.
5. T.D. Perrine, *J. Org. Chem.*, 1951, **16**, 1303.
6. C.D. Cook and R.C. Woodwarth, *J.Am. Chem. Soc.*, 1953, 75, 6242.
7. K. Dimroth, A. Berndt, H. Perst and C. Reichardt, *Org. Synth.* Collective Volume, 1973, **5**, 1130.

1.106 POTASSIUM HYDROXIDE, KOH

Potassium hydroxide has been used since long for the hydroxylation of both α-and β-anthraquinone sulfonic acid by alkali fusion to yield 1, 2-dihydroxyanthraquinone (alizarin).

The hydroxylation of aromatic compounds takes place on heating with solid potassium hydroxide of compounds with low electron density in the ring. Thus, nitrobenzene is converted with o-nitrophenol on heating with potassium hydroxide for 2 hr. at 60–70° in 45% yield[1]. In a simlar way, quinoline gives carbostyryl (2-hydroxyquinoline) on heating with potassium hydroxide for 3 hr at 225°C in 70% yield[2]

NO_2 — KOH Δ, 2 hr. / 60–70°C → NO_2, OH

Nitrobenzene o-Nitrophenol (45%)

KOH, 3 hr. / 225°C → N OH

Quinoline 2-Hydroxyquinoline

Aromatic aldehydes on fusion (140–150°C) with a mixture of sodium and potassium hydroxide gives[3] the corresponding carboxylic acid. However, fusion at higher temperature (190–95°C), there is simiultaneous oxidation and demethylation to give protocatechuic acid.

CH_3O, HO, CHO — KOH-NaOH fusion / 140–150°C, 5-10 min → CH_3O, HO, COOH

Vanillin Vanillic acid 85–95%

KOH-NaOH fusion / 190–95°C, 50–30 min → HO, HO, COOH

Protocatechuic acid 60–65%

Heptadecyl alcohol on heating (240–250°C) with three times its weight of potassium hydroxide yields 76% heptadecanoic acid[4].

Under milder conditions (25–80°C, 10-60 min), benzyl alcohol is converted into benzoic acid in 75% yield on heating with excess of powdred potassium hydroxide in aqueous tert. bytyl alcohol and carbon tetrachloride[5].

References

1. A. Wohl, *Chem. Ber.,* 1899, 32, 3486.
2. A.E. Tschitschibabin, *Chem. Ber.,* 1923, **56**, 1879.
3. I.A. Pearl, *Org. Synth.* Collective Volume, 1963, **4**, 974.
4. A. Heiduschka and J. Ripper, *Chem. Ber.,* 1923, 56.
5. C.V. Meyers, A.M. Malte and W.S. Matthews, *J. Am. Chem. Soc.,* 1969, **91**, 7510.

1.107 POTASSIUM HYPOCHOLORITE, KOCl

This reagent is prepared *in situ* from potassium hydroxide and chlorine. It acts like sodium hypochlorite.

Potasium hypochlorite oxidises methyl ketones to carboxylic acids. Thus mesityl oxide is converted into β,β-dimethyl acrylic acid in 49-53% yield[1].

$$(CH_3)_3CCOCH_3 \xrightarrow[\text{reflux 3-4 hr.; } H_2SO_4]{\text{KOCl, } H_2O\text{, dioxane}} (CH_3)_2C{=}CHCOOH$$

Misityl oxide → 49-53% β,β-Dimethyl acrylic acid

Benzylic alcohols are transformed[2] into aldehydes by potassium hypochlorite.

$$C_6H_5CH_2OH \xrightarrow[\text{36-40°C overnight}]{\text{KOCl, } H_2O\text{, MeOH}} C_6H_5CHO$$

Benzyl alcohol → Benzaldehyde 77%

References

1. L.I. Smith, W.W. Prichard and L.J. Spillane, *Org. Synth.* Collective Volume, 1955, **3**, 302.
2. C.Y. Meyers, *J. Org. Chem.,* 1961, **26**, 1046

1.108 POTASSIUM HYPOIODITE

Like sodium hypoiodite, potassium hypoiodite is also prepared *in situ* from iodine and potassium hydroxide. It oxidizes methyl ketones to carboxylic acids. The byproduct of this reaction is iodoform (*see* also sodium hypoiodite).

An interesting example[1] of the use of potassium hypoiodite is the opening of the ring D of benzyl estrone to benzyl marrianolic acid.

$C_6H_5CH_2O$ — Benzyl estrone $\xrightarrow[\text{(2) KOH, MeOH, } H_2O \text{ reflux, 3 hr.}]{\text{(1) } I_2\text{, KOH, } H_2O\text{, MeOH RT, few hr.}}$ Benzyl marrianolic acid (COOH, CH_2COOH, $C_6H_5CH_2O$)

Reference

1. J. Heer and K. Miescher, *Helv. Chem. Acta*, 1945, **28**, 157.

1.109 POTASSIUM MANGANATE, K_2MnO_4

It is prepared[1,2] from potassium permangante and potassium hydroxide at 100°C or 120–140°C. It has only limited use. It is used[1] for synhydroxylation of double bonds and hydroxylation of tertiary carbons in branched carboxylic acids[2]. In fact, it offers no advantages over potassium permanganate.

References

1. R. Rigby, *J. Chem. Soc.,* 1956, 2452.
2. J. Kenyton and M.C.R. Simons, *J. Chem. Soc.,* 1953, 3580.

1.109A. POTASSIUM PERIODATE, KIO_4

In dilute sulphuric acid, it acts like periodic acid in degrading α-hydroxy acids to aldehydes (*see* periodic acid).

1.110 POTASSIUM PERMANGANATE, $KMnO_4$

Potassium permanganate is mostly used in aqueous solutions. However, it can also be used in the presence of organic solvents, such as petrolum ether, benzene, dichloromethane, tert butyl alcohol, acetone and pyridine. The solubility of potassium permanganate in organic solvents can be achieved by using phase transfer catalysts[1], such as quaternary ammonium salts that form quaternary permanganate which are soluble in solvents immiscible with water. In place of phase transfer catalysts, crown ethers[1] can also be used.

Potassium permanganate is a very powerful oxidising agent. Its reactivity depends on whether it is used in acidic, neutral or basic conditions. In fact, the results of some oxidations with potassium permanganate differ depending on the pH of the reactions[2]. Potassium permanganate is used as a catalyst for oxidation with other oxidants, such as sodium periodate (*see* sodium periodate).

Following are given some of the oxidations affected by potassium permanganate:

1. Potassium permangnate is used in the cleavage of alkenes and cycloalkenes.

$KMnO_2$, $MgSO_4$, H_2O / Me_2CO, –15 to 20°C, 1 h → RT

CHO CHO

55-56%

Reference 3

Reference

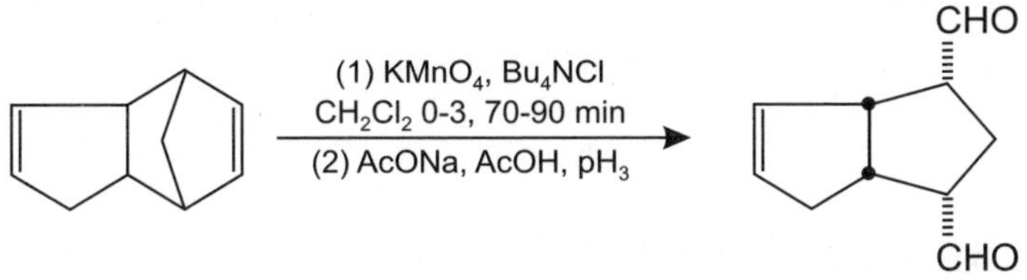

(1) $KMnO_4$, Bu_4NCl
CH_2Cl_2 0-3, 70-90 min
(2) AcONa, AcOH, pH_3

CHO
CHO

81% 4

$C_8H_{17}CH{=}CH_2 \xrightarrow[\text{AcOH, 0°C, 4 hr.}]{KMnO_4,\ H_2O,\ C_6H_6} C_8H_{17}CO_2H$ 85% 5

$C_8H_5CH{=}CHC_6H_5$ (*trans*) $\xrightarrow[C_6H_6,\ \text{RT}]{KMnO_4,\ \text{Crownether}} C_6H_5CO_2H$ 97% 6

$CF_3CCl{=}CCl_2 \xrightarrow[(2)\ H_2O,\ H_2SO_4,\ SO_2]{(1)\ KMnO_4,\ Me_2CO,\ \text{RT, 30 min}} CF_3CO_2H$ 46% 7

$CF_3CCl{=}CClCF_2 \xrightarrow[(2)\ SO_2,\ H_2SO_4]{(1)\ KMnO_4,\ KOH,\ 60°C,\ \text{reflux 8-10 hr}} CF_3CO_2H$ 83% 8

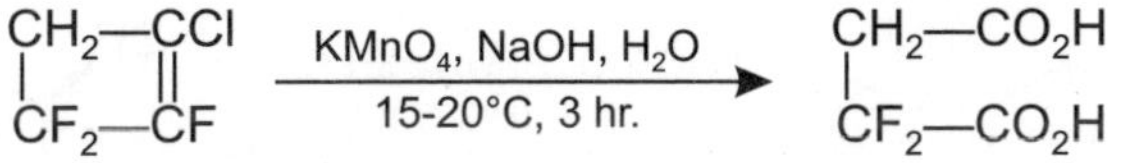

$CH_2{-}CCl$ / $CF_2{-}CF$ (ring) $\xrightarrow[15\text{-}20°C,\ 3\ \text{hr.}]{KMnO_4,\ NaOH,\ H_2O}$ $CH_2{-}CO_2H$ / $CF_2{-}CO_2H$ 9

2. Oxidative degradation of methyl ketones and unsaturated carboxylic acids[11,12].

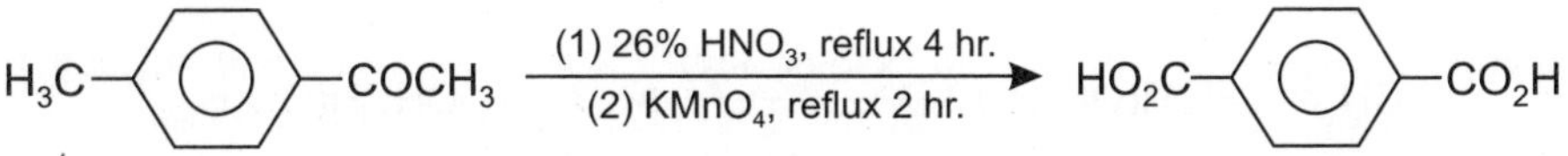

$H_3C{-}C_6H_4{-}COCH_3 \xrightarrow[(2)\ KMnO_4,\ \text{reflux 2 hr.}]{(1)\ 26\%\ HNO_3,\ \text{reflux 4 hr.}} HO_2C{-}C_6H_4{-}CO_2H$

84-88%

$HO_2C(CH_2)_7CH{=}CH(CH_2)_7CO_2H \xrightarrow[5°C,\ 10\ \text{min}]{KMnO_4,\ KOH}$

$HO_2C(CH_2)_7CH(OH){-}CH(OH)(CH_2)_7CO_2H$

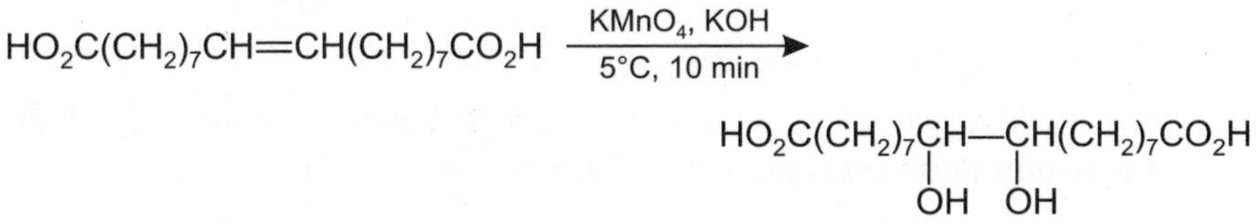

$CH_3(CH_2)_7CH{=}CH(CH_2)_7CO_2H \xrightarrow[20°C,\ 20\ \text{hr.}]{KMnO_4,\ NaIO_4,\ K_2CO_3,\ H_2} CH_3(CH_2)_7CO_2H$ 100% + $HO_2C(CH_2)_7CO_2H$ 100%

When $KMnO_4$ is used along with $NaIO_4$, it is known as **Lemieus-Rudloff reagent.**

3.Dehydrogenation of primary amies to aldehydes[13] ketones and[14] imines[15].

$$C_6H_5CH_2NH_2 \xrightarrow[\text{H}_2\text{O, 50-75°C; steam distillation}]{\text{KMnO}_4\text{, ZnSO}_4\text{, t-BuOH}} C_6H_5CHO \quad 61\%$$

$$\text{Cyclohexylamine} \xrightarrow[\text{70°C, distillation}]{\text{KMnO}_4\text{, CuSO}_4\text{, H}_2\text{O, t-BuOH}} \text{Cyclohexanone} \quad 75\%$$

$$(C_6H_5)_2CHNH_2 \xrightarrow[\text{H}_2\text{O, t-BuOH; 70°C, reflux 8 hr.}]{\text{KMnO}_4\text{, CaSO}_4\text{, 2H}_2\text{O}} C_6H_5C{=}N{-}N{=}C(C_6H_5)_2$$

4. Dehydrogenative coupling.

   Diarylmethanes give tetraalkylethanes on oxidation with potassium permanganate[15a]

$$(C_6H_5)_2CH_2 \xrightarrow{\text{KMnO}_4\text{, −33°C}} (C_6H_5)_2CHCH(C_6H_5)_2 \quad 69\%$$

5. Hydroxylation of dienes[4] and unsaturated acetals[15b]. Potassium permanganate and triethylbenzylammonium chloride as a PTC gives 83% yield of exo, *cis*-diol resulting from hydroxylation of the double bond in the cyclohexene ring[4].

$$\text{Dicyclopentadiene} \xrightarrow[\text{(2) 3\% NaOH, RT, 18 hr}]{\text{(1) KMnO}_4\text{,PhCH}_2\text{NEt}_3\text{Cl, CH}_2\text{Cl}_2\text{, 0-3°C, 1.5 hr.}} \text{exo-}cis\text{-diol (OH, HO)} \quad 83\%$$

$$\underset{\text{Acerolein diethyl acetal}}{CH_2{=}CHCH(OC_2H_5)_2} \xrightarrow[\text{5°C, 3 hr; 100°C, 1 hr}]{\text{KMnO}_4\text{, H}_2\text{O}} \underset{\text{OH}\quad\text{OH}}{CH_2{-}CHCH(OC_2H_5)_2} \quad 67\%$$

6. Syn hydroxylation of alkenes

   Syn hydroxylation[16] is achieved by dilute aqueous solutions (1–2%) of potassium permanganate at low temperatures (0–5°C)

$$\text{1,2-Dimethylcyclopentene} \xrightarrow[\text{−5°C, 1 hr.}]{\text{KMnO}_4\text{, H}_2\text{O, Me}_2\text{CO}} \text{cis-diol (OH OH)} \quad 45\%$$

7. Potassium permanganate is a versatilic oxigizing agent. It is used in the following oxidations:

(*a*) Acetylene carboxylic acids to diketo carboxylic acids[17]

$$\underset{\text{Stearolic acid}}{CH_3(CH_2)_7C{\equiv}C(CH_2)_7CO_2H} \xrightarrow[\text{pH 7.0–7.5, 25°C}]{\text{KOH, KMnO}_4} \underset{\substack{\text{9,10-Diketostearic acid}\\ \text{92-96\%}}}{CH_3(CH_2)_7COCO(CH_2)_7CO_2H}$$

(*b*) Aldehydes to carboxylic acids

Potassium permanganate is frequently used to prepare carboxylic acids from both aliphatic[18] and aromatic[19] aldehydes

$$C_6H_{13}CHO \xrightarrow{\text{KMnO}_4\text{, H}_2\text{SO}_4\text{, H}_2\text{O, 15-20°C}} \underset{\text{76-78\%}}{C_6H_{13}COOH}$$

(*c*) Alkenes to carboxylic acids

$$\underset{\substack{\text{2,3-Dichlorohexafluoro-}\\ \text{-2-butene}}}{CF_3C(Cl){=}C(Cl){-}CF_3} \xrightarrow[\substack{\text{60-95°C (8-10 hr)}\\ \text{40°C, SO}_2}]{\text{KMnO}_4\text{, KOH, H}_2\text{O}} \underset{\substack{\text{Trifluoroacetic acid}\\ \text{87\%}}}{CF_3COOH}$$

(*d*) Internal alkenes to α-hydroxy ketones[20]

$$(C_4H_9)(H)C{=}C(H)(C_4H_9) \xrightarrow[\text{25°C}]{\text{KMnO}_4\text{, Me}_2\text{CO, H}_5\text{O, AcOH}} \underset{\text{73\%}}{C_4H_9CH(OH){-}C({=}O)C_4H_9}$$

(*e*) Terminal alkynes to carboxylic acids

$$C_8H_{17}C{\equiv}CH \xrightarrow[\substack{\text{Alliquat 336, cooling}\\ \text{with ice, 5 hr.}}]{\text{KMnO}_4\text{, H}_2\text{O, C}_5\text{H}_{12}\text{, AcOH}} \underset{\text{70\%}}{C_8H_{17}CO_2H}$$

(*f*) Aromatic side chain to carboxyl groups

Not only methyl group, practically all other alkyl groups attached to aromatic rings are oxidised to carboxylic acids. Some examples[21,22] of oxidation of side chain are given below.

$$R{-}C_6H_4{-}CH_3 \xrightarrow[\text{reflux}]{\text{KMnO}_4\text{, NaOH, H}_2\text{O}} R{-}C_6H_4{-}CO_2H$$

substituent R	yield %
o–Br	59
m–Br	63
p–Br	76%

CH_3 (4-methylpyridine) $\xrightarrow{KMnO_4,\ H_2O,\ Reflux}$ COOH (isonicotinic acid) 60-70%

(*g*) Oxidation of benzene rings to carboxylic acids. Naphthalene is oxidised to phthalic acid[23]

Naphthalene $\xrightarrow[\text{reflux 2 hr}]{KMnO_4,\ NaOH}$ Phthalic acid (CO_2H, CO_2H) 40–41%

α-Nitronaphthalene (NO_2) $\xrightarrow[\text{reflux, 1.5 hr}]{KMnO_4,\ NaOH}$ Phthalonic acid ($COCO_2H$, CO_2H) 74%

(*i*) Benzylic alcohols to aldehydes[24].

$$C_6H_5CH_2OH \xrightarrow[C_6H_6,\ 70°C,\ 2\ hr]{KMnO_4,\ \text{molecular sieves } 4Å} C_6H_5CHO$$

Benzylalcohol → Benzaldehyde 80%

(*j*) Hydroxy carboxylic acids and their esters to Keto acids[25] and their ester[26].

$$C_6H_5CH(OH)CO_2R \xrightarrow[-2\ to\ -4°,\ 1.5\ hr]{KMnO_4,\ NaOH,\ H_2O} C_6H_5COCO_2R$$

35-54%

Mandelic acid, R = H → Phenylglyoxalic acid R = H

Ethylmandalate, R = C_2H_5 → Ethylphenylglyoxalate R = C_2H_5

In case of ethyl mandalate[26], NaH_2PO_4 and $MgSO_4$ are used along with $KMnO_4$

(*k*) Oxidation of hydroxy aldehydes to keto acids[27]

$$C_6H_5CH(OH)(CH_2)_3CHO \xrightarrow[RT]{KMnO_4,\ NaOH,\ H_2O} C_6H_5CO(CH_2)_3CO_2H$$

δ-Hydroxyvaleraldehyde → γ-Benzoylbutyric acid 90%

(*l*) Iodoso compounds to iodoxy compounds

o-Iodosobenzoic acid is converted into o-iodoxybenzoic acid in 81% yield by heating with potassium permanganate in dilute sodium hydroxide solution[28].

$$\text{o-Iodosobenzoic acid (C}_6\text{H}_4\text{(IO)(COOH))} \xrightarrow{KMnO_4/\text{dil NaOH}} \text{o-Iodoxybenzoic acid (C}_6\text{H}_4\text{(IO}_2\text{)(COOH)) (81\%)}$$

(*m*) Lactones to dicarboxylic acids[29]

$$\text{3,5-Dimethoxyphthalide} \xrightarrow[0°C,\ 12\ hr.,\ R.T.\ 24\ hr.]{KMnO_4,\ NaOH} \text{3,5-Dimethoxyphthalic acid (90\%)}$$

(*o*) Mercaptoles to disulfones

Potassium permanganate oxidises acetone 2-chloroethylmercaptole to the corresponding disulfone (sulfonal)[30]

$$(CH_3)_2C(SCH_2CH_2Cl)_2 \xrightarrow[C_6H_6,\ 25\text{-}30°C,\ 2\ hr.]{KMnO_4,\ H_2SO_4,\ H_2O} (CH_3)_2C(SO_2CH_2CH_2Cl)_2$$

Sulfonal 81%

(*p*) Methyl ketons to α-ketocarboxylic acid

Normally ketones like acetophenone, p-methylacetophenone are oxidised to the corresponding benzoic acids. However, 2,4-dimethylacetophenone on shaking with aqueous potassium permanganate (1%) at room temperature gives 66–72% yielded 2,4-dimethylphenylglyoxylic acid.[31,32]

$$\text{2,4-Dimethyl acetophenone} \xrightarrow[RT]{1\%\ aq.\ KMnO_4} \text{2,4-Dimethylphenyl glyoxalic acid (66-72\%)}$$

Another example of interest is the oxidation of 2,5-dimethylpropiophenone to 2,5-dimethylbenzoylacetic acid[33].

$$\text{2,5-Dimethylpropiophenone} \xrightarrow[\text{RT}]{1\%\ \text{aq. } KMnO_4} \text{2,5-Dimethylbenzoyl acetic acid}$$

(2,5-Dimethyl propiophenone: H_3C, CH_3, $COCH_2CH_3$ on benzene ring → 2,5-Dimethylbenzoyl acetic acid: H_3C, CH_3, $COCH_2COOH$)

(*q*) Methylene group to keto group

Methylene groups adjacent to aromatic rings are oxidised to keto group by potassium permanganate[33a;33b]

$$\underset{\text{Ethylbenzene}}{C_6H_5CH_2CH_3} \xrightarrow{KMnO_4\ \text{aq.}} \underset{\text{Acetophenone}}{C_6H_5COCH_3}$$

(*r*) Oxidation of nitro compounds

The oxidation of primary and secondary nitro compounds gives[33c] aldehydes and ketones, respectively.

$$CH_3CH_2CH_2CH_2NO_2 \xrightarrow[\text{Steam distillation}]{KMnO_4,\ 0\text{-}5°C} \underset{76\text{-}90\%}{CH_3CH_2CH_2CHO}$$

$$CH_3CH_2\underset{NO_2}{\underset{|}{C}}H—CH_3 \xrightarrow[\text{Steam distillation}]{KMnO_4,\ 0\text{-}5°C} \underset{94\%}{CH_3CH_2COCH_2}$$

(*s*) Oxidation of primary alcohols to aldehydes[24] and carboxylic acids[6]

Potassium permanganate is rarely used to convert alcohols into aldehydes, sice aldehydes are oxidised to carboxylic acids. However, when adsorbed on molecular sieves, potassium permanganate in benzene at 70° oxidises[24] benzyl acohol and cinnamyl alcohol to the aldehydes in 80 and 94% yields, respectively

$$C_6H_5CH_2OH \xrightarrow[C_6H_6,\ 70°C,\ 2\ \text{hr.}]{KMnO_4,\ \text{molecular sieves 4Å}} \underset{80\%}{C_6H_5CHO}$$

$$C_6H_5CH{=}CHCH_2OH \xrightarrow[C_6H_6,\ 70°C,\ 2\ \text{hr.}]{KMnO_4,\ \text{molecular sieves 4Å}} \underset{90\%}{C_6H_5CH{=}CHCHO}$$

Primary alcohols are oxidised to carboxylic acids by potassium permanganate in presence of crown ether[6].

$$C_6H_5CH_2OH \xrightarrow[\text{RT, 2 hr.}]{KMnO_4,\ \text{dicyclohexano-18-cown-6, ether}} \underset{100\%}{C_6H_5CO_2H}$$

(*t*) Primary amines to nitro compounds

The oxidation of some aliphatic amines to nitro compounds can be affected with by potassium permanganale

$$(CH_3)_3C{-}NH_2 \xrightarrow[\text{RT, 2 hr.}]{KMnO_4,\ H_2O} (CH_3)_3C{-}NO_2$$

tert-Bulylamine → 83% 2-Methyl-2-nitropropane

$$CH_3C(CH_3)_2{-}CH_2C(CH_3)_2{-}NH_2 \xrightarrow[\text{45°C, 8 hr}]{KMnO_4,\ H_2O} H_3CC(CH_3)_2{-}CH_2{-}C(CH_3)_2{-}NO_2$$

4-Amino-2,2,4-trimethylpentane → 69-82% 4-Nitro-2,2,4-trimethylpentane

(*u*) Oxidation of pyridine homologues

Pyridine homologues are oxidised[22] to the corresponding carboxylic acids.

$$\gamma\text{-}CH_3C_5H_4N \xrightarrow[\text{reflux}]{KMnO_4,\ H_2O} \gamma\text{-}HO_2CC_5H_4N$$

γ-Methylpyridine → Pyridine γ-carboxylic acid

(*v*) Oxidation of secondary, alcohols to ketones.

Aqueous potassium permanganate is generally not used to convert secondary alcohols to ketone, since the formed ketones are oxidised to carboxylic acids. However, such oxidations are sussessfully accomplished in organic solvents with solid or supported permanganates.[24]

$$CH_3CH(OH)C_6H_{13} \xrightarrow[C_6H_6,\ 70°C,\ 7\ hr]{KMnO_4,\ \text{molecular sieves}} CH_3COC_6H_{13}\quad 92\%$$

$$C_6H_5CH(OH)C_6H_5 \xrightarrow[C_6H_6,\ 70°C,\ 7\ hr]{KMnO_4,\ \text{molecular sieves}} C_6H_5COC_6H_5\quad 100\%$$

Oxidation of secondary alcohols with aqueous potassium permangante in presence sodium carbonate give carboxylic acid[37].

Cyclohexanol —($KMnO_4$, Na_2CO_3, H_2O; 15-33°C)→ $HO_2C(CH_2)_4CO_2H$ 70% Adipic acid

(*w*) Tertiary amimes to amides

Some tertiary amines are oxidised[38] to amides by potassium permanganate in acetone and pyridine.

p-Bromobenzyl diphenylamine, $Br-C_6H_4-CH_2N(C_6H_5)_2$ —($KMnO_4$, Me_2CO, C_5H_5N; reflux, 1.75 hr)→ $Br-C_6H_4-CON(C_6H_5)_2$ 91% Diphenyl-p-bromo-benzamide

(*x*) Selenides to the corresponding selenone[39]

$C_6H_5SeCH_3$ Methyl phenyl selenide —($KMnO_4$, CH_2Cl_2; 20°C, 3 min)→ $C_6H_5-SeO_2-CH_3$ 75% Methyl phenyl selenone

(*y*) Preparation of saccharin.

Saccharin is prepared by the oxidation of the sodium salt of tolune-o-sulfonamide[40].

Sod. salt of toluene-o-sulfonamide (CH_3, SO_2NHNa) —($KMnO_4$; 60°C, 6 hr)→ [CO_2H, SO_2NHNa] → Saccharin (CO, NNa, SO_2) 80%

References

1. V.K. Ahluwalia and Renu Aggarwal, *Organic Synthesis, Special Techniques*. Narosa Publishing House. 2006, Page 1 to 114 and the references cited there in.
2. N.A. Khan and M.S. Newman, *J. Org. Chem.*, 1952, **17**, 1063.

3. K. Wiesner, E.W.K. Jay, T.Y.R. Tsai, C. Demerson, L. Jay, T. Kanno, A. Vilim and C.S. Wu, Can. *J.Chem.,* 1972, **50**, 1925.
4. T. Ogino and K.Mochizuki, *Chem. Lett.,* 1979, 443.
5. A.P. Krapcho, J.R. Larson and J.M. Eldridge, *J.Org. Chem,* 1977, **42**, 3749.
6. D.J. Sam and H.F. Simmons, *J.Am. Chem. Soc.,* 1972, **94**, 4024.
7. J. Burdon and J.C. Tatlow, *J. Appl. Chem,* 1958, **8**, 293
8. A.L. Henne and P.Trott, *J.Am. Chem. Soc.*, 1947, **69**, 1820
9. T. Sujiu and M. Minato, *Tetrahedron Lett.*, 1987, **28**, 3683.
10. C.F. Koelsch, *Org. Synth.* Collective Volume, 1955, **3**, 791.
11. W.J. Gensler and H.N. Scheim, *J. Am. Chem. Soc.*, 1955, **77**, 4846.
12. R.U. Lemiux and V.E. Rudloff, Can *J. Chem,* 1955, **33**, 1701.
13. C.C. Hach, C.V. Banks and H.Diehl, *Org. Synth.* Collective Volume, 1963, **4**, 229.
14. J.J. Postowsky and B.P. Lungowkin, *Chem. Bev.,* 1935, **68**, 852.
15. H.L. Riley and N.A.C. Friend, *J.Chem. Soc.,* 1932, 2342.
15a. E.M. Kaiser, *J.Am. Chem. Soc.* 1967, **89**, 3659.
15b. H. Rapoport and H.N. Reist, *J.Am. Chem. Soc.,* 1955, **77**, 490.
16. P.D. Barlett and A. Bavely, *J. Am. Chem. Soc.,* 1938, **60**, 2416.
17. H. Goldstein and R. Jaunin, *Helv. Chem. Acta,* 1951, **34**, 2222.
18. R. Runoff, *Org. Synth.* Collective Volume, 1943, **2**, 315.
19. R.L. Shriner and E.C. Kleiderver, *Org. Synth.* Collective Volume, 1943, **2**, 538.
20. N.S. Srinivasan and D.G. Lee, *Syntheses*, 1979, 520.
21. L.A. Bigelow, *J. Am. Chem. Soc.,* 1922, **44**, 2010.
22. R.L. Malan and P.M. Dean, *J.Am. Chem. Soc.,* 1947, **69**, 1797.
23. J.H. Gardner and C.A. Naylor, Jr., *Org. Synth.* Collective Volume, 1943, **2**, 523.
24. S.L. Regen and C.Koteel, *J.Am. Chem. Soc.,* 1977, **99**, 3837.
25. B.B. Corson, R.A. Dodge, S.A. Harris and R.K. Hazen, *Org. Synth.* Collective Volume, 1932, **1**, 241.
26. J.W. Cornsworth, *Org. Synth.* Collective Volume, 1963, **4**, 467.
27. C.W. Smith, D.Y. Norton and S.A. Ballard, *J. Am. Chem. Soc.,* 1951, **73**, 5273.
28. C. Hartman and V.Meyer, *Chem. Ber.,* 1893, **26**, 1727.
29. H. Brockman, F. Kluge and H. Muxfeldl., *Chem. Ber.,* 1957, **90**, 2302.
30. R. Criegee and M. Lederer, *Justus Liebig Ann. Chem.,* 1953, **583**, 1,29.
31. A. Claus and R. Wollner, *Chem. Ber.,* 1885, **18**, 1856.
32. A. Claus, *Chem. Ber.,* 1886, **19**, 230.
33. A. Claus, and E. Fickert, *Chem. Ber.,* 1886, **19**, 1856.
33a. W. Winkler, *Chem. Ber.,* 1948, **81**, 256.
33b. K.E. Crook and S.M. Mc Elvian, *J. Am. Chem. Soc.,* 1930, **52**, 4006.
33c. H. Shechter and F.T. Williams. *Jr. J. Org. Chem.,* 1962, **27**, 3699.
34. A. Calder, A.R. Forrester and S.P. Hepburn, *Org. Synth.* Collective Volume, 1988, **6**, 803.
35. N. Kornblum and R.J. Clutter, *J. Am. Chem. Soc.,* 1954, **76**, 4494.
36. N. Kornblum and W.J. Jones, *Org. Synth.* Collective Volume, 1973, **5**, 845.
37. V.J. Braun and Y. Lemke, *Chem. Ber.,* 1922, **55**, 3526.
38. J. Forrest, S.H. Tucker and M.Whalley, *J.Chem. Soc.,* 1951, 303.
39. A. Krief, W. Dumont, J.-N. Denis, G. Evrad and B. Norbery, *Chem. Commun,* 1985, 569.
40. A.W. Pamfiluv, *Chem. Zentr.,* 1922, **III**, 353.

1.111 POTASSIUM PERSULPHATE

A typical reaction of potassium persulphate is the hydroxylation in the aromatic ring, which take places in the *para* or *ortho* position with respect to the hydroxylic group. This is achieved by alkaline persulphate and is known as **Elbe persulphate oxidation**[1]

$K_2S_2O_8$, 10% NaOH / 20°C, 3-4 hr, overnight

o-Chlorophenol → 2-Chloro-4-hydroxyphenol (62%)

In place of potassium persulphate, sodium or ammonium persulphate can also be used.

The mechanism of persulphate oxidation[1] involves the hydrolysis of the initially formed persulphate(1) to the hydroquinone

Phenoxide + $-KSO_4^-$ → H^+ → (1) $\xrightarrow{H_3O^+}$ Hydroquinone

Some other applications of persulphate oxidation are given below[2].

2-Hydroxypyridine $\xrightarrow[(2)\ H^+]{(1)\ K_2S_2O_8/NaOH,\ 100°C}$ 2,5-Dihydroxypyridine

Salicylic acid $\xrightarrow[(2)\ H^+]{(1)\ K_2S_2O_8/alkali}$ Gentisic acid

Isatin, a cyclic α-keto amide is oxidised with polassium persulphate in presence of H_2SO_4 to 2,3-dioxo-1,4-benzoxazine[3].

Isatin → ($K_2S_2O_8$, H_2SO_4, 0-10°C, few minutes) → 2,3-Dioxo-1,4-benzoxazine 95%

In presence of sulphuric acid, potassium persulphate acts as persulphuric acid[4], H_2SO_5, (*see* persulphuric acid).

Potassium persulphate dehydrogenates primary amines to imines[5] and aromatic primary amines are oxidised to nitroso compounds[6].

Potassium persulphate is also used in conjuction with copper sulphate for the regiospecific oxidation of dimethylanisoles to methoxymethyl benzaldehydes[7] (*see* also copper sulphate).

CH_3O–C_6H_3(CH_3)–CH_3 → ($K_2S_2O_8$, $CuSO_4$, $5H_2O$; MeCN/H_2O (1:1); reflux, 15-30 min) → CH_3O–C_6H_3(CH_3)–CHO

References

1. W. Baker, and N.C. Broun, *J. Chem. Soc*, 1948, 2303.
2. E.J. Bergmann, *J.Am. Chem. Soc.*, 1958, **80**, 3717.
3. G. Reissenweber and D. Mangold, Anglew, *Chem.*, Int. Ed. Engl, 1980, **19**, 222.
4. J. Meisenhecimer and E. Hesse, *Chem.* Ber., 1919, 52, 1161.
5. R.G.R. Bacon and D. Stewart, *J. Chem. Soc.*, C 1966, 1384.
6. W.D. Langley, *Org. Synth.* Collective Volume, 1955, **3**, 334.
7. F.M. Hauser and S.R. Ellenberger, *Synthesis*, 1987, 723.

1.112 POTASSIUM RUTHENATE, K_2RuO_4

It is prepared[1] in situ from ruthenium trichloride and aqueous potassium persulphate.

It oxidises[1] aromatic aldehydes to carboxylic acids.

m-Nitrobenzaldehyde → (K_2RuO_4, H_2O; RT, 2 hr) → *m*-Nitrobenzoic acid 99%

Primary and secondary alcohols are oxidised to carboxylic acids and ketones, respectively.

$$C_6H_5CH_2OH \xrightarrow[\text{RT, 1.5 hr.}]{K_2RuO_4,\ K_2S_2O_8} \underset{98\%}{C_6H_5CO_2H}$$

$$RR'CHOH \xrightarrow[\text{RT, 1.5 hr.}]{K_2RuO_4,\ K_2S_2O_8} RR'C{=}O$$

Potassium ruthernate dehydrogenates[1] primary amines to nitriles.

$$RCH_2NH_2 \xrightarrow[\text{RT, 1.5 hr.}]{K_2RuO_4,\ K_2S_2O_8} RCN$$

Reference

1. H. Schröder and W.P. Griffith, *Chem. Commun.,* 1979, 58.

1.113 POTASSIUM SUPEROXIDE, KO_2

It oxidises Ketones containing α-hydrogens to caboxylic acids at room temperature in presence of phase transfer system with Aliquat 336 (methyltrioctylammonium chloride[1]).

$$\underset{\text{Cyclohexanone}}{C_6H_{10}O} \xrightarrow[\text{RT, 1–8 hr}]{KO_2,\ C_6H_6,\ (C_8H_{17})_3\ NMeCl\ \text{(aliquat 336)}} \underset{\substack{92\% \\ \text{Adipic acid}}}{\begin{array}{l} CH_2CH_2COOH \\ | \\ CH_2CH_2COOH \end{array}}$$

Reference

1. M. Lissel and E.V. Dehmlow, *Tetrahedron Lett*, 1978,3687.

1.114 PYRIDINIUM CHLOROCHROMATE (PCC) $C_5H_5N.HCl.CrO_3$

It is commercially available and is prepared[1] by adding chromium trioxide to a solution of pyridine in hydrochloric acid.

$$\begin{array}{c} \text{aq. HCl} \\ CrO_3,\ \text{pyridine} \end{array} \longrightarrow \underset{\text{PCC}}{C_5H_5N^+H\ \ ClCrO_3^-}$$

It resembles Collins reagent in its oxidising properties.

At room temperature, primary alcohols give aldehydes[2,3]

$$C_5H_{11}C{\equiv}C{-}CH_2OH \xrightarrow[25°C,\ 1\text{–}2\ hr]{PCC} C_5H_{11}C{\equiv}CCHO \quad 84\%$$

$$PhCH{=}CH{-}CH_2OH \xrightarrow[C_6H_{14},\ RT,\ 2\ hr]{PCC,\ Al_2O_3} PhCH{=}CHCHO \quad 84\%$$

Secondary alcohols are oxidised[4] to ketones

PCC oxidises[5] methylene groups to carbonyl especially when the group is flanked by two aromatic rings. Thus, Diphenylmethane yields benzophenone

$$C_6H_5{-}CH_2{-}C_6H_5 \xrightarrow[benzene,\ reflux/8\ hr]{PCC,\ celite} C_6H_5{-}CO{-}C_6H_5$$

Diphenyl methane — 88% Benzophenone

The oxidation of boranes (prepered by the reaction of alkene with boron hydrides) gives aldehydes[3].

$$C_6H_{13}CH{=}CH_2 \xrightarrow[Et_2O,\ 0°C,\ 2\ hr.]{(Me_2CHCHMe)_2BH} C_6H_{13}CH_2B(CHMeCHMe_2)_2$$

$$C_6H_{13}CH_2B(CHMeCHMe_2)_2 \xrightarrow[40°C,\ 2\ hr]{[O]\ PCC,\ CH_2Cl_2} C_6H_{13}CH_2CHO \quad 71\%$$

PCC being a stronger oxidant converts ethers into esters or lactones[7].

$$CH_2{=}CHOC_2H_5 \xrightarrow[RT,\ 1\ hr]{PCC,\ CH_2Cl_2} CH_3COOC_2H_5$$

Ethylvinyl ether — Ethylacetate (75%)

Glucal (ring bearing CH_2OR, OR, RO) $\xrightarrow[RT,\ 40\ hr.]{PCC,\ CH_2Cl_2}$ Lactone (ring bearing CH_2OR, OR, RO, =O)

$R = C_6H_5CH_2$ Glucals — 60% Lactones

The *bis* (trimethylsilyl) ether of hydroquinone is oxidised[8] by PCC to *p*-benzoquinone

$$(CH_3)_3SiO{-}C_6H_4{-}OSi(CH_3)_3 \xrightarrow[25°C,\ 2\ hr]{PCC,\ CH_2Cl_2} O{=}C_6H_4{=}O$$

Bis (trimethylsilyl) ether of hydroquinone — 99% *p*-Benzoquinone

The reaction of 1-methyl-4-cycloocten-1-ol with PCC gives 100% yield of the ketooxide[9]. In this case a peculiar transannular oxidation takes place.

H_3C OH — PCC, CH_2Cl_2, reflux, 40 hr → CH_3, O

1-Methyl-4-cycloocten-1-ol — 100%, Keto oxide

References

1. E.J. Corey and G. Schmidt, *Tetrahedron Lett.*, 1979, 399.
2. E. Compaigne, W.M. Budde and G.F. Schaeffer, *Org. Synth.* Collective Volume, 1963, **4**, 31.
3. Y.-S. Cheng, W.-L. Liu and S.–H. *Chem. Synthesis*, 1980, 223.
4. E.J. Corey and J.W. Suggs, *Tetrahedron Lett.* 1975, 2647.
5. R. Rathore, N. Saxena and S. Chandrasekaran, *Synth. Commun.,* 1986, **16**, 1493.
6. G.. Piancatelli, A. Scettri and M. Daurid, *Tetrahedron Lett.,* 1977, 3483
7. P. Rollin and P. Sinay, *Carbohydrate Res.,* 1981, **98**, 139.
8. J.P. Willis, K.A.Z. Gogins and L.L. Miller, *J. Org. Chem.,* 1981, **46**, 3215.
9. M.F. Schecht and H.-J.Kim, *Tetrahedron Lett.,* 1986, **27**, 4889.

1.115 PYRIDINIUM DICHROMATE, $(C_2H_5NH)_2Cr_2O_7$

It is prepared[1] by adding pyridine to a cold solution of chromium trioxide in water at lower temperature (< 30°C) and by diluting the mixture with acetone. Precaution must be taken in preparing and handling of this reagent, since it can explode even in aquous media[2].

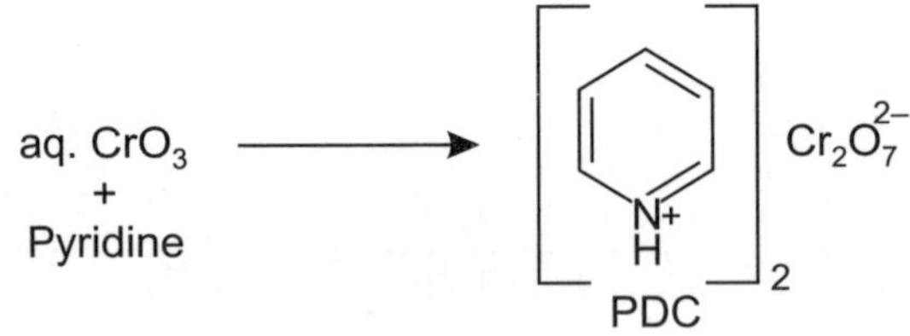

Pyridinium dichromate in dichloromethane oxidises primary alcohols to aldehydes[1]. However, oxidation in dimethylformamide at 25°C gives carboxylic acids[1].

$$\underset{\text{Decanol}}{CH_3(CH_2)_8CH_2OH} \xrightarrow[CH_2Cl_2,\ 25°C]{(C_5H_5NH)_2.Cr_2O_7} \underset{\text{Decanal 98\%}}{CH_3(CH_2)_8CHO}$$

$$\text{Decanol} \xrightarrow[DMF,\ 25°C]{(C_5H_5NH)_2.Cr_2O_7} CH_3(CH_2)_8COOH$$

$$\text{Cyclohexylmethanol (C}_6\text{H}_{11}\text{CH}_2\text{OH)} \xrightarrow[\text{DMF, 25°C}]{(C_5H_5NH)_2.Cr_2O_7} \text{Cyclohexane carboxylic acid (C}_6\text{H}_{11}\text{COOH) } 84\%$$

Allylic secondary alchols are oxidised more rapidly than their saturated analogues; the product obtainted is the ketone in high yields[1]

$$CH_2{=}CHCH(OH)C_5H_{11} \xrightarrow[CH_2Cl_2,\ 25°C,\ 9\ hr]{1.5\ (C_5H_5NH)_2,\ Cr_2O_7} CH_2{=}CHCOC_5H_{11}\quad 80\%$$

References

1. E.J. Corey and G. Schmidt, *Tetrahedron, Lett.,* 1979, 399.
2. J. Salmon, *Chem. Ber.*, 1982, 703.

1.116 QUINOLINIUM CHLOROCHROMATE, $C_9H_7N.\ HCl.\ CrO_3$

It is prepared[1] by adding qunoline to a stirred mixture of chromic acid and concentrated hydrochloric acid in water. This reagent oxidises primary alcohols in preference to secondary alcohols[1].

All the pyridine and pyridine analogue-containing hexavalent chromium reagents give very high yields under mild conditions mostly at room temperature or lower than room temperature. At times, these reagents are selective oxidants.

Reference

1. J. Singh, P.S. Kalsi, G.S. Jawanda and B.R. Chhabra, Chem. Ind. (London), 1986, 751.

1.117 RANEY NICKEL, Ni(R)

It converts secondary alcohols into ketones in 70-98% yield (refluxing with Raney Ni in benzene for 1-24 hr for with azotropic removal of water.

$$\text{Norborneol (OH)} \xrightarrow[\text{Paraffin oil, 150-180°C, 6 hr}]{\text{Raney Ni, KOH (catalyst)}} \text{Norbornanone (=O) } 95\%$$

$$CH_3CH(OH)C_6H_{13} \xrightarrow[4\ hr]{Raney\ Ni,\ C_6H_6,\ 125\text{-}135°C} CH_3COC_6H_{13}\quad 93\%$$

Another example is the conversion of cholesterol into 3-cholestanone[3] in 80% yield.

Raney Ni, cyclohexanone, reflux, 24 hr

Cholesterol → 3-Cholestanone 80%

$CH_3CH(OH)C_6H_{13}$ —Raney Ni, C_6H_6, 125-135°C, 4 hr→ $CH_3C(=O)C_6H_{13}$ 93%

Raney Ni and hydrogen are used for removal of hydroxy group from aromatic compounds via the formation of tosyloxy group.

β-Naphthol —TsCl, Me_2CO, K_2CO_3, reflux→ (OTs) —Raney Ni/H_2, EtOH→ Naphthalene

References

1. M.E. Krafft, W.J. Crooks, III, B. Zorc and S.E. Milezanowski, *J. Org., Chem.,* 1988, **53**, 3158.
2. K. Alder and H. Wirtz, Justus Leibigs *Ann. Chem.,* 1956, **601**, 138.
3. E.C. Kleiderer and E.C. Kornfield, *J. Org. Chem,* 1948, **13**, 455.

1.118 RED LEAD, Pb_3O_4

It is used in degradative oxidation of α-hydroxy acids to ketones[1]. Thus, 2-(p-methoxybenzyl-p-methoxymandelic acid is cenverted into desoxyanision (p-methoxybenzyl p-methoxyphenyl ketone) in 98% yield an heating with red lead.

$CH_3O-C_6H_4-C(OH)(COOH)-CH_2-C_6H_4-OCH_3$ —Pb_3O_4, Δ→ $CH_3O-C_6H_4-CO-CH_2-C_6H_4-OCH_3$

2-(p-Methoxybenzoyl)-p- -methoxymandelic acid

Desoxyanision

Red lead acts as lead tetraacetate *in situ*[2]. Thus, acenaphthene on heating with red lead and acetic acid (60-70°C) gives 80% yield of 1-acetoxyacenaphthene, the latter compound on heating with sodium hydroxide gives the corresponding alcohol in 70–74% yield.

Pb_3O_4, AcOH, 60–70°C; NaOH, MeOH, 60–70°C

Acenaphthene → 1-Acetoxy acenaphenene ($OCOCH_3$) → 1-Hydroxy acenaphthene (OH)

Red lead in acetic acid cleaves vicinal diols to carbonyl compounds[3]. In this reaction also red lead acts as lead tetraacetate *in situ*.[2]

Red lead, Pb_3O_4 is also used in the preparation of lead tetracetate and tetrakis (trifluoroacetate)[4] (*see* also lead dioxide).

References

1. E. Rohrann, R. Y. Jones and H. A. Shonle, *J. Am. Chem. Soc,* 1944, **66**, 1856.
2. J. Cason, *Org. Synth.,* Collective Volume, 1955, **3**, 3.
3. L. Vargha and M. Remenyi, *J. Chem. Soc.,* 1951, 1068.
4. R.E. Partch, *J.Am. Chem. Soc.,* 1967, **89**, 3662.

1.119 RUTHENIUM TETROXIDE, RuO_4

It is usually prepared *in situ* from black ruthenium dioxide (RuO_2 or $RuO_2.XH_2O$) or ruthenium trichloride ($RuCl_3$ or $RuCl_3.3H_2O$) by oxidation with sodium hypochlorite[1], sodium bromate[2], periodic acid[3], sodium periodate[4] or peroxyacetic acid. This reagent, in fact, acts as a catalyst, since it is reoxidisded after reaction with substrates by one the oxidants mentioned above. Appropriate precautions must be taken since the reaction mixture generates obnoxious smell. Following are given some of the applications of ruthenium tetroxide:

1. Oxidative cleavage of alkene is effected by ruthenium tetroxide, generated in situ. Some examples are given below[5,4,3].

C_6H_5, C_6H_5 → RuO_4, $NaIO_4$, Me_2CO, 20-25°C, 4 hr → CO_2H

80-83%

$$(E)\text{-}C_4H_9CH{=}CHC_4H_9 \xrightarrow[CCl_4,\ MeOH,\ RT\ 2\ hr]{RuCl_3,\ H_2O,\ NaIO_4} C_4H_9CO_2H\ \ 88\%$$

$$C_8H_{17}CH{=}CH_2 \xrightarrow[RT,\ 1\ hr]{RuO_2,\ AcO_2H,\ C_2Cl_3F_3} C_8H_{17}CO_2H\ \ 92\%$$

2. Oxidation of aldehydes to carboxylic acids.

   The oxidation of aldehydes to acids by pure ruthenium tetroxide gives very low yields[2].

3. Diarylacelyenes are converted into α–diketones[6] by ruthenium tetroxide and sodium hypochlorite or periodate.

$$C_6H_5C{\equiv}CC_6H_5 \xrightarrow[CCl_4,\ RT,\ 1\ hr]{RuO_2,\ NaOCl\ or\ NaIO_4} C_6H_5COCOC_6H_5\ \ 83\%$$

4. Oxidation[7] of ethyl N-acetyl-α-pipecolinate

$$\xrightarrow[RT,\ 14\ hr]{RuO_2,\ NaIO_4,\ AcOEt,\ H_2O}$$

N, CO_2Et, $COCH_3$ → O, N, CO_2Et, $COCH_3$

Ethyl N-acetyl-
-α-pipecolinate

95%
Ethyl 1-acetyl-2-piperidone-
-6-carboxylate

5. A oxidation of lactams into imides[2]

$$\xrightarrow[10\text{-}15°C]{RuO_4,\ CCl_4}$$

N H O → O N H O

64%

6. Oxidative cleavage of the benzene ring.

   Phenyl groups can be oxidised[4] to caboxyls by ruthenium tetroxide with sodium periodate or sodium hypochlorite as reoxodants.

$$C_6H_{11}\text{–}C_6H_5 \xrightarrow[CCl_4,\ RT,\ 24\ hr.]{RuCl_3.H_2O,\ NaIO_4,\ MeCN} C_6H_{11}\text{–}CO_2H$$

Phenylcyclohexane

64%
Cyclohexanecarboxylic acid

In a similar way, β-phenylpropionic acid is converted into succinic acid[1].

$$C_6H_5CH_2CH_2CO_2H \xrightarrow[\text{RT, 3 hr}]{\text{RuCl}_3\text{, 2.25\% NaOCl}} HO_2CCH_2CH_2CO_2H + C_6H_5CO_2H$$

β-Phenylpropionic acid → 94%, Succinic acid + 6%, Benzoic acid

Napthaline is oxidised by ruthenium tetroxide to phthalic acid[8] and α-nitronaphthalene to 3-nitrophthalic acid.

Naphthalene $\xrightarrow[\text{H}_2\text{O, CCl}_4\text{, RT, 60 hr}]{\text{RuO}_4\text{, 5.25\% NaOCl}}$ Phthalic acid (70%)

α-Nitronaphthalene $\xrightarrow[\text{RT, 7 day}]{\text{RuO}_4\text{, NaOCl, H}_2\text{O, CCl}_4}$ 3-Nitrophthalic acid (63%) + Phthalic acid (7%)

7. Primary alcohols are oxidised to aldehyldes[2]. However, aldehydes may undergo further oxidation to carboxylic and

$$C_6H_5CH_2OH \xrightarrow[\text{10-15°C, RT overnight}]{\text{RuO}_4\text{, CCl}_4} RCHO$$

90%

It is most appropriate to isolate the aldehydes as soon as formed.

8. Secondary alcohols are oxidised[10] to ketones.

Cyclohexanol $\xrightarrow[\text{25°C, 1 hr}]{\text{RuO}_2\text{, NaIO}_4\text{, NaOH}}$ Cyclohexanone 85%

9. Sulphides are oxidised[11] to sulfones and sulfoxides

$$C_6H_5SC_6H_5 \longrightarrow C_6H_5S(=O)C_6H_5 + C_6H_5S(=O)_2C_6H_5$$

32% 42%

References

1. S. Wolfe, S.K. Hasan and J.R. Compbell, *Chem. Commun.,* 1970, 1420.
2. L.M. Berkowitz and P.N. Rylander, *J. Am. Chem. Soc.,* 1958, **80**, 6682.
3. C. Guizard, H. Cheradame, Y. Brunel and C.G. Beguin, *J. Fluorine Chem.,* 1979, **13**, 175.
4. P.H.J. Carlsen, T. Kasiuri, V.S. Martin and K.B. Sharpless. *J. Org. Chem.,* 1981, **46**, 3936
5. S. Sarel and Y. Yanuka, *J. Org. Chem.,* 1959, **24**, 2018.
6. H. Gopal and A.J. Gordon, *Tetrahedron Lett.,* 1971, 2941.
7. S. Yoshifuja, K. –I. Tanaka, T., Kawai and Y. Nitta, *Chem. Pharm. Bull,* 1986, **34**, 3873.
8. J. Böeseken and G. Slooff, Rec. *Trav. Chem.* Pays-Bas 1930, **49**, 100.
9. U.A. Spitzer and D.G. Lee, *J. Org. Chem.,* 1974, **39**, 2468.
10. D.G. Lee, D.T. Hall and J.H. Cleland, Can. *J. Chem.,* 1972, **50**, 3741.
11. C. Djerassi and R.R. Engle, *J.Am. Chem. Soc.,* 1953, **75**, 3838.

1.120 SELENIUM, Se

Selenium is known to bring about chemical dehydrogenations[1] like sulphur (*see* sulphur).

It is mostly used in the structure elucidation of steroids. In fact all steroids on selenium dehydrogenation at 360°C yield Diel's hydrocarbon (3'-methyl-1,2-cyclopentenophenanthrene).

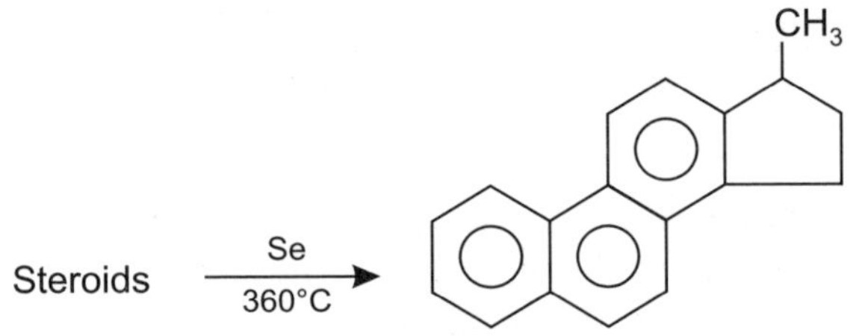

Diel's hydrocarbon

Compared to sulphur, selenium requires higher temperature. Consequently, side reactions, such as rearrangements occur[1].

Reference

1. V. Prelog, A. Komzak and K. Moor, Helv., *Chem. Acta,* 1942, **25**, 1654.

1.121 SELENIUM DIOXIDE, SeO_2

Selenium dioxide is a very selective oxidant[1,2]. Another related selective oxidant is **selenious acid**, which is obtaned[1,2] by the evaporation of an aqueous solution of selenium dioxide. Following are given some of the important applications of selenium dioxide:

(*i*) Selenium dioxide is used as a catalyst for anti hydroxylation with hydrogen peroxide. Thus, cyclohexene is converted[3] into DL-trans-1,2-cyclohexanediol by hydrogen peroxide in presence of selenium dioxide.

(*ii*) It is also used as a catalyst[4] for the oxidation of sulphides to sulfones. In this case, the double bond (if present into sulphides) remains uneffected. This is because the affinity of sulphur to oxygen is higher than that of carbon-carbon multiple bonds.

$$C_6H_5SCH_2CH{=}CH_2 \xrightarrow[< RT,\ 1\ hr]{30\%\ H_2O_2/SeO_2,\ MeOH} C_6H_5S(=O)_2CH_2CH{=}CH_2$$

100%

(*iii*) Aldehydes are converted into carboxylic acids by hydrogen peroxide in presence of selenium dioxide as acatalyst. Thus, acrolein is oxidised[5] to acrylic acid in 90% yield with 12% hydrogen peroxide in tert. amyl alcohol at 40°C for 3 hr. In a simlar way methacrolein is converted[5] into methacrylic acid by 90% hydrogen peroxide in presence of selenium dioxide at 60°C.

$$\underset{\text{Acrolein}}{CH_2{=}CH{-}CHO} \xrightarrow[SeO_2,\ 40°C,\ 3\ hr]{12\%\ H_2O_2/\text{tert. amylalcohol}} \underset{\text{90\% Acrylic acid}}{CH_2{=}CH{-}COOH}$$

$$\underset{\text{Methacrolein}}{CH_2{=}C(CH_3){-}CHO} \xrightarrow[60°C,\ 1.75\ hr]{90\%\ H_2O_2,\ SeO_2} \underset{\text{74\% Methacrylic acid}}{CH_2{=}C(CH_3){-}COOH}$$

(*iv*) Alkenes on oxidation with hydrogen peroxide in presence of selenium dioxide in aqueous solvents give allylic alcohols. One such example[6] is given below.

$$\text{β-Pinene} \xrightarrow[40\text{-}50°C]{SeO_2,\ H_2O_2,\ t\text{-}BuOH,\ H_2O} \textit{trans}\text{-pinocarveol}\ (49\text{-}55\%)$$

β-Pinene

49-55%

trans-pinocarveol

(*v*) Selenium dioxide is also capable of oxidising alkynes to vicinal dicarbonyl compounds. Thus diphenylacetylene is oxidised to benzil with selenium dioxide in presnece of small ammount of sulphuric acid

$$\underset{\text{Diphenylacetylene}}{Ph{-}C{\equiv}C{-}Ph} \xrightarrow{SeO_2/H_2SO_4} \underset{\text{Benzil (84\%)}}{Ph{-}C(=O){-}C(=O){-}Ph}$$

It is to be noted that only internal alkynes give vicinal dicarbonyl compounds. Terminal alkynes are oxidised to phenylglyoxylic acid

$$\underset{\text{Phenylacetylene}}{Ph—C{\equiv}C—H} \xrightarrow{SeO_2/H_2SO_4} \underset{\text{Phenylglyoxylic acid}}{Ph—\overset{O}{\overset{||}{C}}—\overset{O}{\overset{||}{C}}—OH}$$

(*vi*) Aromatic compounds are oxidised by selenium dioxide to quinones. Thus, anthracene on oxidation gives anthraquinone[8] in 76% yield.

Anthracene $\xrightarrow{SeO_2,\ 165\text{-}170°C}$ Anthraquinone (76%)

(*vii*) Methyl homologues of aromatic compounds are oxidised to aldehydes by selenium dioxide. Thus, toluene gives benzaldehyde[9-11].

Toluene (C_6H_5–CH_3) $\xrightarrow[\text{reflux}]{SeO_2,\ EtOH}$ Benzaldehyde (C_6H_5–CHO)

In aromatic heterocycles, the methyl groups in α-position with respect to nitrogen are oxidised expecially easely. Thus, selective oxidation of 8-ethyl-2-methlquinoline givs[10] 8-ethyl-2-formyl quinoline.

8-Ethyl-2-methyl quinolene (8-CH_2CH_3, 2-CH_3) $\xrightarrow[\text{reflux 15-20 hrs}]{SeO_2,\ EOH}$ 8-Ethyl-2-formyl quinolene (8-CH_2CH_3, 2-CHO)

(*viii*) Selenium dioxide, converts aryl and aralkyl allyl ethers into aldehydes and the correspoding alcohols. Thus, cinnamyl methyl ether is converted into cinnamaldehyde and methanol.

$$\underset{\text{Cinnamyl methyl ether}}{C_6H_5CH{=}CHCH_2OCH_3} \xrightarrow[\text{reflux, 1 hr.}]{SeO_2,\ AcOH,\ \text{dioxane}} \underset{\text{Cinnamaldehyde 60\%}}{C_6H_5CH{=}CHCHO} + \underset{\text{Methanol}}{CH_3OH}$$

Similarly, allyl benzyl ether is converted into benzyl alcohol and acrolein

$$\underset{\text{Allylbenzyl ether}}{CH_2{=}CHCH_2OCH_2C_6H_5} \xrightarrow[\text{reflux, 1 hr.}]{SeO_2,\ AcOH,\ \text{dioxane}} \underset{\text{Benzyl alcohol}}{C_6H_5CH_2OH} + \underset{\text{Acrolein}}{CH_2{=}CH{-}CHO}$$

(*ix*) Oxidation of ketones to α-dicarbonyl compounds. Methyl and methylene groups adjacent to carbonyl groups are easily oxidised to carbonyls to yield α-keto aldehydes or α-diketones. The oxidation is brought about by selenium dioxide or selenious acid. Some examples[13-15] are given below.

$$p\text{-}CH_3OC_6H_4COCH_2C_6H_5 \xrightarrow[\text{OR } 1.1\ SeO_2,\ 70\%\ AcOH,\ 89°C,\ 12\ hr.]{1.1\ SeO_2,\ \text{dioxane},\ H_2O,\ \text{reflux 5-8 hr}} \underset{83\text{-}100\%}{p\text{-}CH_3OC_6H_4COCOC_6H_5}$$

$$\text{Cyclohexanone} \xrightarrow[\text{cooling, 3 hr, RT, 11 hr}]{H_2SeO_2,\ H_2O,\ \text{dioxane}} \underset{60\%}{\text{Cyclohexane-1,2-dione}}$$

2-Acetylmesitylene on oxidation with selenium dioxide gives[16] about 83% yield mesityl glyoxal.

$$\underset{\text{2-Acetylmesitylene}}{2,4,6\text{-}(CH_3)_3C_6H_2COCH_3} \xrightarrow[\text{reflux, 5 hr}]{SeO_2,\ \text{dioxane},\ H_2O} \underset{\text{83\% Mesitylglyoxal}}{2,4,6\text{-}(CH_3)_3C_6H_2COCHO}$$

The above oxidation is applicable to acetophenones[17-19] and deoxybenzoins[13,16].

$$R{-}C_6H_4{-}COCH_3 \xrightarrow[\text{reflux, 5 hr}]{SeO_2, H_2O,\ \text{dioxane}} R{-}C_6H_4{-}COCHO$$

R = H, Br

R = H (69-72% yield)
R = Br (86% yield)

$$\underset{\text{Deoxybenzoin}}{C_6H_5{-}\overset{O}{\overset{\|}{C}}{-}CH_2{-}C_6H_5} \xrightarrow[\text{reflux, 5 hr}]{SeO_2, H_2O,\ \text{dioxane}} \underset{\text{Benzil}}{C_6H_5{-}\underset{O}{\underset{\|}{C}}{-}\underset{O}{\underset{\|}{C}}{-}C_6H_5}$$

The conversion of methylene group to carbonyl is especially easy when the group is flanked by two aromatic rings or two carbonyl groups. Thus, diphenyl methane yields benzophenone[8].

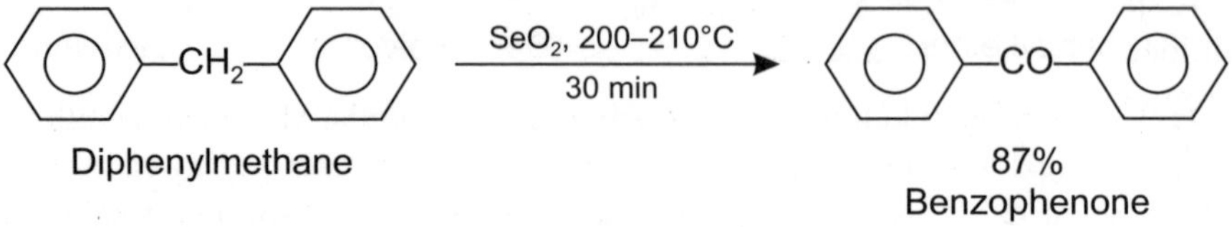

$$C_6H_5{-}CH_2{-}C_6H_5 \xrightarrow[\text{30 min}]{SeO_2,\ 200\text{–}210°C} C_6H_5{-}CO{-}C_6H_5$$

Diphenylmethane

87%
Benzophenone

Some other examples are given below:

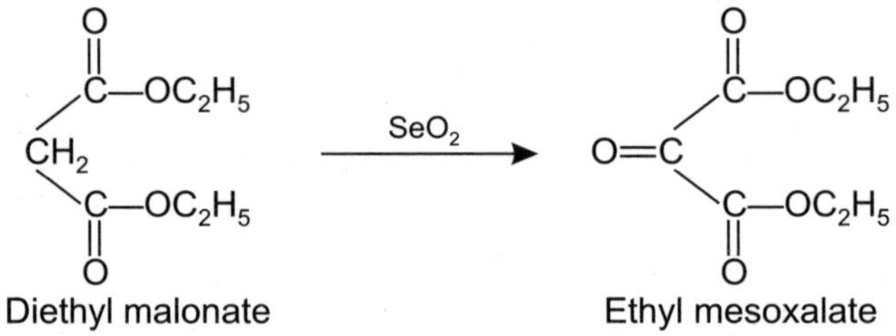

$$CH_2(COOC_2H_5)_2 \xrightarrow{SeO_2} O{=}C(COOC_2H_5)_2$$

Diethyl malonate → Ethyl mesoxalate

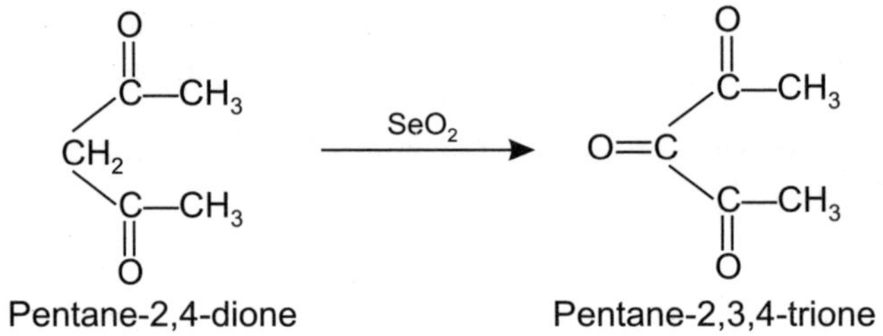

$$CH_3COCH_2COCH_3 \xrightarrow{SeO_2} CH_3CO{-}CO{-}COCH_3$$

Pentane-2,4-dione → Pentane-2,3,4-trione

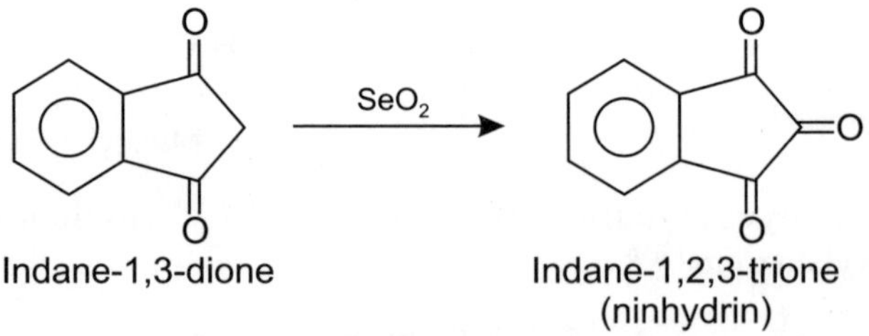

Indane-1,3-dione $\xrightarrow{SeO_2}$ Indane-1,2,3-trione (ninhydrin)

(*x*) Dehydrogenation[20]. Selenium dioxide is an oxidising agent, but in certain cases it converts carbonyl compounds to α,β-unsaturated compounds by removing hydrogen.

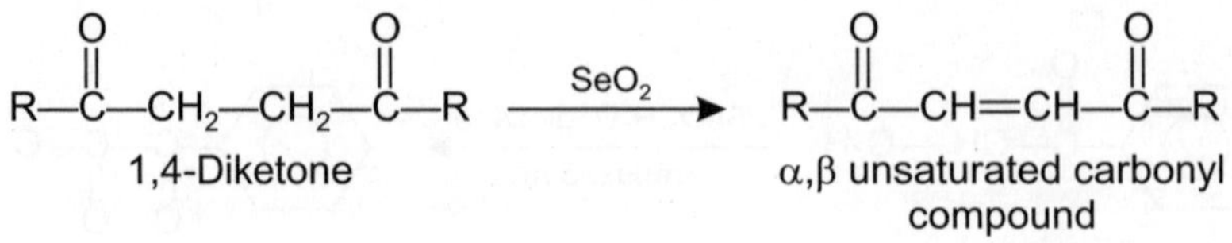

$$R{-}CO{-}CH_2{-}CH_2{-}CO{-}R \xrightarrow{SeO_2} R{-}CO{-}CH{=}CH{-}CO{-}R$$

1,4-Diketone → α,β unsaturated carbonyl compound

$$\underset{\text{1,2-Diarylethane}}{Ar{-}CH_2{-}CH_2{-}Ar} \xrightarrow{SeO_2} \underset{\text{1,2-Diarylethylene}}{R{-}CH{=}CH{-}R}$$

A special feature of oxidations with selenium dioxide is the formation of red colloided selenium, which is difficult to separate by distillation. However, it can be removed by treatment of the reaction product with potassium cyanide[23], mercury[22] or deactivated nickel[23].

References

1. A.R. Ronzio and T.D. Wargh, *Org. Synth.* Collective Volume, 1955, **3**, 438.
2. C. C. Hach, C.V. Banks and H. Diehl, *Org. Synth.* Collective Volume, 1963, **4**, 229.
3. M. Mugdan and D.P. Young, *J. Chem. Soc.,* 1949, 2988.
4. J. Drabowicz, P. Lyzwa and M. Mitolajczyk, Phosphorus Sulfur, 1983, **17**, 169.
5. C.W. Smith and R.T. Holm, *J. Org. Chem.,* 1957, **22**, 746.
6. C.M. Coxon., E. Dansted and M.P. Hartshorn, *Org. Synth.* Collective Volume, 1988, **6**, 946.
7. Sonoda, Yamamoto, Murai and Tsutsumi, *Chem., Lett.,* 1972, 229.
8. J.J. Postowsky and B.P. Lugowkin, Chem., Ber., 1935, **68**, 852.
9. H. Kaplan, *J. Am. Chem. Soc.,* 1941, **63**, 2654.
10. R.A. Glenn and J.R. Barley, *J. Am. Chem. Soc.,* 1941, **63**, 639.
11. A. Burgher and L.R., Modlin, Jr., *J.Am. Chem. Soc.,* 1940, **62**, 1079.
12. K. Kariyone and H. Yazawa, *Tetrahedron Lett.,* 1932, 2342.
13. E.R. Bockstahler and D.L. Wright, *J. Am. Chem. Soc.,* 1949, **71**, 3760.
14. E.J. Corey and J.P. Schaefer, *J. Am. Chem. Soc.,* 1960, **82**, 918.
15. C.C. Hach, C.V. Banks and H. Diehl, *Org. Synths.,* Collective Volume, 1963, **4**, 229.
16. A.R. Gray and R.C. Fuson, *J.Am. Chem. Soc.,* 1934, **56**, 739.
17. H.A. Riley and A.R. Gray, *Org. Synth.* Collective Volume, 1943, **2**, 509.
18. T. Suga, K. Kihara and T. Matsuura, *Bull Chem., Soc. Jpn.,* 1965, **38**, 1141.
19. M.B. Floyd, M.T. Du, P.F. Fabio, L.A. Jacob and B.D. Johnson, *J. Org., Chem.,* 1985, 50, 5022.
20. R. I. Fryer, G.A. Archer, B. Brust, W.Zally and L.H. Siernabach, *J.Org. Chem.,* 1965, **30**, 1308
21. H. Watonable, *Chem. Pharm. Bull.,* 1957, **5**, 426.
22. J. Jacques, G. Oursson and C. Sandris, *Bull. Soc. Chim.* Fr., 1955, 1293.
23. M. Heller and S. Bernstein, *J. Org. Chem.,* 1961, **26**, 3876.

1.122 SILVER

Catalytic dehydrongenations of primary alcohols over silver are carried out[1,2] in an insufficient amount of air or oxygen at 300–380°C and give aldehydes in 70–100% yeild. A typical industrial dehydrogenation is the conversion of methallyl alcohol into methacrolein.

$$\underset{\text{Methallyl alcohol}}{CH_2{=}C(CH_3)CH_2OH} \xrightarrow{\text{Ag, 350°C, air}} \underset{\text{Methacrolin 95\%}}{CH_2{=}C(CH_3)CHO}$$

In place of silver, copper can also be used (*see* copper)

Another example is the dehydrogenation of cinnamyl alcohol into cinnamaldehyde[1].

$$\underset{\text{Cinnamyl alcohol}}{C_6H_5CH{=}CHCH_2OH} \xrightarrow{\text{Air/Aq, 300°C}} \underset{\text{Cinnamaldehyde 60-70\%}}{C_6H_5CH{=}CHCHO}$$

The well knwon cannizzaro reaction (conversion of aromatic aldehydes into aromatic carboxylic acids and aromatic alcohols) takes place in presence of alkali. This reactions catyalyed by silver[3].

Reference

1. C. Moureu and G. Mignonac, C.R. Hebd. *Sciences Acid. Sci.,* 1920, **170**, 258.
2. R. R. Davies and H.H. Hodgson, *J. Chem., Soc.,* 1943, 282.
3. I. A. Pearl, *J. Org. Chem.,* 1947, **12**, 79.

1.123 SILVER CHROMATE, Ag_2CrO_4

It is prepared by adding an aqueous solution of silver nitrate to solution of potassium chromate in water.

Silver chromate is used with iodine for the preparation of α-iodoketones from alkenes′.

$$Ag_2CrO_4 + I_2 \longrightarrow AgO{-}Cr(=O)_2{-}OI + AgI$$

$$C_6H_5CH{=}CH_2 + AgO{-}Cr(=O)_2{-}O{-}I \longrightarrow C_6H_5C(H)(O{-}Cr(=O)_2{-}OAg){-}CH_2I \longrightarrow C_6H_5COCH_2I + AgHCrO_3$$

$$\underset{\text{Cyclohexene}}{\text{Cyclohexene}} \xrightarrow[CH_2Cl_2,\ 0°C,\ 25\ min]{Ag_2CrO_4,\ I_2,\ C_5H_5N} \underset{\text{2-Iodocyclohexanone}}{\text{2-Iodocyclohexanone}}$$

Reference

1. G Cardillo and M. Shimizu, *J. Org. Chem.,* 1977, **42**, 4268.

1.124 SILVER DICHROMATE-PYRIDINE COMPLEX (TETRAPYRIDINE SILVER DICHROMATE), $Ag_2Cr_2O_7.4C_5H_5N$

It is prepared by adding a warm solution of potassium dichromate to silver nitrate solution in water and pyridine. It oxidises allylic and benzylic alcohols to aldehydes and acyloins to α-diketones[1].

Reference

1. H. Firouzabadi, A. Sardarian and H. Charibi, *Synth. Commun.*, 1984, **14**, 89.

1.125 SINGLET OXYGEN

As already stated (*see* oxygen), the excited state of oxygen is called singlet oxygen. In singlet oxygen there are two odd electrons possessing antiparallel spins [↑.0:0.↓]

Singet oxygen in formed[1] by the irradiation of gaseous oxygen with ultraviolet light in presence of sensitizers (like benzophenone, methylene blue, rose bengal, chlorophyll *etc.*) that absorb the light of the particular wavelength. It is also generated chemically by the treatment of hydrogen peroxide in alkaline medium (equation 1) with sodium hypochlorite[2], bromine[3] (equation 2) or organic peroxy acids[3] (equation 3).

$$H_2O_2 + NaOH + NaOCl: \quad \bar{O}{-}O{-}\overset{+}{H} + \overset{+}{Cl} \longrightarrow O_2 + HCl \qquad (1)$$

$$H_2O_2 \xrightarrow[C_6H_5Cl,\ 25°C]{Br_2,\ KOH} \bar{O}{-}O{-}H + Br{-}Br \xrightarrow[-HBr]{-Br^-} \ddot{O}{=}\ddot{O} \qquad (2)$$

$$RC(=O)O{-}\bar{O} + H{-}O{-}O{-}H \longrightarrow R{-}C(=O){-}O{-}H + \bar{O}H + O_2 \qquad (3)$$

It can also be obtained by the decomposition of aromatic endoperoxides, such as 9,10-diphenylanthracene endoperoxide[4] (equation 4) and also by the decomposition of triphenyl phosphite ozonide[5] (equation 5).

$$\text{9,10-Diphenyl anthracene endoperoxide} \xrightarrow{\text{heat}} \text{9,10-diphenylanthracene} + O_2 \qquad (4)$$

9,10-Diphenyl anthracene endoperoxide

$$(C_6H_5O)_3P \xrightarrow{O_3} (C_6H_5O)_3P(O_3) \longrightarrow (C_6H_5O)_3PO + O_2 \quad (5)$$

Triphenyl phosphite

Follwing are given some of the reactions in which singlet oxygen is used.

(*i*) Formation of Hydroperoxides

Singlet oxygen reacts with alkenes to form hydroperoxides. The hydroperoxide group enters in the α-position with respect to double bond[6] (equation 6).

$$(H_3C)(CH_3{-}H)C{=}C(CH_3)_2 \xrightarrow{O_2} (H_3C)(H_2C{=})C{-}C(CH_3)_2{-}O{-}O{-}H \quad (6)$$

The hydroperoxides on reduction give the corresponding alcohol (equation 7).

$$-\overset{|}{\underset{|}{C}}-\overset{|}{C}{=}\overset{|}{C} \xrightarrow[\text{sensitiser}]{O_2,\,h\nu} -\overset{|}{C}{=}\overset{|}{C}-\overset{OOH}{\underset{|}{C}}- \xrightarrow{\text{Redn.}} -\overset{|}{C}{=}\overset{|}{C}-\overset{OH}{\underset{|}{C}}- \quad (7)$$

Allylic alcohol

The reaction of alkenes with singlet oxygen is beleved to proceed by a concerted mechanism[7] (equation 8).

$$-\overset{|}{\underset{H}{C}}-C{=}C \;\; O{=}O \longrightarrow -\overset{|}{C}{=}C-\overset{OOH}{\underset{|}{C}}- \quad (8)$$

(*ii*) Formation of Endooxides

Conjugated dieines on reaction with singlet oxygen form oxides or endooxides by 1,4-addition (Diels-alder reaction)[8] (equation 9)

Butadiene + O=O ⟶ Endooxide

Cyclopentadiene + O=O ⟶ Endooxide (9)

(*iii*) Formation of dioxetanes

Compounds containing double bonds react with singlet oxygen to form dioxetanes, which being unstable disingtegrate to give dicarbonyl compounds (equation-10)[9].

O_2 → Dioxetane → CHO CHO (10)

(*iv*) Oxidation of alkenes to epoxides

Alkenes react with singlet oxygen to form epoxides (equation-11)[10]

$$R{-}C{=}C{-}R + O_2 \longrightarrow R{-}\overset{O}{C{-}C}{-}R \quad (11)$$

(*v*) Oxidation of Secondary alcohols to Ketones

Secondary alcohols on reaction with singlet oxygen give the corresponding ketones via the formation of hydroperoxides (equation 12)[11]

$$>C{-}OH \xrightarrow[\text{sensitiser}]{O_2,\ h\nu} >C(O{-}OH){-}OH \longrightarrow >C{=}O + H_2O_2 \quad (12)$$

Sec. alcohol Ketone

(*vi*) Conversion of tertiary amines to secondary amines.

Tertiary amines on reaction with singlet oxygen undergo oxidative degradation leading, to the formation[12] of secondary amines (equation-13)

$$>N{-}CH_3 \xrightarrow{O_2} >NH + CO_2 + H_2O \quad (13)$$

Tert. amine Sec. amine

(vii) Introduction of oxygen function (Hydroxyl group)

Singlet oxygen has been used for the specific introduction of an oxygen function (hydroxyl group) at the site of the double bond. This has been explored in the synthesis[13] of a number of natural products (equation–14)

$\xrightarrow[\text{sensitiser}]{O_2,\ h\nu}$ OOH $\xrightarrow[\text{aq. } CH_3OH]{Na_2SO_3}$ OH (14)

α-Pinene *trans*-pinocarveyl hydroperoxide *trans*-pinocarveol

References

1. C.S. Foote, *Acc. Chem. Res.*, 1968, **1**., 104.
2. C.S. Foote, S. Wexler and W. Ando. *Tetrahedron Lett.*, 1965, 4 , 111.
3. E. Mckeown and W.A. Waters, *I. Chem. Soc.* B. 1966,1040.
4. H.H. Wasserman and J.R. Scheffer, *J. Am. Chem., Soc.*, 1967, **89**, 3073.
5. N.J. Turro, Y. Ito, M. –F. Chow, W. Adam; O. Rodriquez and F. Yany, *J. Am. Chem. Soc.*, 1977, **99**, 5836.
6. A.G. Schultz and R.H. Schlessinger, *Tetrahedron Lett.*, 1970, 2731; G. O. Schenk, K. Gollnick and O.A. Neumüller, *Justius Llibigs Ann. Chem.*, 1957, **603**, 43.
7. C.S. Foote, *Pure and Appl. Chem.*, 1971, **27**, 635; K. Yamaguchi, S. Yabushita, T. Fueno and H.W. Hawk, *J. Am. Chem. Soc.*, 1981, **103**, 5043.
8. C.N. Skold and R.H. Schlessingev, *Tetrahedron Lett.*, 1970, 791; A. Windaus and J. Bruken., *Justus Liebigs Ann. Chem.*, 1928, **460**, 225.
9. S. Mazur and C.S. Foote, *J. Am. Chem. Soc.*, 1970, **92**, 3225; H. H. Wasserman and S. Terao, *Tetrahedron Letters*, 1975, 1735
10. N. Shimizu and P.D. Barlett, *J. Am. Chem. Soc.*, 1976, **98**, 4193.
11. G. O. Schenck, H.-D. Becker, K. -H. Schulte-Elle and C.H. Krauch, *Chem. Ber.*, 1963, **96**, 509.
12. J.H.E. Linder, H.J. Kuhn and K. Gollnick, *Tetrahedron Lett.*, 1972, 1705.
13. H.H. Wasserman and J.L. Ives, *Tetrahedron*, 1981, **37**, 1825.

1.126 SODIUM BISMUTATE, $NaBIO_3$

It cleaves[12] vicinal diols to dicarbonyl compounds. However, the yield is low (23%). Lead tetraacetate is a better reagent and gives about 70% yield (*see* lead tetraacetate).

OH OH

$NaBiO_3$, H_3PO_4, H_2O, Et_2O
30°C, 2 hr.

CHO CHO

cis-1,2-Cyclohexanediol

Adipicaldehyde
23%

The *trans* isomer gives 49% yield of adipicaldehyde.

The treatment of 17β, 17aβ-dimethyl-D-homoandrostane-3β, 17α, 17aα-triol 3β-acetate with sodium bismutate yields 3β-acetory – 17, 17a-dimethyl-17, 17a-secohomoandrostane-17,17 a-dione[2].

CH_3 OH OH CH_3 AcO H

$NaBiO_3$, 70% AcOH
RT, 18 hr.

CH_3 O O CH_3 AcO H 87%

17β, 17aβ-dimethyl-
D-homoandrostane-
-3β, 17α, 17aα-triol
3β-acetate

3-Acetoxy-17,17a-dimethyl-
-17, 17a-secohomoandrostane-
-17, 17a-dione

References

1. W. Regby, *J.Chem. Soc.*, 1950,1907.
2. M. Uskoković, M. gut, E.N. Trachtenberg, W. Klyne and R.I. Dorfman, *J. Am. Chem. Soc.*, 1960, **82**, 4965.

1.127 SODIUM BROMATE, $NaBrO_3$

It oxidises acyloins to α-diktones[1]

$$C_6H_5COCH(OH)C_6H_5 \xrightarrow[100°C,\ 5\text{-}6\ hr]{NaBrO_3,\ NaOH,\ H_2O} C_6H_5COCOC_6H_5$$

Acyloin → α-diketone 84-90%

Secondary alcohols are oxidised to ketones[2] in presence of catalytic amounts to ceric ammonium nitrates or ceric sulphate.

References

1. D. A. Balland W. M. Dehn, *Org. Synth.* Collective Volume, 1932,**1**, 89.
2. S. Kanemoto, H. Tomiokd, K. Osmina and H. Nozaki, *Bull. Chem. Soc. Jpn.*, 1986, **59**, 105.

1.128 SODIUM BROMITE, $NaBrO_2.H_2O$

It oxidises secondary alcohols in preference to primary alcohol. Thus, menthol on treatment with sodium bromite in aqueous acetic acid at room temperature affords 86% of menthone after 5.5 hr.

Menthol $\xrightarrow[RT,\ 5.5\ hr.]{NaBrO_2.H_2O,HOAc,\ H_2O}$ 86%, Menthone

Similarly, cyclohexanol yields cyclohexanone[1]

Cyclohaxanol $\xrightarrow[RT,\ 5.5\ hr.]{NaBrO_2.H_2O,\ AcOH}$ 100% Cyclohaxanone

Sodium bromate oxidises[1] sulphides to sulfoxides

$$\underset{\text{Diphenyl sulphide}}{C_6H_5SC_6H_5} \xrightarrow[\text{RT, 30 min}]{NaBrO_2.H_2O,\ H_2O} \underset{\text{Diphenyl sulfoxide}}{C_6H_5S(=O)C_6H_5}$$

Reference

1. T. Kageyama and M. Okawara, Synthesis, 1983, 815.

1.129 SODIUM CHLORATE, $NaClO_3$; POTASSIUM CHLORATE, $KClO_3$; SILVER CHLORATE, $AgClO_3$; BARIUM CHLORATE, $Ba(ClO_3)_2$

These reagents oxidise organic compounds only in the presence of catalyst (usually osmium tetroxide or vanadium pentoxide) since, such oxidations do not occur without catalyst, it is believed that the real oxidants are osmium tetroxide and vanadium pentoxide. The function of the chlorates is only in reoxidation. Thus, osmium tetroxide is used for syn hydroxylation of olefins. Since osmium tetroxide is very expensive, usually only a catalytic amounts are used in presence of an oxidising agent like sodium chlorate, potassium chlorate, silver chlorate or barium chlorate. The oxidising agent reoxidises hexavalent osmium to octavalent osmium.

Hydroxylation of double bond[1,2] of olefins or unsaturated carboxylic acid is accomplished stereoselectively by osmium tetraoxide in presence of a chlorate (of sodium, potassium or barium).

Hydroquinone can be oxidised[3] to p-benzoquinone by sodium chlorate/ vanadium pentoxide in 92-96% yield.

$$\underset{\text{Hydroquinone}}{HO-C_6H_4-OH} \xrightarrow[H_2O,\ 40°C,\ 4\ hr.]{NaClO_3/V_2O_5,\ H_2SO_4} \underset{\substack{92\text{-}96\% \\ p\text{-Benzoquinone}}}{O=C_6H_4=O}$$

These reagents are also useful in converting furfural to fumaric acid.[4,5]

References

1. G. Braux. *J. Am. Chem. Soc.,* 1929, **51**, 228.
2. K. A. Hofmann, O. Ehrhart and O–Schneider, *Chem. Ber.,* 1913, **46**, 1657.
3. H.M. Underwood, Jr., and W. L. Walsh, *Org. Synth.* Collective Volume, 1943, **2**, 533.
4. N.A. Milas, *J. Am. Chem. Soc.,* 1927, **49**, 2005.
5. N. A. Milas, *Org. Synth.* Collective Volume. 1943, **21**, 302.

1.130 SODIUM CHLORITE, $NaClO_2$

It is commercially available. It oxidises[1] aldehydes to carboxylic acids at room temperature. In this oxidation, the byproducts, hypochlorous acid and chlorine dioxide must be removed (to prevent side reactions) by hdrogen peroxide, sulphuric acid, dimethyl sulfoxide or resorcinol.

$$C_6H_5CHO \xrightarrow[\text{resorcinol, pH 4, RT, 18 hr}]{NaClO_2,\ \text{t-BuOH}} \underset{88\%}{C_6H_5COOH}$$

In a similar way vanillin in oxidised[1] to vanillic acid in 81% yield. Even unsaturated aldehydes are oxidised[1,2] to the coresponding carboxylic acids.

$$\underset{\text{Cinnamaldehyde}}{C_6H_5CH{=}CHCHO} \xrightarrow[\text{OR } NaClO_2,\ 35\%\ H_2O_2,\ \text{MeCN},\ H_2O,\ 10°C,\ 1\ \text{hr}]{NaClO_2,\ \text{t-BuOH},\ H_2O,\ \text{resorcinol, acetate buffer, RT, 45 min}} \underset{\text{72-90\% Cinnamic acid}}{C_6H_5CH{=}CHCOOH}$$

Oxidation of aldehydes to acids with sodium chlorite are very selective and much cheapter than those with silver oxide.

Another application of sodium chlorite in the oxidation of aldehyde-bisulphite compound to carboxylic acids in the presence of dimethyl sulfoxide and acitic anhydride[3].

References

1. B. O. Lindgren and T. Nilsoon, *Acta Chem. Scand.*, 1973, **27**, 888.
2. E. Dalcanale and F. Montanari; *J. Org. Chem,* 1986, **51**, 567.
3. P. G. M. Wuts and C. L. Bergh, *Tetrahedron Lett.,* 1986, **27**, 3995.

1.131 SODIUM DICHROMATE, $Na_2Cr_2O_7$

Sodium dichromate in presence of dilute sulphuric acid has been conventionally used for the oxidation of primary alcohols to aldehydes.

$$3RCH_2OH + Na_2Cr_2O_7 + 4H_2SO_4 \longrightarrow 3RCHO + Cr_2(SO_4)_3 + Na_2SO_4 + 7H_2O$$

In this case, care must be taken to prevent overoxidation of the aldehyde to carboxylic acids. Lower boiling aldehydes can be removed by concomitant steam distillation and thus escape further oxidation. It has been found[1] that neutral sodium dichromate even under the above conditions does not cause much overoxidation. Thus benzyl alcohol is oxidised to benzaldehyde at pH 5.6 at a rate 18 times faster than that of oxidation of benzaldehyde to benzoic acid.

It is best to carry out dichromate oxidation in aqueous media in presence of 'Phase transfer catalyst', and the oxidation is carried out in a two phase system. A better modification is to carry out the oxidation with tetrabutylammonium dichromate prepared *in situ* from sodium dichromate and tetrabutylammonium bisulphate in 3M sulphuric acid and methylene chloride; this oxidation takes place at room temperature[2] in few minutes. A typical example[2] is given below:

$$C_6H_5CH_2OH \xrightarrow[\text{3M } H_2SO_4,\ CH_2Cl_2,\ \text{RT 1 min}]{Na_2Cr_2O_7,\ Bu_4NHSO_4} C_6H_5CHO \quad 100\%$$

In acidic conditions, oxidation of primary alcohols gives the esters. one such example[3] is given below

$$\underset{\text{Butyl alcohol}}{C_4H_9OH} \xrightarrow[20\text{–}35°C]{Na_2Cr_2O_7,\ H_2SO_4,\ H_2O} \underset{\substack{41\text{-}47\% \\ \text{Butylbutyrate}}}{C_3H_7COOC_4H_9}$$

Seccondary alcohols are oxidised to ketones. Thus, cyclohexanol on oxidation gives cyclohexanone.

Cyclohexanol → Cyclohexanone

Reagent	*% yeld*	*Retence*
$Na_2Cr_2O_7.2H_2O$ (100% excess) H_2SO_4, Et_2O, 0°C	98	4
$Na_2Cr_2O_7.2H_2O$ (equivalent amount), H_2SO_4, Et_2O, 25°C	92	4
$Na_2Cr_2O_7$. $2H_2O$, H_2SO_4, DMSO, 70°C, > 30 min	89	5
$Na_2Cr_2O_7.2H_2O$, H_2SO_4, Et_2O, 25°C, 2.25 hr	92	6

Sodium dichromate in sulphuric acid oxidises aromatic amines to quinones. Thus, 3-Chloro-4-aminophenol on sirring with sodium dichromate and sulphuric acid at temperatures below 35°C gives 58–53% yield of chloro-p-benzoguinone[7].

$$\text{3-Chloro-4-amino phenol} \xrightarrow[< 30°C,\ 1\ hr.]{Na_2Cr_2O_7/H_2SO_4} \text{Chlorobenzoquinone (58-63\%)}$$

Some aromatic compounds are oxidised to quinones by sodium dichromate in presence sulphuric acid and a PTC.

$$\text{Anthracene} \xrightarrow[Bu_4NHSO_4,\ 70°C,\ 2\ min]{Na_2Cr_2O_7,\ H_2SO_4} \text{Anthraquinone (82-92\%)}$$

Sodium, dichromate in sulphuric acid oxidises aromatic side chain to carboxyl groups. Some examples[8,9] are given below.

R_1	R_2	R_3	Conditions	Product
H	NO_2	H	$Na_2Cr_2O_7/H_2SO_4$, reflux, 1 hr.	*p*-Nitrobenzoic acid 82–80%
NO_2	NO_2	NO_2	$Na_2Cr_2O_7/H_2SO_4$, 45-55°C, 3-4 hr.	3,4,5-Trinitrobenzoic acid 57-69%

Similarly, homologues of napthalene are oxdised into naphthalenecarboxylic acids on heating in an autoclave at 250°C for 18 hr with an aqueous solution of sodium dichromate. Thus, β-methylnaphthalene yields[10] β-naphthoic acid in 93% yield and 2,3-dimethylnapthalene yield[11] 2,3-napthalenedicarboxylic acid in 87-93% yield[11]. Similarly, methyl anthracences and methylphenantherenes are converted[10] into the corresponding carboxylic acids in 90% yield.

The five-membered ring in a derivative of acenaphthene is cleaved to form a dicarboxylic acid that forms an anhydride[12].

Sodium dichromate in dilute sulphuric acid brings about degradation of benzene rings to carboxyls particularly in presence of electron donating groups. Thus, m-trifluromethyl aniline is converted into trifluoroacetic acid in 90-95% yield on heating[13] with sodium dichromate and dilute sulphuric acid for 30–40 min., at 70–170°C.

Hydroxylamines on treatment with sodium dichromate in sulphuric acid yields[14] nitroso compound.

$$\underset{\text{Phenyl hydroxylamine}}{C_6H_5NHOH} \xrightarrow[H_2O,\ 0°C,\ 3\ min]{Na_2Cr_2O_7/H_2SO_4} \underset{\text{Nitrosobenzene}}{C_6H_5NO}$$

The above method is of special interest, since nitroso compounds cannot be prepared by the reduction of nitro compounds.

Methylene groups adjacent to aromatic ringes can be oxidised to keto groups by sodium dichromate[15,16].

Oxidation of hydroqunine to p-benzoquinone can be effected[4] with sodium dichromate and sulphuric acid.

$$\underset{\text{Hydroquinone}}{HO{-}C_6H_4{-}OH} \xrightarrow[H_2O,\ 20\text{-}30°C,\ 30\text{-}45\ min]{Na_2Cr_2O_7,\ H_2SO_4} \underset{\substack{76\text{-}81\% \\ p\text{-Benzoquinone}}}{O{=}C_6H_4{=}O}$$

The oxidation of methyl ketone gives a carboxylic acid with one less carbon by sodium dichromate. Thus, 2-acetyl fluorene an heating with sodium dichromate–acetic acid gives the carboxylic acid[17]. In this case the methylene groups is also oxidised to

$$\underset{\text{2-Acetylfluorene}}{COCH_3} \xrightarrow[100°C,\ 10\ hr]{Na_2Cr_2O_7.2H_2O,\ AcOH,\ Ac_2O} \underset{67\text{-}74\%}{COOH}$$

References

1. D.G. Lee and U.A. Spetzer. *J. Org. Chem.*, 1970, **35**, 3589.
2. D. Pletcher and S. J. D. Tait. *J. Chem Soc.*, Perkin Trans. 2, 1979, 788.
3. Y. R. Robertson, *Org. Synth.* Collective Volume, 1932, **I**, 138.
4. H. C. Brown, C.P. Garg and K. –T. Liu, *J. Org. Chem.*, 1971, **36**, 387.
5. Y. S. Rao and R. Filler, *J. Org. Chem.*, 1974, **39**, 3305.
6. H. C. Brown and C. P. Garg, *J. Am. Chem. Soc.*, 1961, **83**, 2952.
7. R. E. Harman, *Org. Synth.* Collective Volume, 1963, **4**, 148.

8. O. Kamm and A. O. Matthews, *Org. Synth.* Collective Volume, 1932, **1**, 392.
9. H.T. Clarke and W.W. Hartman. *Org. Synth.* Collective Volume., 1932, **1**, 543.
10. L. Friedman, D. L. Fishel and H. Shechter, *J. Org. Chem.,* 1965, **30**, 1453.
11. L. Friedman, *Org. Synth.* Collective Volume, 1973, **5**, 810.
12. D.V. Nightingale and W.S. Wagner, *J. Org. Chem.,* 1960, **25**, 32.
13. R. Wächter, *Angew. Chem.* 1955, **67**, 305.
14. G. H. Coleman, C.M. Mccloskey and F.A. Stuart, *Org. Synth.* Collective Volume, 1955, **3**, 668.
15. E. H. Huntress, E. B. Herhberg and I. S. Cliff, *J. Am. Chem. Soc.,* 1931, **53**, 2720.
16. C.F.H. Allen and J.A. Van Allan, *Org., Synth.* Collective Volume, 1955, **3**, 1.
17. Y. Rieveshl and F. E. Ray, *Org. Synth.* Collective Volume, 1995, 3, 420

1.132 SODIUM HYPOBROMITE, NaOBr

It is prepored *in situ* from bromine and aqueous sodium hydroxide under cooling. It is mostly used in the degradation to acids. Two such examples[1,2] are given below.

$CH_3O-C_6H_4-C_6H_4-COCH_3$ $\xrightarrow[\text{dioxane, 35-40°C, 45 min, }H^+]{Br_2/NaOH/0°}$ $CH_3O-C_6H_4-C_6H_4-COOH$

4-Acetyl-4′-methoxy biphenyl → 4-(4′-Methoxyphenyl) benzoic acid

Pregnenolone acetate $\xrightarrow[\text{−5°C to 0°C; 2.5 hr., 90°C, HCl}]{\text{NaOBr } (Br_2, NaOH),\ H_2O, \text{ dioxane}}$ 3-Acetoxyetienic acid (91-95%)

In certain cases, not only methyl ketones, but other alkyl ketons are degraded to carboxylic acids. Some examples[3,4] are given below.

$C_6H_5COCH_2CH_3 \xrightarrow[22-25°C]{NaOBr} C_6H_5COOH$

Propiophenone → Benzoic acid (96%)

C_4H_9 … N … COC_3H_7 $\xrightarrow[\text{(2) NaHSO}_3\text{, H}_2\text{SO}_4]{\text{(1) Br}_2\text{, NaOH, H}_2\text{O, RT, 15 hr}}$ C_4H_9 … N … CO_2H

5-Butyl-2-butryl pyridine

79%
5-Butyl-α-picolinic acid

References

1. W. S. Johnson, C.D. Gutsche and R.D. Offenhauer, *J. Am. Chem. Soc,* 1946, **68**, 1648.
2. J. Staunton and E. J. Ei Senbraun, *Org. Synth.* Collective Volume, 1973, **5**, 8.
3. S. L. Friees, *J. Am. Chem. Soc.,* 1949, **71**, 2571.
4. E. Hardegger and E. Nikles, *Helv. Chim. Acta,* 1957, **40**, 2428.

1.133 SODIUM HYPOCHLORITE, NaOCl

It is commercially available as dilute aqueous solution (5.25%) under the name clorox or bleach. It can also be prepared[1] *in situ* by passing chlorine into an aqueous solution of sodium hydroxide. In presence to tetrabutyl ammonium hydrogen sulphate, it is converted into tetrabutylammonium hypochlorite, which is soluble in organic solvents[2,3].

Sodium hypochlorite is used in the following oxidations:

1. Sodium hypochlorite oxidises α-amino acids to give aldehydes. Thus, valine and sodium hypochlorite and 10 mol % sodium hydroxide on steam distillation gives isobutyraldehyde in 79% yield[4].
2. Methyl ketones are oxidised by sodium hypochlorite to the corresponding carboxylic acids. Thus, isopropenyl methyl ketone gives methacrylic acid in 41% yield[5].

$$CH_2{=}C(CH_3)COCH_3 \xrightarrow[< 20°C,\ > 1\ hr,\ H_2SO_4]{3N\ NaOCl,\ 6N\ NaOH} CH_2{=}C(CH_3)COOH$$

Isopropenyl methylketone

41%
Methacrylic acid

3. Dehydrogenation of primary amines to nitriles. Octylamine is dehydrogenated to octanenitrile in 60% yield by treatment with sodium hypochlorite and tetrabulytammonium hydrogen sulphate[6] in ethyl acetate at room temperature for 35 min.
4. Primary amino group bonded to secondary carbon are oxidised to ketones. Thus, cyclohexylamine gives cyclohexanone

$$C_6H_{11}NH_2 \xrightarrow[AcOEt,\ RT,\ 2\ hr]{NaOCl,\ Bu_4NHSO_4} C_6H_{10}{=}O$$

Cyclohexyl amine

98%
Cyclohexanone

References

1. D. T. Mowry and E. L. Ringwald, *J. Am. Chem. Soc.,* 1950, **72**, 12037.
2. Krishman, D. Y. Kuhn and G. A. Hamilton, *J. Am. Chem. Soc.,* 1977, **99**, 8121.
3. S. Cacchi and F. La Torre, *Chem.* Ind. (London), 1986, 286.
4. K. Langheld, *Chem. Ber.,* 1909, **42**, 2360.
5. T. White, *J. Chem. Soc.*, 1943, 238.
6. G. A. Lee and H. H. Freedman, *Tetrahedron Lett.*, 1976, 1641.

1.134 SODIUM HYPOIODITE, NaOI

It is prepared[1] *in situ* from iodine and sodium hydroxide and is used for the conversion of methyl ketones (CH_3COR), acetaldehyde, ethanol and secondary methyl carbinols (CH_3CHOHR) into iodoform. The reaction is known as **Lieben Iodoform reaction**[2]. It is best carried out by dissolving the compound in dioxane, adding dilute sodium hydroxide and then adding a slight excess of iodine in potasium iodide solution, warming and finally adding water.

$$RCOCH_3 + NaOI \longrightarrow RCOCI_3 + 3NaOH$$

$$RCOCI_3 + NaOH \longrightarrow CHI_3 + RCOO^-Na^+$$

The mechanism of the reaction is given below:

abstraction of proton

abstraction of proton

Resonance stabilized carboxylate

Iodoform

References

1. K. J. Morgan, J. Bardwell and C.F. Culles, *J. Chem. Soc.*, 1950, 3190.
2. A. Lieben, Ann. (sppl), 1870, **7**, 208; R.C. Fuson and B.A. Bull, *Chem. Rev.*, 1934, **15**, 275.

1.135 SODIUM IODATE, $NaIO_3$

Like potassium iodate, sodium iodate is used for the oxidation of aromatic polyhydroxy compounds to quinones[1,2].

An interesting example[2] is given below.

OH, OH — $NaIO_3$, H_2O, aqueous EtOH, RT, 20 min → O, O

N H — N H

References

1. T. W. Evans and W. M. Dehn, *J. Am. Chem. Soc.*, 1930, **52**, 3647.
2. J.M. Bruce, *J. Chem. Soc.*, 1959, 2366.

1.135a SODIUM METAPERIODATE $NaIO_4$

It is commercially available. It is Known[1] that periodic acid oxidises 2,3-dimethyl indole to 2-formyl-3-methylinidole (*see* periodic acid). However, sodium pariodate cleaves[1] the double bond to give 85% yield of o-acetamidoacetophenone in 85% yield.

CH_3, CH_3, N H — $NaIO_4$, H_2O, MeOH, RT, 8 hr → $COCH_3$, $NHCOCH_3$

2,3-Dimethyl indole — o-Acetamidoacetophenone (85%)

Cyclic olefines are oxidised[2] with osmium tetraoxide in presence of sodium metaperiodate to give dialdehyde in 77% yield.

OsO_4, $NaIO_4$, Et_2O, H_2O, 24-26°C, 2 hr. → CHO, CHO

Cyclohexene — Hexane-1,6-dialdehyde (70%)

Another example of the cleavage of a double bond is by using the sequence of hydroxylation by osmic acid and subsequent fission of the diol by sodium periodate. Thus, 1-allyl-1-ethyl-7-methoxy-2-tetralone is converted into 1-ethyl-1-formylmethyl-2-tetralone by the following sequence of reactions[3]:

$NaClO_4/OsO_4$, THF, H_2O, RT, 22 hr

CH_3O, O, H_5C_2, $CH_2CH{=}CH_2$

1-Allyl-1-ethyl-7-methoxy--2-tetralone

CH_3O, O, H_5C_2, $CH_2CH(OH){-}CH_2OH$

$NaIO_4$, THF, H_2O

CH_3O, O, H_5C_2, CH_2CHO

1-Ethyl-1-formylmethyl--2-tetralone
94%

An interesting example is the fission of vicinol diols is the conversion of ethyl-2-acetamido-2-deoxy-1-thio-α-D-galactofuranoside into 4-formylthioacetal-furanoside in 80% yield.

H, O, H, OH, H, HO—HC, SC_2H_5, HO H_2CH, $NHCOCH_3$

Ethyl 2-acetamido--2-deoxy-1-thio--α-D-galaclofuranoside

$NaIO_4/H_2O$, 0-3°C, 30 min

H, O, H, OH, H, O=CH, SC_2H_5, H, $NHCOCH_3$

>80%
4-Formylthioacetal furanoside

Triphenylphosphine ylides are cleaved[4] by sodium periodate to triphenylphosphine oxide and carbonyl compound.

$$C_6H_5COCH{=}P(C_6H_5)_3 \xrightarrow[\text{reflux, 1 hr}]{NaIO_4/H_2O} C_6H_5COCHO + Ph_3P{=}O$$

Triphenyl phosphine ylide — 100%

Unsaturated carboxylic acids can be oxidised with $KMnO_4$ and sodium periodate. Thus, undecylenic acid gives 73% sebacic acid[4a].

$$CH_2{=}CH(CH_2)_8CO_2H \xrightarrow[H_2O,\ 20°C,\ 20\ hr]{KMnO_4/NaIO_4} HO_2C(CH_2)_8CO_2H$$

Sebacic acid
73%

Acetylenic sulphides are oxidised[5] to sulfoxide by sodium periodate in acetonitrile-water

$$C_6H_5C{\equiv}CSCH_3 \xrightarrow[-10°C,\ -5\ to\ -10^{10},\ 12\ hr]{NaIO_4,\ MeCN,\ H_2O} C_6H_5C{\equiv}C\overset{O}{\overset{\|}{S}}CH_3$$

46%

Sodium periodate is used as a catalyst[6] in the oxidation of 3-hydroxycyclobutanecarboxylate to 3-cyclobutanonecarboxylate in 78% yield by ruthenium tetroxide.

•

HO / CO_2Et $\xrightarrow[\text{water-}CCl_4,\ \text{RT, overnight}]{RuO_4/NaIO_4}$ O / CO_2Et

Ethyl 3-hydroxycyclobutane carboxylate

Ethyl 3-cyclobutanone carboxylate (78%)

A number of sulphides are oxidised to sulfoxides[7,8] by sodium periodate. Some examples are given below.

$$C_6H_5SCH_3 \xrightarrow{NaIO_4,\ H_2O,\ 0°C,\ 3\text{-}12\ hr} C_6H_5\overset{O}{\overset{\|}{S}}CH_3$$

88-91%

$$C_6H_5SC_6H_5 \xrightarrow{NaIO_4,\ H_2O,\ 0°C,\ 3\text{-}12\ hr} C_6H_5\overset{O}{\overset{\|}{S}}C_6H_5$$

98%

The cyclic mercaptal of acetaldehyde (2-methyl-1,3-dithiane) is oxidised by sodium meta iodate to a mixture of diastereomeric *cis* and *trans* monosulfoxides[9]

CH_3CH (S, S ring) $\xrightarrow[-7\ to\ 0°C,\ -5°C,\ 1\ hr,\ 0°C,\ overnight]{NaIO_4,\ MeOH,\ H_2O}$

2-Methyl-1, 3-dithiane

CH_3CH (S=O, S ring) 7% + CH_3CH (O=S, S ring) 80%

Like sulphides, selenides can be oxidised to the corresponding selenoxides[10].

$$C_6H_5SeCH_3 \xrightarrow{NaIO_4,\ MeOH,\ H_2O,\ 0°C} C_6H_5\overset{O}{\overset{\|}{S}}eCH_3 \quad 79\%$$

Using silica 'doped' with 10% $NaIO_4$, sulphides can be oxided to sulfoxides and sulfones depending on the reaction conditions. This oxidation is carried out in solid state using MW irradiation[11].

$$R{-}S{-}R_1 \xrightarrow[MW,\ 1\text{–}3\ min]{20\%\ NaIO_4\text{–Silica (3.0 eq)}} R{-}SO_2{-}R_1 \quad 72\text{–}93\%$$

$$R{-}S{-}R_1 \xrightarrow[MW,\ 0.5\text{–}2.5\ min]{20\%\ NaIO_4\text{–Silica (1.7 eq)}} R{-}\overset{O}{\overset{\|}{S}}{-}R_1 \quad 76\text{–}85\%$$

$R = R_1$ = Ph, $PhCH_2$, n-Bu
R = $PhCH_2$; R_1 = Ph;
R = Ph, n—$C_{12}H_{25}$; R_1 = Me

References

1. L. J. Dolby and D.L. Booth, *J. Am. Chem., Soc.* 1966, **88**, 1049.
2. R. Pappo, D.S. Allen, Jr., R.U. Lemieux and W.S. Johnson. *J. Org. Chem.,* 1956, **21**, 478.
3. K. Wiesner, K.K. Chan and C. Demerson, *Tetrahedron Lett.,* 1965, 2893.
4. H.–J. Bestman, R. Armsen and H. Wagner, *Chem., Ber.,* 1969, **102**, 2259.

4a. R. U. Lemieux and V.E. Rudloff, *Can. J. Chem.,* 1955, **33**, 1701.

5. G.A. Russel and L.A. Ochrymowycz, *J. Org., Chem.,* 1970, **35**, 2106.
6. J.A. Caputo and R. Fuchs, *Tetrahedron Lett.,* 1967, 4729.
7. N.J. Leonard and C.R. Johnson, *J. Org. Chem.,* 1962, **27**, 282.
8. C.R. Johnson and J.E. Keiser, *Org. Synth.* Collective Volume. 1973, **5**, 791.
9. B. Riegel, R. B. Moffett and A.U. McIntosh, *Org. Synth.* Collective Volumes, 1955, **3**, 234.
10. M. Cingerini, S. Colonna and R. Giovini, *Chem.* Ind (London), 1969,1737.
11. R.S. Verma, R.K. Saini and H.M. Meshram, *Tetrahedron Lett,* 1997, **38**, 6525.

1.136 SODIUM PERBORATE, $NaBO_3.4H_2O$

It is used for oxidation of primary amines to azocompounds[1] or nitro compounds[2]

$$C_6H_5NH_2 \xrightarrow{NaBO_3.4H_2O} C_6H_5N{=}NC_6H_5$$

The reagent also oxidise sulfides to sulfoxides and sulfones[2]

$$\underset{\text{Diphenyl sulphide}}{C_6H_5SC_6H_5} \xrightarrow[NaBO_3]{AcOH,\ 50\text{-}55°C} \underset{\text{Diphenyl sulfoxide}}{C_6H_5\overset{O}{\overset{\|}{S}}C_6H_5} \text{ and or } \underset{\text{Diphenyl sulphone}}{C_6H_5\overset{O}{\overset{\|}{\underset{O}{\underset{\|}{S}}}}C_6H_5}$$

In the above reaction, using small excess of sodium perborate gives exclusively diphenyl sulfoxide (71% yield). However, using large excess of sodium perborate gives exclusively diphenyl sulphone (98%).

Referencess

1. P. Santurri, F. Robbins and R. Slubbings, *Org. Synth.* Collective Volume, 1973, **5**, 341.
2. A. Mckillop and J.A. Tarbin, *Tetradron Lett.*, 1983, **24**, 1505.

1.137 SODIUM PERMANGANATE MONOHYDRATE, $NaMnO_4 . H_2O$

It is commercially available and is used for the following oxidations:

1. It oxidises alkenes to carboxylic acids,[1]

NaMnO$_4$, CO$_2$, isooctane; < 50°C, 4 hr → cis-cyclopentane-1,3-dicarboxylic acid (CO_2H, CO_2H) 65-68%

2. It oxidises alcohols to carbonyl compounds[2]

$$CH_3(CH_2)_6CH_2OH \xrightarrow[69°C,\ 5\ hr]{NaMnO_4,\ H_2O,\ C_6H_{14}} CH_3(CH_2)_6CO_2H \quad 67\%$$

$$\underset{\text{Benzyl alcohol}}{C_6H_5CH_2OH} \xrightarrow[69°C,\ 6\ hr]{NaMnO_4.H_2O,\ hexane} C_6H_5CO_2H \quad 81\%$$

Cyclohexanol (OH) → NaMnO$_4$.H$_2$O, hexane; reflux, 1.5 hr → cyclohexanone (O) 100%

$$CH_3CH(OH)C_6H_{13} \xrightarrow[69°C,\ 2.5\ hr]{NaMnO_4.H_2O,\ C_6H_{14}} CH_3COC_6H_{13} \quad 95\%$$

3. Sulfinic acids are converted into sulfonic acids[3]

$$C_{12}H_{25}-C_6H_4-SO_2H \xrightarrow[35\text{-}40°C,\ RT,\ overnight]{NaMnO_4.NaOH} C_{12}H_{25}-C_6H_4-SO_3Na \quad 70\%$$

4. Sulfoxides are oxidised to sulfones[4]

$$(C_6H_5CH_2)_2S{=}O \xrightarrow[20°C,\ 7\ min]{NaMnO_4,\ dioxane,\ H_2O} (C_6H_5CH_2)_2S({=}O)_2 \quad 95\text{-}100\%$$

Sulphides are not oxidised with sodium permanganate in dioxane[4].

References

1. S.F. Birch, W. J. Oldham and E.A. Johnson, *J.Chem. Soc.*, 1947, 818.
2. F.M. Menger and C. Lee, *Tetrahedron lett.*, 1981, **22**, 1655
3. W.E. Truice and J.F. Lyons, *J. Am. Chem., Soc.*, 1951, **73**, 126
4. H.B. Henbest and S.A. Khan, *Chem. Commun;* 1968, 1036.

1.138 SODIUM PEROXIDE, Na_2O_2

It is a pale yellow solid (m.p. 460°C). It is rearly used. However, it oxidises aldoximes to carboxylic acids[1] and ketones esters (lactones)[2].

References

1. T.L. Ho and Y.A. Ohal, *Synthesis*, 1976, 807.
2. M. Matsumoto and H. Kobayashi, *Heterocycles*, 1986, **24**, 2443.

1.139 SODIUM RUTHENATE, Na_2RuO_4

It is prepared[1] *in situ* from a solution of ruthenium tetroxide in carbon tetrachloride and 1M sodium hydroxide solution. The mixture is shaken for 2 hr. at room temperature; the reagent remains in aqueous solution which is bright-orange in colour.

It oxidises primary alcohols[1] to carboxylic acids and secondary alcohols to ketones.

$$C_6H_5CH_2OH \xrightarrow[25°C,\ 1\ hr]{Na_2RuO_4,\ NaOH,\ H_2O} \underset{97\%}{C_6H_5COOH}$$

$$R{-}CHOH{-}R' \xrightarrow[25°C,\ 1\ hr]{Na_2RuO_4,\ NaOH,\ H_2O} R{-}CO{-}R'$$

Reference

1. D.Y. Lee, D.T. Hall and J.H. Cleland, *Can. J. Chem.*, 1972, **50**, 3741.

1.140 SULPHUR, S_X

Sulphur is dehydrogenating agent that converts compounds containing six memberd ringis into their aromatic analogues[1,2]. Such dehydrogenations (known as chemical dehydrogenations occur at lower temperatures compared to catalytic dehydrogenation), (*e.g.*, with Pd/C, Pt/C) one such example is given below.

C_6H_5 (dihydronaphthalene) $\xrightarrow[250–270°C]{S}$ C_6H_5 (naphthalene)

91-94%

An important application of sulphur is the **Willgerodt reaction.** It involves heating ketones with ammonium polysulphide (prepared by saturating a concentrated aqueous solution of ammonia with hydrogen sulphide). The reaction is carried out in sealed tubes or autoclaves at 160–220°C. The reaction products, mixutres of amides, thioamides and some times acids, were hydrolysed with dilute aqueous sodium or potassium hydroxide to give carboxylic acids[3] having the same number of carbon atoms as the starting ketones. The **Kindler modification** avoids heating under pressure or using autoclaves. In this procedure the ketone is refluxed with a low boiling seecondary amine, such as morpholine (instead of ammonium sulphide) with sulphur. The formed thiomorpholide is subsequently hydrolysed to acids. The Willgerodt-Kindler reaction is mostly used to convert alkyl aryl ketones (which are synthesised by Friedel Crafts reaction) into arylalkanoic acids with or without the isolation of the intermediate thiomorpholide[4].

$C_6H_5COCH_2CH_3$ Propiophenone → (S, HN(morpholine)O; reflux, 6 hr) → [$C_6H_5COCH_2CSN$(morpholine)O] thiomorpholide → (10% NaOH/EtOH, reflux, 6-8 hr.) → $C_6H_5CH_2CH_2COOH$ β-Phenylpropionic acid 65%

Another example of Willgerodt-Kindler reaction is the conversion[5] of 3-acetyl pyridine into the corresponding carboxylic acid.

3-Acetyl pyridine ($COCH_3$) → (S, HN(morpholine)O; reflux, 12 hr) → 3-pyridyl-CH_2CSN(morpholine)O 80% → (10% KOH/EtOH, reflux 72 hr) → 3-pyridyl-CH_2COOH 74% 3-Pyridylacetic acid

A typical oxido-reductive reaction takes place when p-aminotoluene is heated with sodium polysulphide to give[6] p-aminobenzaldehyde.

References

1. E.B. Hershberg and L.F. Fieser, *Org. Synth.* Collective Volume, 1943, **2**, 423.
2. R. Weiss, *Org. Synth.* Collective Volume, 1955, **3**, 729.
3. L.F. Fieser and Y.W. Kilmer, *J. Am. Chem., Soc.*, 1940, **62**, 1554.
4. E. Schwenk and D. Papa, *J. Org., Chem.*, 1946, **11**, 798.
5. M. Hartmann and W. Bosshard, *Helv. Chem. Acta.*, 1941, **24**E, 28.
6. E. Compaigne, W.M. Budde and G.F. Schaeffer, *Org. Synth.* Collective Volume, 1963, **4**, 31.

1.141 SULFURYL CHLORIDE, SO_2Cl_2

It is used in the oxidation[1] of sulphides of sulfoxides

$$\underset{\text{Diphenyl sulphide}}{C_6H_5SC_6H_5} \xrightarrow[\substack{-30 \text{ to } -40°C \\ 9 \text{ min, EtOH}}]{SO_2Cl_2,\ CH_2Cl_2} \underset{\substack{95\% \\ \text{Diphenyl sulfoxide}}}{C_6H_5-\overset{\overset{O}{\|}}{S}-C_6H_5}$$

Reference

1. V.J. Traynelis, Y. Yoshikaua, S.M. Tarka and J.R. Livingston, Jr., *J. Org. Chem.*, 1973, **38**, 3986.

1.142 SULPHURIC ACID, H_2SO_4

It rarely acts as an oxidant. However, it can be used for dehydrogenative coupling of aromatic rings. Thus, 1-amino-2-methylnaphthalene on refluxing with 66% sulphuric acid for 50 hr. or in presence of ferric oxide for 8 hr. yields[1] 4,4[1]-bis (1-amino-2-methyl) binaphthyl.

Reference

1. H.E. Fierz-David, L. Blangey and H. Dübendorfer, *Helv. Chem., Acta*, 1946, **29**, 1661.

1.143 TETRABUTYLAMMONIUM CHLOROCHROMATE, $(C_4H_9)_4NCrO_3Cl$

It is prepared[1] by adding a solution of chromium trioxide in concentrated hydrochloric acid to a solution of tetrabutylammonium hydrogensulphate in 5 N hydrochloric acid at 0°C. The dry reagent is used as chloroform for the oxidation of alcohols to carbonyl compounds[2].

$$C_6H_5CH_2OH \xrightarrow[\text{CHCl}_3\text{, RT, 18 hr; reflux, 4 hr}]{Bu_4NCrO_3Cl} C_6H_5CHO \quad 68\%$$

$$C_6H_5CH(OH)CH_3 \xrightarrow[\text{CHCl}_3\text{, reflux 2 hr}]{Bu_4NCrO_3Cl} C_6H_5COCH_3$$

References

1. S. Cacchi, *Synthesis*, 1979, 356.
2. E. Santaniello, F. Milani and R. Casati, *Synthesis*, 1983, 749.

1.144 TETRABUTYLAMMONIUM CHROMATE, $(C_4H_9)_4NHCrO_4$

It is obtained by mixing solutions of chromium trioxide and tetrabutylammonium chloride in water. It is used in dry state in dichloromethane and chloroform for the oxidation[1,2] of alcohols to carbonly compounds.

$$C_6H_5CH_2OH \xrightarrow[\text{CCl}_4\text{, 60°C, 1 hr}]{CrO_3\text{, }Bu_4\ NCl\text{, }H_2O} C_6H_5CHO \quad 81\%$$

In the above case, the reagent is prepared and used *in situ* in a mixture of water and carbon tetrachloride.

$$\text{Cyclohexanol} \xrightarrow[\text{RT, 2 hr}]{CrO_3\text{, }Bu_4NCl\text{, }CH_2Cl_2} \text{Cyclohexanone} \quad 70\%$$

References

1. G. Cardillo and M. Shimizu, *J. Org. Chem.*, 1977, **42**, 4268.
2. T. Brunelet, C. Jouitteau and G. Gelbard, *J. Org. Chem.*, 1986, **51**, 4016.

1.145 TETRABUTYLAMMONIUM PERIODATE, $(C_4H_9)_4$ NIO_4

It is prepared *in situ* from tetrabutylammonium hydrogen sulphate and sodium periodate. Since it is soluble in chloroform and other organic solvents, it is useful in two phase systems. Its oxidising properties are similar to that of periodic acid and alkaline periodates.

It is useful in degradative oxidation[2] of acids to aldehydes. Thus, phenylacetic acid gives benzaldehyde in 50–85% yields when refluxed with the reagent in dioxane.

Degradative oxidation of α-hydroxy carboxylic acids yield aldehydes or ketones.

$$C_4H_9CH(OH)CO_2H \xrightarrow[\text{reflux, 3 hr}]{Bu_4NIO_4,\ CHCl_3} C_4H_9CHO \quad 90\%$$

Benzylic halides are oxidised to aromatic aldehydes with tetrabutyl ammonium periodate.

$$C_6H_5CH_2Br \xrightarrow[\text{reflux, 3 hr}]{Bu_4NIO_4,\text{ dioxane}} C_6H_5CHO$$

Benzyl bromide 85%

Sulphides containing sulfur in the ring are oxidised to cyclic sulfoxodes[1]

$$\text{Tetrahydrothiophene} \xrightarrow[\text{reflux, 2 hr}]{Bu_4NiO_4,\ CHCl_3} \text{Tetramethylene sulfoxide (S=O)}$$

Tetrahydro thiophene

90% Tetramethylene sulfoxide

References

1. E. Santaniello and T.–L. Tan, *J. Am Chem. Soc.*, 1948, **70**, 602.
2. E. Santaniello, F. Ponti and A. Manzocchi, *Tetrahedron Lett.*, 1980, **21**, 2655.

1.146 TETRABUTYLAMMONIUM PERMANGANATE $(C_4H_9)_4NMnO_4$

It is prepared[1] by the addition of slight excess of concentrated aqueous solution of tetrabutylammonium bromide to a well stirred solution of potassium permanganate in water. The formed reagent is filtered, washed with water and dried in vacuo. The reagent should be handeled with caution, since ignition of crystalline material during handling (weighing *etc*) has been reported[2].

It oxidises aldehydes to carboxylic acids[1].

Pipronal $\xrightarrow[\text{reflux, 3 hr}]{Bu_4NMnO_4,\ \text{dioxane}}$ Pipronalic acid (100%)

Primary alcohols and secondary alcohols are oxidised to carboxylic acids[3] and ketones[1] respectively.

$$CH_3(CH_2)_6CH_2OH \xrightarrow[\text{RT, 3 hr}]{Bu_4NMnO_4,\ C_6H_6} CH_3(CH_2)_6CO_2H \quad 47\%$$

$$C_6H_5CH_2OH \xrightarrow[\text{RT}]{Bu_4NMnO_4,\ C_5H_5N} C_6H_5CO_2H \quad 98\%$$

$$C_6H_5CH(OH)C_6H_5 \xrightarrow[\text{RT}]{Bu_4NMnO_4,\ C_5H_5N} C_6H_5COC_6H_5 \quad 97\%$$

References

1. T. Sala and M.V. Sargent, *Chem. Commun.,* 1978, 253.
2. J.A. Morris and D.C. Mills, *Chem.,* Ind. (London), 1978, 446.
3. A.W. Herriott and D. Picker, *Tetrahedron Lett.*, 1974, 1511.

1.147 THALLIUM NITRATE $Tl(NO_3)_3$

It is used as an oxidising agent[1]. Some of the reactions are given below.

(*i*) Oxidation of alkenes[2]

Thallium nitrate reacts with alkenes in methanol solution to give carbonyl compounds or 1,2-glycol dimethyl ethers.

$$CH_3(CH_2)_7CH{=}CH_2 \xrightarrow{Tl(ONO_2)_3/MeOH} CH_3(CH_2)_7COCH_3 \ (28\%) + CH_3(CH_2)_7CH(OMe){-}CH_2(OMe) \ (52\%)$$

Similarly, styrene forms 1,2-dimethoxyderivative

$$C_6H_5CH{=}CH_2 \xrightarrow[\text{MeOH}]{Tl(ONO_2)_3} C_6H_5CH(OCH_3){-}CH_2OCH_3$$

However, if the above reaction is carried out in dilute nitric acid, rearragement takes place giving aryl acetaldehydes

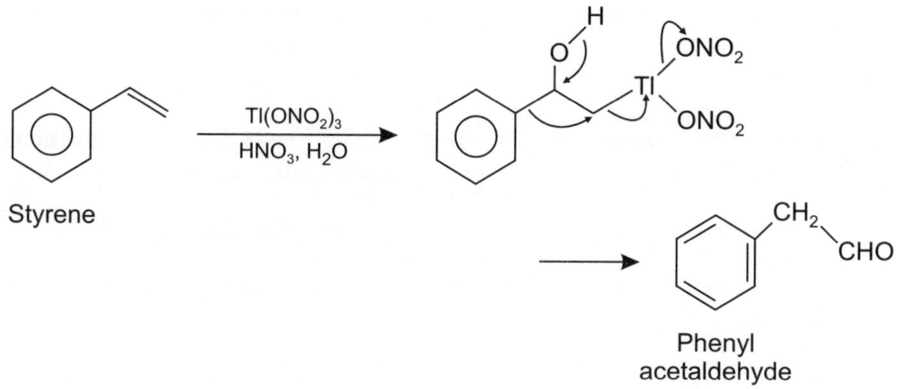

$$\text{Styrene} \xrightarrow[HNO_3,\ H_2O]{Tl(ONO_2)_3} \text{PhCH(OH)CH}_2\text{Tl(ONO}_2)_2 \longrightarrow C_6H_5CH_2CHO$$

Styrene

Phenyl acetaldehyde

Oxidation of cycle alkene proceeds by an oxidative rearrangement to ring contracted products[1]. Thus, cyclohexene gives dimethyl acetal of cyclopentanecarboxaldehyde.

Cyclohexene $\xrightarrow[MeOH]{Tl(ONO_2)_3}$ (cyclohexyl cation with $Tl(ONO_2)_2$) $\xrightarrow{CH_3OH}$ 2-methoxycyclohexyl-$Tl(ONO_2)_2$ $\longrightarrow$ $\overset{+}{C}HOCH_3$-cyclopentane $\xrightarrow{CH_3OH}$ $CH(OCH_3)_2$-cyclopentane

Cyclopentane carboxaldehyde dimethylacetal

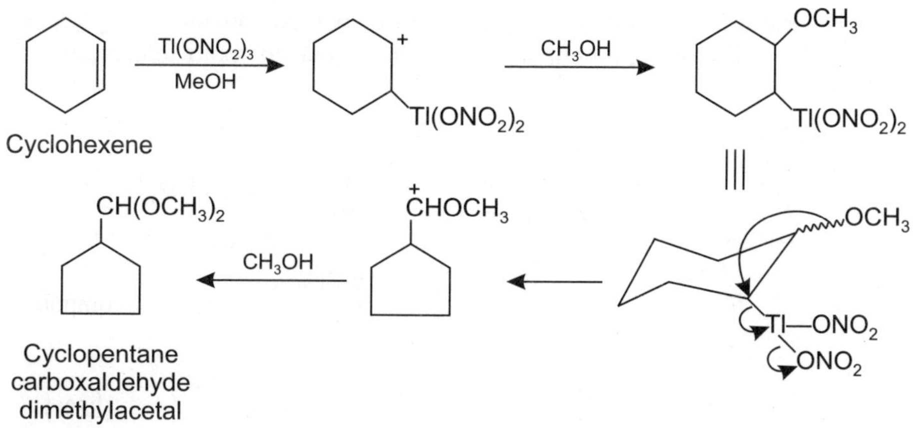

(*ii*) Oxidation of Alkynes

Dialkylalkynes are oxidised to acyloins or their alkyl ether by thallium nitrate in aqueous or aquous-alcoholic solution, respectively[3]. Only symmetrical alkynes give acceptable yields (70-90%).

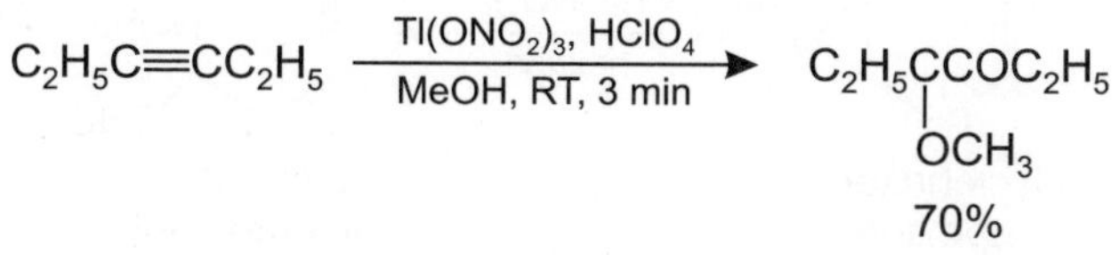

$$C_2H_5C{\equiv}CC_2H_5 \xrightarrow[MeOH,\ RT,\ 3\ min]{Tl(ONO_2)_3,\ HClO_4} C_2H_5\underset{\substack{|\\ OCH_3}}{C}COC_2H_5$$

70%

Alkylarylacetylenes undergo oxidative rearrangement to esters of alkyl arylacetic and[3]

$$C_6H_5C{\equiv}CCH_3 \xrightarrow[\text{reflux, 2 hr}]{Tl(ONO_2)_3,\ MeOH} C_6H_5\underset{\substack{|\\CH_3}}{C}H{-}CO_2CH_3$$

80%

Diarylacetylenes are converted in 55–90% yields into α-diketones by refluxing for 2–7 hr. with thallium nitrate in glyme solution, containing perchloric acid[3].

$$C_6H_5C{\equiv}CC_6H_5 \xrightarrow[\text{glyme, reflux, 3 hr}]{Tl(ONO_2)_3,\ 70\%\ HClO_4} C_6H_5COCOC_6H_5$$

Terminal acetylenes on oxidiation[3] of thalium nitrate in presence of perchloric acid give carboxylic acids with one less carbon

$$RC{\equiv}CH \xrightarrow[\text{glyme, RT, } H_2O\text{, 1 hr}]{Tl(ONO_2)_2,\ HClO_4} RCO_2H$$

$R{=}C_6H_{13}$ 80%

(*iii*) Oxidation of Methyl Ketones

Aryl methyl ketones are oxidised with thallium nitrate in methanolic solution in the persence of perchloric acid to give the correspoding carboxylic acid. Thus, acetophenone and its substituted derivatives are converted[4] into methyl arylacetates in 61–94% yields by treatment with thallium nitrate at room temperature for 2–18 hr.

$$ArCOCH_3 \rightleftharpoons Ar\overset{\overset{OH}{|}}{C}{=}CH_3 \xrightarrow[HNO_3]{Tl(ONO_2)_2,\ MeOH} Ar{-}\underset{\underset{OCH_3}{|}}{\overset{\overset{O-H}{|}}{C}}{-}CH_2{-}\overset{\overset{ONO_2}{|}}{Tl}{-}ONO_2 \xrightarrow{-HNO_3,\ -TlNO_3} ArCH_2CO_2CH_3$$

61–94%

(*iv*) Oxidation of cyclohexene oxide[2]

$$\text{Cyclohexene epoxide} \xrightarrow[\text{hexane}]{Tl(ONO_2)_3} \text{2-Nitrato cyclohexanol (OH, } ONO_2\text{)}$$

Cyclohexene epoxide

2-Nitrato cyclohexanol

(*v*) Oxidation of α,β-unsaturated ketones

Benzalacetophenone on oxidation with thallium nitrate gives benzil

$$\text{Benzalacetophenone} \xrightarrow[H_2O]{Tl(ONO_2)_3} Ph{-}\overset{O}{\overset{\|}{C}}{-}\overset{O}{\overset{\|}{C}}{-}Ph \ (\text{Benzil})$$

References

1. E.A. Braude, A.G. Brook and R.P. Linstead, *J. Chem. Soc.*, 1957, 3569.
2. V.K. Ahluwalia and R.K. Parashar, *Organic Reaction Mechanisms,* Narosa, 2007, Page 193–195.
3. A. Mc Killop, O.H. Oldenziel, B.P. Swann, E.C. Taylor and R.L. Robey, *J. Am. Chem., Soc.,* 1973, **95**, 1296.
4. A. Mckillop, B.R. Swann and E.C. Taylor, *J. Am. Chem. Soc.,* 1971, **93**, 4919.

1.148 THALLIUM MONOACETATE, $TlOCOCH_3$

Thallium monoacetate in presence of iodine accomplishes stereoselective hydroxylation of alkenes[1]

Cyclohexene → (1) $TlOCOCH_3$, I_2, AcOH, 80°C, 0.5 hr; (2) H_2O, reflux 9 hr or NaOH, EtOH, reflux 2 hr; syn hydroxylation → Cyclohexane syndiol (OH OH), 70-75%

Cyclohexene → (1) $TlOCOCH_3$, I_2, AcOH, reflux 9 hr; (2) NaOH, EtOH, reflux 3 hr; anti hydroxylation → Cyclohexane antidiol (OH, OH), 65–70%

Reference

1. R.C. Combil and P.S. Rutledge, *Org. Synth.* Collective Volume, 1988, **6**, 348.

1.149 THALLIUM TRIACETATE $Tl(OCOCH_3)_3.1.5H_2O$

Cyclohexene on treatment with thallium triacetate gives[1] a mixture of diastereomers in varying amounts depending on the reaction conditions and byproducts as a result of rearrangements.

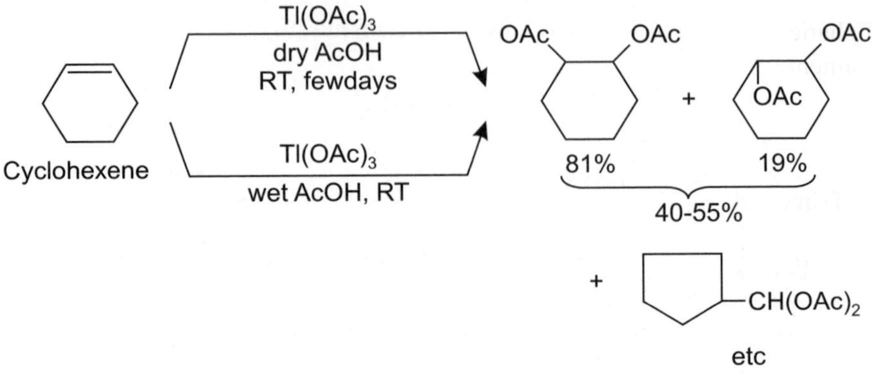

$Tl(OAc)_3$, dry AcOH, RT, fewdays
Cyclohexene
$Tl(OAc)_3$, wet AcOH, RT

OAc OAc + OAc OAc
81% 19%
40-55%
+ $CH(OAc)_2$
etc

Thallium triacetate is used for epoxidation of alkenes[2]. Thus, the oxidation of isobutylene with thallium triacetate gives 82% of isobutylene oxide[2].

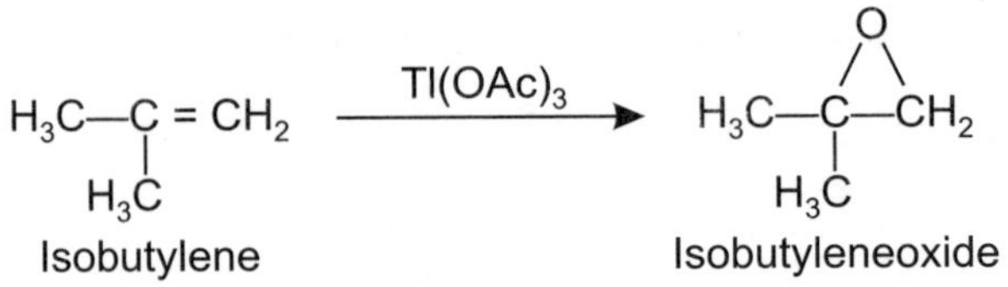

$$H_3C-C(CH_3)=CH_2 \xrightarrow{Tl(OAc)_3} H_3C-C(CH_3)(-O-)CH_2$$

Isobutylene Isobutyleneoxide

References

1. C.B. Anderson and S. Winstein, *J. Org. Chem.*, 1963, **28**, 605.
2. W. Kruse and T.M. Bednarski, *J. Org. Chem.*, 1971, **36**, 1154.

1.150 TRICHLOROISOCYANURIC ACID, $C_3Cl_3N_3O_3$

It converts ethers into esters[1]. Another related reagent, sodium N-chloro-p-toluenesulfonamide **(chloramine-T)**, p–$CH_3C_6H_4SO_2NClNa.3H_2O$, in presence of osmium tetroxide converts alkenes into vicinal hydroxy sulfonamino compounds[2].

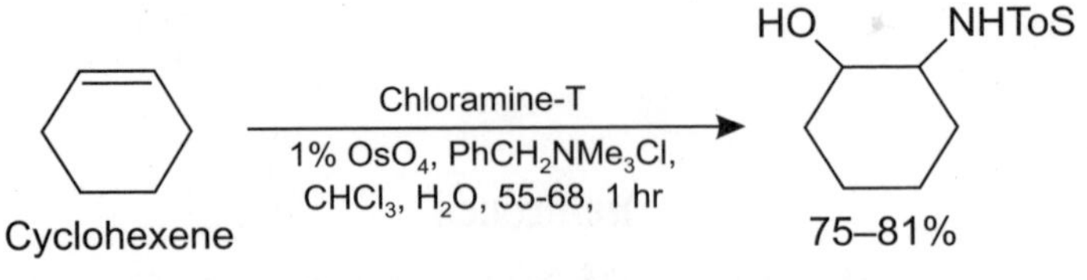

Chloramine-T
1% OsO_4, $PhCH_2NMe_3Cl$, $CHCl_3$, H_2O, 55-68, 1 hr
Cyclohexene
HO NHToS
75–81%

References

1. E.C. Juenge and D.A. Beal, *Tetrahedron Lett.*, 1968, 5819.
2. E. Herranz and K.B. Sharpless, *Org. Synth.*, 1983, **61**, 85.

1.151 TRIMETHYLAMINE OXIDE, $(CH_3)_3NO$

Aliphatic primary halides, especially the iodides are converted into aldehydes by treatment with trimethylamine oxide[1].

$$C_7H_{15}CH_2I \xrightarrow[\text{40-50°C, 1 hr; reflux, 20 min}]{(CH_3)_3NO,\ CHCl_3} C_7H_{15}CHO \quad 41.5\text{–}43\%$$

Trimethylamine oxide is used to convert alkylboranes into alcohols[2,3]

$$\underset{\text{Tributyl borane}}{(C_4H_9)_3B} \xrightarrow[\text{exothermic reaction 0°C}]{Me_3NO} \underset{\text{Dibutylborinate}}{(C_4H_9)_2BOC_4H_9} \xrightarrow[25°C]{Me_3NO} \underset{\text{Butylboronate}}{C_4H_9B(OC_4H_9)_2}$$

$$\underset{\text{Butylboronate}}{C_4H_9B(OC_4H_9)_2} \xrightarrow[\text{reflux, } CHCl_3\text{, 24 hr}]{Me_3NO} \underset{\text{Tributylborate}}{B(OC_4H_9)_3} \xrightarrow[\text{decanol } (\textit{trans}\text{ esterification})]{\text{distil with}} C_4H_9OH$$

N-Methylmorpholine Oxide, is used as a reoxidant in hydroxylation of alkenes with osmium tetroxide[4].

Cyclohexene $\xrightarrow[\text{Me}_2\text{CO, EtOH; reflux, 2 hr; synhydroxylation}]{\text{OsO}_4\text{, N-methylmorpholine N-oxide}}$ cyclohexane-1,2-diol (OH OH) 70–75%

The oxidation of phosphites to phosphates is very important in the chemistry of nucleoside phosphite. One example[5] is given below:

Dinucleoside methyl phosphite (R = pMeC$_6$H$_4$Ph$_2$CO; 3′-OSiMe$_2$(CMe$_3$)) $\xrightarrow[\text{CH}_2\text{Cl}_2\text{, 25°C, 7 hr}]{\text{N-Methylmorpholine oxide}}$ dinucleoside methyl phosphate (P=O; OSiMe$_2$(CMe$_3$)) 80%

R = pMeC$_6$H$_4$Ph$_2$CO

Pyridine oxide hydroxylates aromatic rings to phenols[6]. The hydroxylation takes place, probably by an intramolecular transfer of oxygen, on irradiation of 2-benzylpyridine-N-oxide complex with boron trifluoride.

BF_3, hν
$C_6H_6NCH_2Cl_2$

2-Benzylpyridine N-oxide → 43%

An important observation is that the presence of oxygen on the nitrogen of the pyridine ring reverses the direction of electrophilic substitutions in the pyridine. Normally, electrophilic attack on pyridine occur in β-positions. However, attack on pyridine oxide occurs at α-and γ-positions. After the introduction of the electrophiles, the pyridine oxide is converted into pyridine by mild reduction, for example treatement[6] with salts of iron or titanium.

References

1. V. Franzen, *Org. Synth.* Collective Volume, 1973, **5**, 872.
2. J.A. Soderquist and M.R. Najafi, *J. Org. Chem.*, 1986, **51**, 1130.
3. G.W. Kabalka and H.C. Hedgecock, Jr., *J. Org. Chem.,* 1975, **40**, 1776.
4. V. Van Rheenen, D.Y. Cha and W.M. Hartley. *Org. Synth.* Collective Volume, 1988, **6**, 342.
5. Y. Hayakawa, M. Uchiyama and R. Noyori, *Tetrahedron Lett.,* 1986, **27**, 4191.
6. G. Serva-Errante and P.G. Sammes, *Chem. Commun,* 1975, 573.

1.152 TRIPHENYLMETHYLPHOSPHONIUM PERMANGANATE, $(C_6H_5)_3PCH_3MnO_4$

It is prepared[1] by adding an aquous solution of potassium permanganate to a solution of triphenylmethylphosphonium bromide in water. The formed reagent is filtered and dried (vacuo over phosphorus pentoxide). It should be handeled with care, since explosion may occur when organic substrates are added to solid oxidant. It is used for syn hydroxylation of alkenes[1].

$Ph_3PMeMnO_4$, CH_2Cl_2
–70°C, 30 min

Cyclooctene → cis 80%

For hydroxylation, a dilute solution of the alkene in dichloromethane is added to a prepared solution of triphenylmethylphosphonium permanganate in dichloromethane at –70°C.

Reference

1. W. Reischl and E. Zbiral, *Tetrahedron*, 1979, **35**, 1109.

1.153 TRIPHENYLMETHYL TETRAFLUOBRORATE, $(C_6H_5)_3C^+BF_4^-$

It oxidises secondary alcohols in preference to primary alchohol[1].

$$R_2CHOH \xrightarrow[-HBF_6]{Ph_3CBF_4} R_2CHOCPh_3 \xrightarrow[\text{catalyst}]{Ph_3CBF_4} R_2C{=}O + Ph_3CH$$

$$CH_3CH(OH)C_6H_{13} \xrightarrow[CH_2Cl_2,\ 25°C,\ 10\ hr]{2Ph_3CBF_4} CH_3COC_6H_{13}\ (84\%)$$

$$\text{cyclohexanol} \xrightarrow[CH_2Cl_2,\ 25°C,\ 13.5\ hr]{2Ph_3CBF_4} \text{cyclohexanone}\ (66\%)$$

Reference

1. M.E. Jung and R.W. Broun, *Tetrahedron Lett.*, 1978, 2771.

1.153a TRIPHENYL PHOSPHATE OZONIDE, $(C_6H_5O)_3PO_3$

It is prepared[1] by the ozonolysis of triphenyl phosphite in dichloromethane solution at –78°. As an oxidant, it converts trimethylsilyl ethers of enols to α, β, unsaturated ketones into unsaturated hydroxy ketones[2].

$$\text{OSi(CH}_3)_3\text{ diene} \xrightarrow[(2)\ Et_3N,\ Ph_3P,\ Et_2O,\ RT,\ 1\ hr]{(1)\ (PhO)_3PO_3,\ CH_2Cl_2,\ -70°,\ 3\ hr} \text{HO-enone}\ (79\%)$$

References

1. M.V. Bhatt, G.V. Rao, and K.S.R. Rao, *Chem. Commun*, 1971, 822.
2. C. Iwata, Y. Takemoto, A. Nakamura and T. Imanishi, *Tetrahedron Lett.,* 1985, **26**, 3227.

1.154 TUNGSTEN HEXAFLUORIDE, WF_6

It resembles molybdenum hexafluoride (MoF_6) and molydenum oxychloride (MoOCl) in its oxidation properties. This is also used in regeneration of ketones from their derivatives (*see* molybdenum hexafluoride)

Reference

1. G.A. Olah and J. Welch, *Synthesis*, 1976, 809.

1.154a UREA-HYDROGEN PEROXIDE COMPLEX (UHP)

A novel oxidant[1], it is commercially available and can also be easily prepared. It is useful for the oxidation of hydroxylated aldehydes and ketones to hydroxylated phenols. The oxidation is conducted in solid phase.

Salicyaldehyde —(UPH, 55°C; 1.5 hr)→ Catechol (85%)

p-Hydroxybenzaldehyde —(UPH, 55°C; 3 hr)→ Hydroquinone (83%)

o-Hydroxy acetophenone —(UPH, 85°C; 1 hr)→ Catechol (86%)

p-Hydroxyacetophenone —(UPH, 85°C; 1 hr)→ Hydroquinone (80%)

2-Hydroxy-4-Nitro benzaldehyde —(UPH, 85°C; 25 min)→ 3,4-Dihydroxy nitrobenzone (83%)

p-Methoxybenzaldehyde —(UPH, 85°C; 45 min)→ Hydroquinol monomethyl ether (85%)

Benzaldehyde is oxidised to benzoic acid (95%)

UPH is represented as NH_2CONH_2
|
HOOH

UPH oxidises nitriles to amides. Thus, benzylcyanide and phenyl cyanide are oxidised (85°, 1-1.5 hr) to phenyl acetamide and benzamide, respectively in 80-85% yeild sulphides are oxidised to sulfoxodes or sulfones by UPH. Nitrogen heterocycles are oxidised to the corresponding N-oxides.

Reference

1. R.S. Varma and K.P. Naicker, *Organic Letters,* 1999, **1**, 189.

1.155 ZINC DICHROMATE TRIHYDRATE, $ZnCr_2O_7.3H_2O$

It is obtamied[1] by adding zinc carbonate to a cold solution of chromium trioxide in dilute sulphuric acid.

If oxidises[1] acetylenes to α-diketones. Aromatic compounds are oxidised to quinones[1].

Anthracene —($ZnCr_2O_7$, $3H_2O$; CCl_4, RT, 3.5 hr.)→ Anthraquinone 80%

Zinc dichromate oxidises cyclic ethers to lactones in 70% yield[1]

$$\text{THF} \xrightarrow[\text{CH}_2\text{Cl}_2,\ \text{RT}]{\text{ZnCr}_2\text{O}_7.3\text{H}_2\text{O}} \gamma\text{-butyrolactone (70\%)}$$

Primary alcohols and secondary alcohols are oxidised to carboxylic acids and ketones, respectively

$$C_6H_5CH_2OH \xrightarrow[\text{CH}_2\text{Cl}_2,\ \text{RT, 6 hr}]{\text{ZnCr}_2\text{O}_7.3\text{H}_2\text{O}} \underset{90\%}{C_6H_5COOH}$$

$$C_6H_5CH(OH)C_6H_5 \xrightarrow[\text{CH}_2\text{Cl}_2,\ \text{RT, 6 hr}]{\text{ZnCr}_2\text{O}_7.3\text{H}_2\text{O}} \underset{95\%}{C_6H_5COC_6H_5}$$

Zinc dichromate is also used[1] for the oxidative regenration of carbonyl compounds from their oximes

$$C_6H_5C(=NOH)C_6H_5 \xrightarrow[\text{CH}_2\text{Cl}_2,\ \text{RT}]{\text{ZnCr}_2\text{O}_7.3\text{H}_2\text{O}} \underset{70\text{–}90\%}{C_6H_5C(=O)C_6H_5}$$

Reference

1. H. Firouzabadi, A.R. Sardarian, H. Moosavipour and G.M. Afshari, *Synthesis*, 1986, 285.

2

Oxidation of Specific Types of Organic Compounds

2.1 OXIDATION OF HYDROCARBONS

2.1.1 Oxidation of Alkanes (Hydrocarbons)

Generally speaking alkanes, as a class are characterised by a general inertness to many chemical reagents. In alkanes, the carbon-carbon and carbon-hydrogen bonds are quite strong; they do not break unless alkanes are heated to very high temperatures. Alkanes, as we know are uneffected by most bases, acids and oxidising agents under normal conditions. Under vigorous conditions, both chromic acid and permanganate attack alkanes, but the reaction is of little synthetic value since usually mixture of products are obtained.

It is however, well-known that methane, an alkane can be oxidesed into methyl alcohol and formic acid with oxygen in persence of cataly (for detail *see* oxygen section 1.88).

Aromatic Hydrocarbons

Benzene ring, in the absence of any activating hydroxyl or amino substituents is only very slowly attacked by chromic acid or permanganate However, under certain conditions benzene can be oxidised by oxygen in presensce of V_2O_5 at 400-500°C to malic anhydride, which on hydrolysis gives malic acid. Also toluene can be oxidised in vapour phase in presence of catalysts (MnO_2)Mo_2O_3/ZrO_2) at 500°C into benzaldehyde. Similarly, naphthalene can be oxidised (air/V_2O_5/500°C) into phthalic anhydride. In place of tolune, o-xylene can also be used (for more details *see* oxygen, section 1.88).

Benzene is also converted[1] into *trans*, *trans*-muconic acid by micrococcus spheroids like organism.

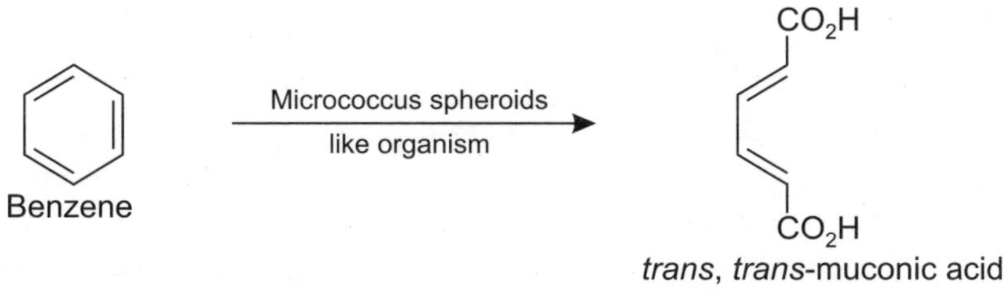

Benzene —(Micrococcus spheroids like organism)→ *trans, trans*-muconic acid

It has been reported[2] that oxygen assisted oxidation of benzene in presence of pseudomonas putida gives *cis* diol.

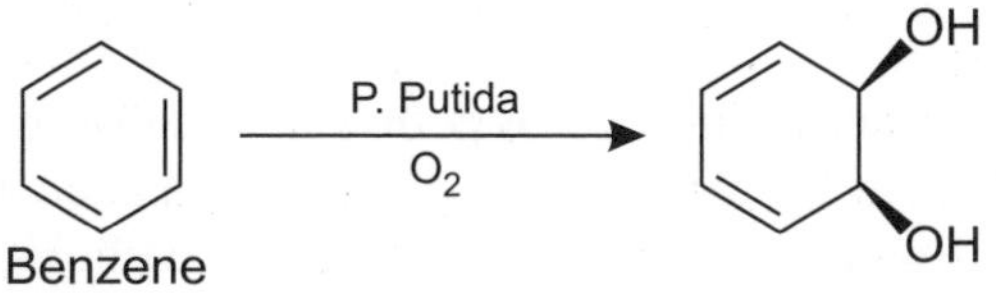

Benzene —(P. Putida, O_2)→

The *cis* diol obtanied above can be converted by four steps into 1,2,3,4-tetra hydroxy compound, conduritol-F, and by five steps into hexahydroxy compound, pinitol[4], an antidiabetic agents[5].

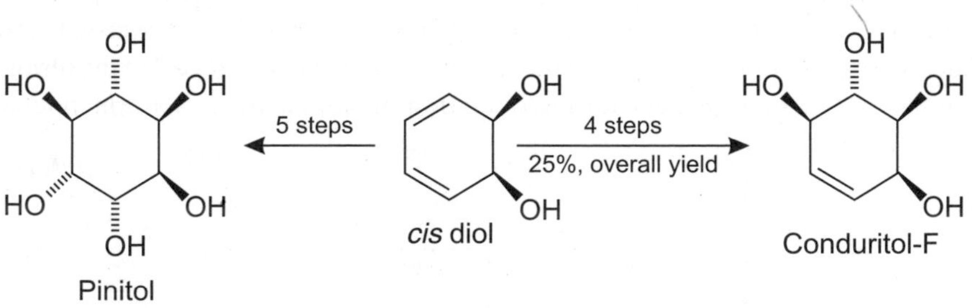

Pinitol ←(5 steps)— *cis* diol —(4 steps, 25%, overall yield)→ Conduritol-F

In a similar way, chlorobenzene and 4-chlorotoluene gave the diols[6,7].

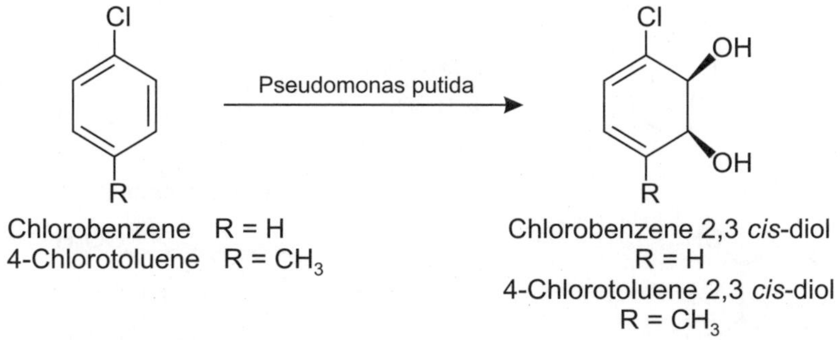

Chlorobenzene R = H
4-Chlorotoluene R = CH_3

—(Pseudomonas putida)→

Chlorobenzene 2,3 *cis*-diol R = H
4-Chlorotoluene 2,3 *cis*-diol R = CH_3

The chlorobenzene 2,3-*cis* diol obtained above is converted in 4 steps (18% overall yield) to 2,3-isopropylidene-L-ribose-γ-lactone[8,9].

Chlorobenzene 2,3-*cis* diol $\xrightarrow[\text{22\% Yield}]{\text{4 steps}}$ 2,3-Isopropylidene--L-ribose-γ-lactone

The use of the *cis* diols derived from functionalized aromatic compounds is an important method for the preparation of molecules used in asymmetric synthesis.

In the absence of activating hydroxyl or amino substituents, benzene rings are only slowly attacked by chromic and or permanganate. However, alkyl side chains are degraded with the formation of benzene carboxylic acids. This is a useful method for the prepation of benzene carboxylic acids. The oxidation of the side chain of benzene derivatives can be affected with aqueous potassium permanganate, or with chromic acid in aqueous solution or with sodium dichromate at elevated temperatures or with nitric acid. In case of aqueous potassium permanganate is best to carry out the oxidation in presence of a crown ether. Under these conditions, permanganate becomes, soluble in benzene and the resulting solutions is excellent reagent for the oxidation of side chain in benzene nucleus.

Reference

Toluene $\xrightarrow[\text{dicyclohexyl-18 crown –6}]{C_6H_6,\ \text{aq. } KMnO_4}$ Benzoic acid (100%) — 10

$C_6H_5C_8H_{17}$ $\xrightarrow[\text{aq. } H_2SO_4]{CrO_3,\ CH_3COOH}$ Benzoic acid (50%) — 11

2,3-Dimethylnaphthalene (CH_3, CH_3) $\xrightarrow[(2)\ H_3O^+]{(1)\ Na_2Cr_2O_7,\ H_2O,\ 250°C}$ naphthalene-2,3-dicarboxylic acid (CO_2H, CO_2H) — 12

$H_3C{-}C_6H_4{-}CH(CH_3)_2$ $\xrightarrow[H_2O,\ \text{reflux}]{HNO_3}$ $H_3C{-}C_6H_4{-}CO_2H$ (87–93%) — 13

The conversion of a methyl group attached to a benzene ring into the formyl group can be achieved by oxidation with chromium (VI) oxide in acetic anhydride[14] in presence of a acid or with a solution of chromyl chloride in carbon disulphide or carbon tetrachloride (the **Etard reaction**)[15,16].

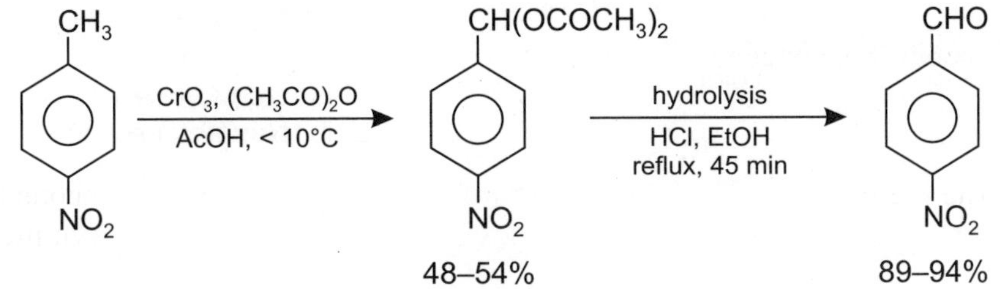

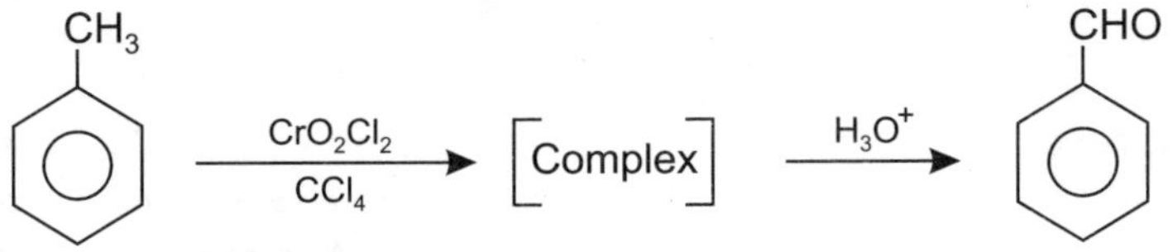

$$p\text{-}NO_2C_6H_4CH_3 \xrightarrow[AcOH,\ <10°C]{CrO_3,\ (CH_3CO)_2O} p\text{-}NO_2C_6H_4CH(OCOCH_3)_2 \xrightarrow[\text{HCl, EtOH, reflux, 45 min}]{\text{hydrolysis}} p\text{-}NO_2C_6H_4CHO$$

48–54%   89–94%

$$C_6H_5CH_3 \xrightarrow[CCl_4]{CrO_2Cl_2} [\text{Complex}] \xrightarrow{H_3O^+} C_6H_5CHO$$

The mechanism[17] of Etard reaction is given below.

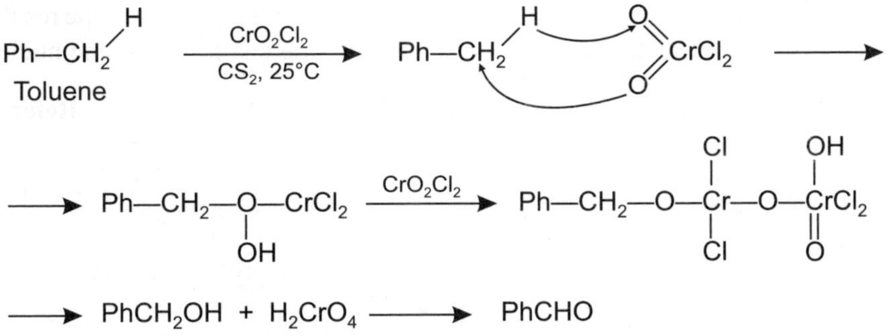

$$\underset{\text{Toluene}}{Ph{-}CH_2{-}H} \xrightarrow[CS_2,\ 25°C]{CrO_2Cl_2} Ph{-}CH_2{-}H \cdots O{=}CrCl_2{=}O \longrightarrow$$

$$\longrightarrow Ph{-}CH_2{-}O{-}CrCl_2(OH) \xrightarrow{CrO_2Cl_2} Ph{-}CH_2{-}O{-}CrCl_2{-}O{-}CrCl_2(OH)(=O)$$

$$\longrightarrow PhCH_2OH + H_2CrO_4 \longrightarrow PhCHO$$

Aromatic compounds can be oxidised to **quinones.** Some examples are given below.

Reference

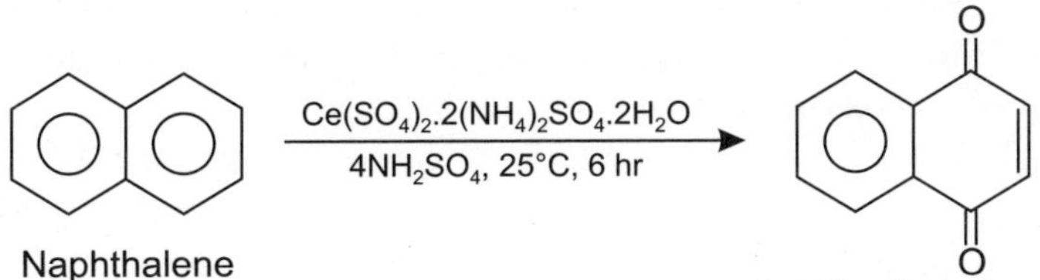

$$\underset{\text{Naphthalene}}{C_{10}H_8} \xrightarrow[4NH_2SO_4,\ 25°C,\ 6\ hr]{Ce(SO_4)_2.2(NH_4)_2SO_4.2H_2O} \underset{\text{1,4-Napthoquinone 90–95\%}}{C_{10}H_6O_2}$$ 18

Reference

2,3-Dimethylnaphthalene $\xrightarrow[\text{20–30°C, 15 min; RT, 3 days}]{CrO_3, AcOH}$ 2,3-Demethyl-Naphthoquinone 19

Anthracene $\xrightarrow[Bu_4NHSO4, 70°C, 2 min]{Na_2Cr_2O_7, H_2SO_4}$ Anthraquinone (82–92%) 20

Chrysene $\xrightarrow[Bu_4NHSO_4, 70°C, 2 min]{K_2Cr_2O_7, H_2SO_4, H_2O}$ 5,6-Chrysenequinone (79%) 21

Oxidative Cleavage of Benzene Ring

The cleavage of benzene ring is very rare. It can only be cleaved to aldehyde by ozonolysis. A classical example is the cleavage of o-xylene[22] to a mixture of glyoxal, methyl glyoxal and diacetyl. An interesting example is the conversion of phenanthrene into o–o′-diformyl biphenyl[23] (diphenylaldehyde) or o-formylbiphenyl-o′-carboxylic acid.

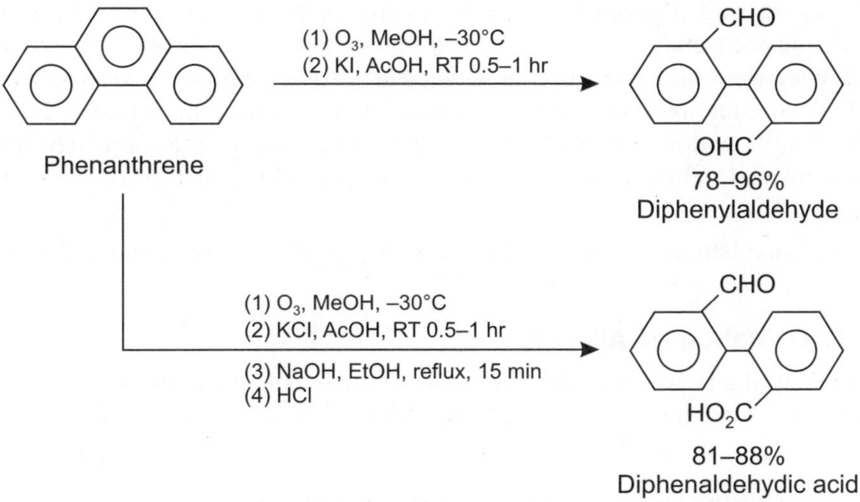

Phenanthrene

(1) O_3, MeOH, –30°C
(2) KI, AcOH, RT 0.5–1 hr

CHO
OHC
78–96%
Diphenylaldehyde

(1) O_3, MeOH, –30°C
(2) KCl, AcOH, RT 0.5–1 hr
(3) NaOH, EtOH, reflux, 15 min
(4) HCl

CHO
HO_2C
81–88%
Diphenaldehydic acid

It is however well-known that naphthalene[24], 1–nitronaphthalene[25], quinoline[26] and oxidised to the corresponding o-dicarboxylic acids.

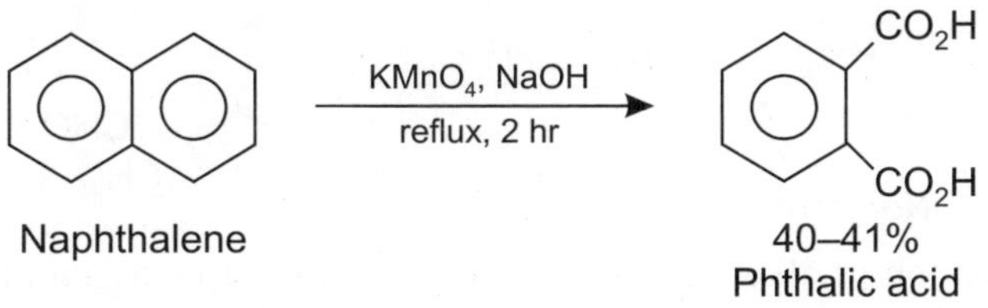

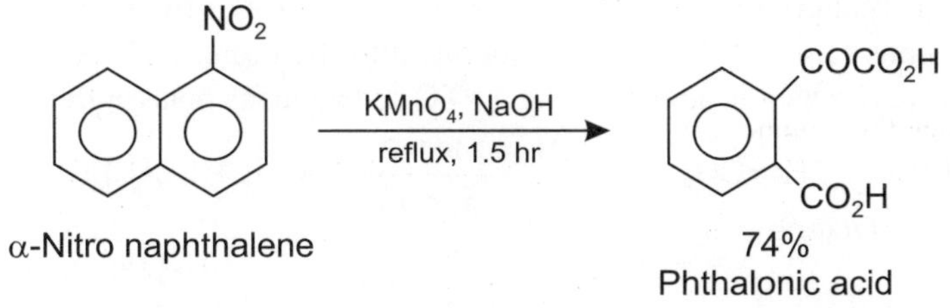

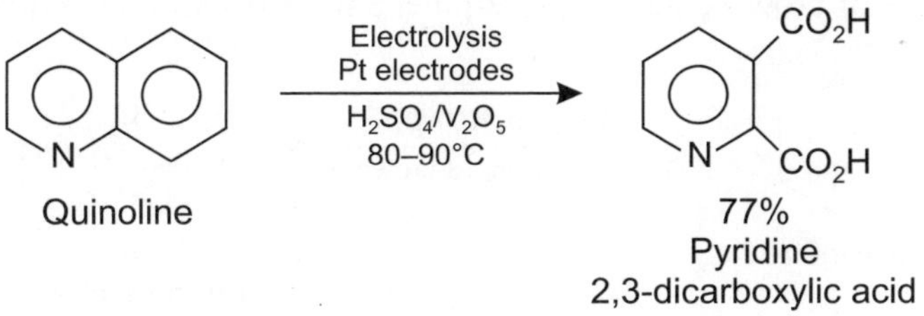

Naphthalene

$KMnO_4$, NaOH
reflux, 2 hr

CO_2H
CO_2H
40–41%
Phthalic acid

NO_2
α-Nitro naphthalene

$KMnO_4$, NaOH
reflux, 1.5 hr

$COCO_2H$
CO_2H
74%
Phthalonic acid

N
Quinoline

Electrolysis
Pt electrodes
H_2SO_4/V_2O_5
80–90°C

CO_2H
N
CO_2H
77%
Pyridine
2,3-dicarboxylic acid

Hydroxylation of Aromatic rings

On heating with potassium hydroxide, hydroxylation of aromatic rings occurs. Thus, nitrobenzene is converted into o-nitrophenol on heating with potassium hydroxide for 2 hr. at 60–70°C in 45% yield[27]. Also, quinoline gives carbostyril (2-hydroxyquinoline) on heating with potassium hydroxide for 3 hr. to 225°C in 77% yield[28]. Hydroxylation of both α-and β-anthraquinone sulfonic acid by alkali fusion to yield alizarin (1,2-dihydroxyanthraquinone) is historically important reaction.

Hydroxylation of aromatic rings with enzymes has already been described (*see* page 207).

2.1.2 Oxidation of Alkenes

Oxidation of alkenes may take place both at the double bond and at the adjacent allylic positions. Important synthetic reactions of each type are known involving various oxidising agents.

2.1.2.1 Oxidation of carbon-carbon double bonds

The carbon-carbon double bonds in alkenes can be fissioned by neutral solution of potassium permanganate. Thus, norbornene on oxidation with neutral potassium permanganate gives the dialdehyde.

$$\text{Norbornene} \xrightarrow[\text{Neutral } KMnO_4]{[O]} \text{(dialdehyde: CHO, CHO)}$$

The oxidation of alkenes with aqueous solution of potassium permanganate can also be affected in presence of an PTC[29] and crown ethers. The catalytic action of the quaternary salt (PTC) is due to the ability of their soluble cations to transfer anions (*e.g.*, MnO_4^-) from aqueous into the organic phase. For example, 1–octene gives heptanoic acid by using a PTC and aqueous potassium permanganate.

$$\underset{\text{1-Octene}}{CH_3(CH_2)_5CH{=}CH_2} \xrightarrow[CH_3(CH_2)_{15}\overset{+}{N}(CH_3)_3Cl]{C_6H_6,\ aq.\ KMnO_4} \underset{\substack{95\% \\ \text{Heptanoic acid}}}{CH_3(CH_2)_4CH_2CO_2H}$$

Another example of carbon-carbon bond fission is oxidation with permanganate is in presence of crown ether. Thus, α-pinene is oxidised to *cis* norpinonic acid.

$$\text{α-Pinene} \xrightarrow[\text{Dicyclohexano-18-crown-6}]{C_6H_6,\ aq.\ KMnO_4} \underset{\substack{90\% \\ cis\text{-norpinonic acid}}}{\text{(O, COOH)}}$$

In place of phase transfer catalyst or crown ether, it is also possible to carry out the oxidations by using permanganate salts, which are soluble in organic solvents. One such example is tetra-*n*-butylammonium permanganate[30].

An interesting modification of the permanganate oxidation procedure utilizes a mixture of potassium permanganate and potassium periodate in aqeuous solution at pH 7 for the oxidation of olefins[28a]. Under these conditions, the α-hydroxy-ketone and 1,2-diol resulting from the reaction of olefin with permanganate are cleaved by the periodate to form carboxylic acid.

$$CH_3(CH_2)_7CH{=}CH(CH_2)_7CO_2H \xrightarrow[K_2CO_3,\ H_2O]{NaIO_4,\ KMnO_4} \xrightarrow{H_3O^+}$$

$$\longrightarrow CH_3(CH_2)_7CO_2H + HO_2C(CH_2)_7CO_2H$$

The fission of the olefinic double bond with chromic acid, particularly when increasing number of phenyl groups are attached to the double bond becomes easier; the product obtained is the carboxylic acid. This procedure, known as **Barbier-Wieland degradation**[30a] is used for degradation of aliphatic acids (particularly in steroid chain) to the next lower homologhue. The various steps involved are given below.

$$RCH_2COOH \xrightarrow{H^+/EtOH} RCH_2COOEt \xrightarrow{2PhMgX}$$

$$RCH_2C(OH)Ph_2 \xrightarrow[-H_2O]{Ac_2O} \underset{\text{Olefin}}{RCH{=}CPh_2} \xrightarrow[CrO_3,\ H_2O,\ HOAc,\ 50°C]{[O]} RCOOH + Ph_2CO$$

As seen in the above B. W. degradation, the intermediate is an olefin ($RCH{=}CPh_2$), which on oxidation with chromic acid gives carboxylic acid.

Another example of the oxidation of the olefinic double bond is by chromyl chloride is methylene chloride to give an aldehyde. Thus 2-Phenyl propene is converted into 2-phenyl acetaldehyde[30b].

$$\underset{\text{2-Phenylpropene}}{C_6H_5C(CH_3){=}CH_2} \xrightarrow[CH_2Cl_2]{CrO_2Cl_2} [C_6H_5(CH_3)C{-}CH_2\ \text{epoxide-}O^+{-}CrCl_2{=}O] \longrightarrow \underset{\text{60\% 2-Phenylacetaldehyde}}{C_6H_5CH(CH_3){-}CHO}$$

In a similar way, oxidation of 2,4,4-trimethylpentene with chromyl chloride followed by treatment with zinc affords 2,4,4-trimethyl pental in 70-78% yield[30c].

$$\underset{\text{2,4,4-Trimethylpentene}}{H_2C{=}C(CH_3){-}CH_2{-}C(CH_3)_2{-}CH_3} \xrightarrow[(2)\ Zn]{(1)\ CrO_2Cl_2/CH_2Cl_2/0\text{–}5°C} \underset{\text{7–78\% 2,4,4-Trimethylpental}}{OHC{-}CH(CH_3){-}CH_2{-}C(CH_3)_2{-}CH_3}$$

Oxidative cleavage of double bonds can also be affected by ozonolysis and further treatment of the ozonide. For details *see* ozone, Section 1.90.

Alkenes can be oxidised with thallium nitrate. The nature of the product obtained depends an the structure of the alkene and the reaction conditions. For details *see* thallium nitrate, section 1.147.

2.1.2.2 Epoxidation

(*a*) Epoxidation with Peracids

A number of peracids having the general formula, RCO_3H have been used to affect epoxidation. Some common per acids are performic (HCO_3H), peracetic acid (CH_3CO_3H), perbenzoic acid ($C_6H_5CO_3H$), trifluoro peracetic acid (F_3CCO_3H), m-chloroperbenzoic acid [*m*-$ClC_6H_4CO_3H$) and monoperphalic acid (o–$HO_2CC_6H_4CO_3H$). Performic and peracetic acids are generally prepared *in situ* by the reaction of hydrogen peroxide on the corresponding carboxylic acids

$$RCO_2H + H_2O_2 \rightleftharpoons RCO_3H + H_2O$$

Peracids react with alkenes to give stable three-membered rings containing one oxygen atom, epoxides or oxiranes. The process is called epoxidation. It is believed that the reaction takes place by transfer of oxygen atom from peracid to the alkene. The reaction involves a one-step process. The epoxidation is stereospecific and leads to the syn addition of the oxygen atom

(EPOXIDATION)

The rate of epoxidation is increased in the presence of electron withdrawing groups in the peroxy acid; thus, trifluoro per acetic acid is more effective than peracetic acid. The rate of epoxidation is also increased if the alkene has electron-donating substituents. Thus, terminal alkenes react slowly with per acids compared to alkyl substituted alkenes like butene-2. Some examples of epoxidations are given below.

Reference 31

Reference

$$CF_2{=}CFCF_3 \xrightarrow[\text{-50 to -40°C, 4.5 hr, RT}]{\text{30\% } H_2O_2\text{, KOH, MeOH}} CF_2{-}CFCF_3 \text{ (epoxide, O bridge) } \quad 23\text{–}51\%$$ 32

$$CF_2{=}CHC_6H_{13} \xrightarrow[CH_2Cl_2\text{, 0°C, 1.8 hr}]{\text{90\% } H_2O_2\text{, maleic anhydride (CHCO)}_2\text{O}} CH_2{-}CHC_6H_{13} \text{ (epoxide) } \quad 80\%$$ 33

$$\text{Cyclohexene} \xrightarrow[CH_2Cl_2\text{, RT, overnight}]{mClC_6H_4CO_3H\text{, KF,}} \text{cyclohexene oxide} \quad > 95\%$$ 34

$$C_6H_5CH{=}CH_2 \xrightarrow[\text{NaOH (pH 7.5), 50°C, 6 hr}]{\text{50\% } H_2O_2\text{, MeCN, MeOH}} C_6H_5CH{-}CH_2 \text{ (epoxide) } \quad 74\%$$ 35

As see in some of the examples cited above, epoxidation has been affected with hydrogen peroxide in alkaline medium, in this case the reactive species is an anion ($-\bar{O}-O-H$)[36,37].

$$>C{=}C< + \bar{O}{-}O{-}H \longrightarrow >C{=}C< \;(:\ddot{O}:{-}O{-}H) \longrightarrow >C{-}C< \text{ (O bridge)} + \bar{O}H$$

In the presence of nitrile, hydrogen peroxide in alkali forms peroxycarboximidic acid, which in the presence of alkene form an epoxide and a carboxamide[35].

The epoxidations with peroxy acids are stereoselective and take place by syn-addition to the double bond; this has been established by x-ray analysis of the formed products. Thus, *cis*-alkene and trans alkene give only the *cis*-epoxide and *trans*-epoxide, respectively. This implies that the reaction is concerted. In fact, the one step mechanism retains the stereochemistry of the starting alkenes.

$$\text{RHC}{=}\text{CHR} \; (\textit{cis alkene}) \xrightarrow{C_6H_5CO_3H} \textit{cis}\text{-epoxide (R, R; H, H)}$$

cis alkene

R(H)C=C(H)R —$C_6H_5CO_3H$→ epoxide

trans alkene

For more applications of peracids for epoxidation *see* Peracids, section 1.97

(*b*) Epoxidation with dimethyldioxirane

Dimethyldioxirane is used for the epoxidations of olefins. Some examples[38,39] are given below.

$C_6H_5(H)C{=}C(H)C_6H_5$ —(dimethyldioxirane, $(H_3C)_2C(O)O$)→ Epoxide

Olefin — Epoxide

$C_6H_5(H)C{=}C(H)CO_2Et$ —(dimethyldioxirane, $(H_3C)_2C(O)O$)→ epoxide

Ethyl *trans*-cinnamate

63%

Ethyl 2,3-epoxydihydro cinnamate

Dimethyldioxirane is also used for allenes to form spiro dioxides[40]

$H_2C = C = CH_2$ + Dimethyl dioxirane → Spirodioxide

Allene — Dimethyl dioxirane — Spirodioxide

(*c*) Epoxidation with tert-butyl hydroperoxide

Tert-butyl hydroperoxide $(CH_3)_3COOH$ is a new versatile reagent. It is used for epoxidations in presence of vanadium (V^{5+}) or molybdenum (Mo^{6+}) catalyst (vanadyl or molybdenyl acetyl acetates have been frequently used). Epoxidation with *t*-butylhydroperoxide is more stereoselective than with peracids.

For more details about epoxodation of alkenes with *t*-butylhydroperoxide *see* (*t*-Butyl hydroperoxide, section 1.19). Even enantioselective epoxidation which is normally achieved by enzymes or microogranisms can be achieved by t-butylhydroperoxide (*see* Section 1.19)

A useful new method for the asymmetric epoxidation of allylic alcohols has been developed by Sharpless (1980). It involves the use of t-butylhydroperoxide using a titanium complex as catalyst, $[Ti(OiPr)_4]$. Thus, the oxidation of allylic alcohols with t-butyl hydroperoxide in presence of either (+) or (–) diethyl tartrate and titanium tetraisopropoxide yields the corresponding asymmetric epoxidation product in high optical yield. In this chiral epoxidation, there is uniformly high asymmetric induction and the absolute configuration of the formed epoxide can be predicted. It has been established that for a given tartarate, the system delivers the epoxide oxygen from the same face of olefin regardless of the substitution pattern. If allyl alcohol is represented as shown below, (so that the CH_2OH groups is at the lower right), oxygen is delivered[41,42] from the top face in the presenc of D–(–) diethyl tartarate and from the bottom face in the presence of L-(+)-diethyl tartarate.

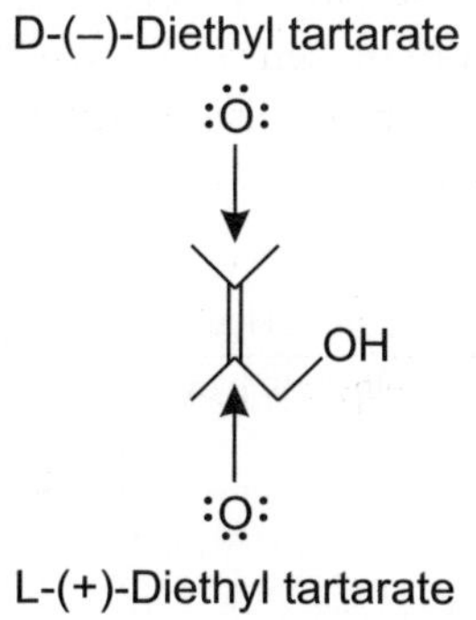

D-(–)-Diethyl tartarate

:Ö:

OH

:Ö:

L-(+)-Diethyl tartarate

A number of examples of highly enantioselective epoxidations of allylic alcohols have been reported. Some examples are given below.

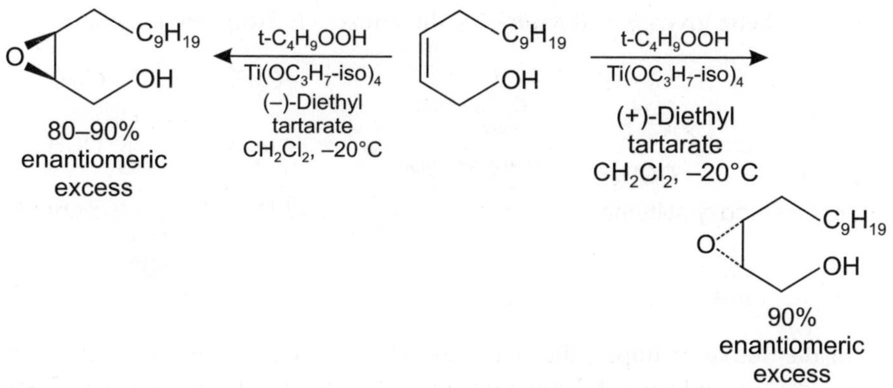

C_9H_{19} OH

t-C_4H_9OOH

$Ti(OC_3H_7\text{-iso})_4$

(–)-Diethyl tartarate

CH_2Cl_2, –20°C

80–90% enantiomeric excess

t-C_4H_9OOH

$Ti(OC_3H_7\text{-iso})_4$

(+)-Diethyl tartarate

CH_2Cl_2, –20°C

90% enantiomeric excess

CH_3 OH → t-C_4H_9OOH, Ti(O i-Pr)$_4$ / CH_2Cl, –20°C, (+)-Diethyl tartarate → CH_3 O H OH

Geraniol 2S,3S-Diastereomer

For more details see chiral oxidants section 1.26

2.1.2.3 Formation of Dioxetanes

Dioxetanes are compounds containing two atoms of oxygen in four membered ring. These are formed by the addition of one molecule of oxygen (singlet oxygen, generated chemically or photochemically) across a double bond. Thus, the oxidation of 1,2-diphenylcyclobutene with singlet oxygen generated by the irradiation of the solution of the compound in presence of a sensitizer give dioxetane, which is an intermediate in the conversion of unsaturated hydrocarbons into diketones[43].

H_5C_6 C_6H_5
1, 2-Diphenylcyclobutene

O_2, hν, CH_2Cl_2 / methylene blue → [C_6H_5 C_6H_5 O—O] → $C_6H_5CO-CH_2-CH_2-COC_6H_5$ 40%

The stereospecific addition of singlet oxygen to cis and trans 1,2-di methoxystilbene give *cis*-and *trans*- 1,2-dimethoxy–1, 2-diphenyl oxetane[44].

$C_6H_5(CH_3O)C=C(C_6H_5)OCH_3$
α, α′-Dimethoxy stilbene

cis
trans

O_2, hν, Et_2O / 15% MOH / Methylene blue / RT, 14 min →

1,2-Dimethoxy-1,2-diphenyl oxetane
cis 83%
trans 75%

In the above example, the oxetanes obtained are stable compounds. In a similar way *cis*-and *trans*-1,2-diethoxyethylene under irradiation with rose bengal as a sensitizer in flurotrichloroethane at –78° gives *cis*-and *trans*- 1,2-diethoxy-dioxetane[42]

EtO, OEt, C=C, H, H — cis and trans → (O_2, hν, CHFCl–CHCl$_2$; –78°C, rose bengal) → cis and trans 1,2-diethroxydioxetane

An intersting example is the synthesis of biadamantylene dioxetane[46]. It is obtained by dissolving biadamantylene in methylene chloride at –78°C followed by triphenyl phosphite ozonide in the same solvent and finally treatment with a mixture of methonol and pyridine. The dioxetane decomposes at 150°C.

Biaadamantylene → (1) $(PhO)_3PO_3$, CH_2Cl_2, –78°C (2) MeOH-C_5H_5N (1:1). –78°C → 91% Biadamantylene dioxetane

In the above example, singlet oxygen in generated chemically from $(PhO)_3PO_3$. For more details *see* singlet oxygen, section 1.125.

2.1.2.4 Formation of Ozonide

Ozone reacts with alkenes to form ozonides, which are not isolated, but converted into aldehydes, alcohols, carboxylic acids. For mechanism of ozonolysis and details about ozonolysis and conversion of ozonides into other productus *see* ozone, section 1.90.

This procedure is used for oxidative cleavage of double bonds.

2.1.2.5 Oxidation of Alkenes to Ketones

Alkenes can be oxidised to ketones of same chain length by using salts of copper, palladium, and mercury as catalysts and air. Some examples are given below.

$$CH_2{=}CH\,R \longrightarrow CH_3COR$$

R	*Reagent*	*% Yield*	*Reference*
C_8H_{17}	O_2, $PdCl_2$, CuCl, H_2O, DMF, RT, 24 hr	63–73	47
C_6H_{13}	30%H_2O_2, $Pd(OAc)_2$, AcOH, 80°C, 6 hr.	96	48
C_6H_{13}	CrO_3, H_2SO_4, H_2O, $Hg(OCoEt)_2$, Me_2CO, 25°C, 4 hr.	82	49

The oxidation of alkenes to ketones by palladium (II) salts and oxygen in presence of copper salt is known as the **Wacker Process.** Terminal alkenes give predominantly the ketone isomer rather than the aldehyde product.

$$R\text{-}CH_2CH=CH_2 \xrightarrow[CuCl_2,\ O_2]{PdCl_2,\ H_2O} R\text{-}CH_2COCH_3$$

Termmal alkenes can also be oxidised to ketones by alkaline potassium permanganate

$$\underset{\text{2-Methylbutene}}{CH_3\text{—}CH_2\text{—}\overset{CH_3}{\overset{|}{C}}{=}CH_2} \xrightarrow[(2)\ H_3O^+]{(1)\ KMnO_4,\ OH^-,\ \Delta} \underset{\text{Ethylmethyl ketone}}{H_3CCH_2\text{—}COCH_3} + CO_2 + H_2O$$

2.1.2.6 Hydroxylation of Alkenes

During hydroxylation of alkenes, there is addition of two hydroxyl groups across the double bond; the products obtained are vicinal diols. The addition can occur either in the *syn* or the *anti* mode. In open chain alkenes (with the exception of terminal alkenes for which stereochemistry is irrevalent), *syn* hydroxylation of *cis* alkenes give erythro (or meso) diols and trans alkene give threo (or DL) diols. However, *anti*-hodroxylation of *cis* alkenes gives threo (or DL diols), whereas anti hydroxylation of trans alkenes gives erythro (or meso) diols. *Syn* hydroxylation of cycloalkenes give *cis* diols, whereas *anti* hydroxylation gives trans diols. All these are summarised in Table.1.

Nature of alkene	***Mode of addition***	***Nature of the diol obtained***
cis (symmetrical)	syn	meso
	anti	DL
trans (symmetrical)	syn	DL
	anti	meso
cis (unsymmetrical)	syn	DL-erythro
	arti	DL-threo
trans (unsymmetrical)	syn	DL-threo
	anti	DL-ethryo
Cyclic (3–7 membered)	syn	*cis*
	anti	*trans*

Syn Hydroxylation of Alkenes

Syn Hydroxylation of alkenes can be achieved by potassium permanganate or with osmium tetroxide (osmic acid).

Syn Hydroxylation with potassium permanganate.

Alkenes on treatment with dilute potassium permanganate solution in presence of alkali give 1,2-diols. In this reaction, the purple colour of potassium permanganate is decolourised. This change of colour is the basis of a test for the presenece of unsaturation (double or triple bonds) and is known as **Baeyers test.**

Following are given some examples of syn hydroxylation with potassium permanganate:

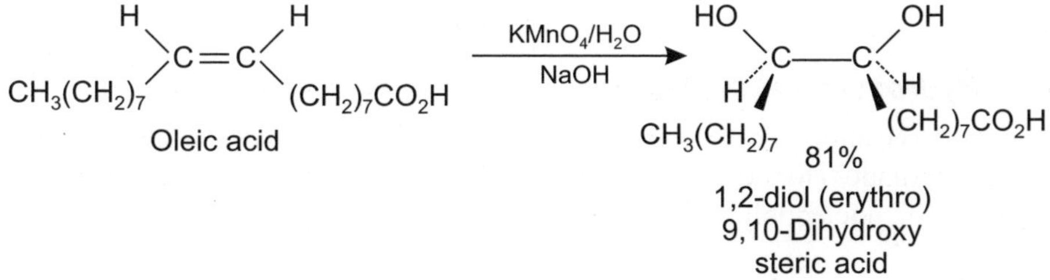
H
H
C=C
$CH_3(CH_2)_7$
$(CH_2)_7CO_2H$
Oleic acid
$KMnO_4/H_2O$
NaOH
HO
OH
C
C
H
H
$CH_3(CH_2)_7$
$(CH_2)_7CO_2H$
81%
1,2-diol (erythro)
9,10-Dihydroxy
steric acid

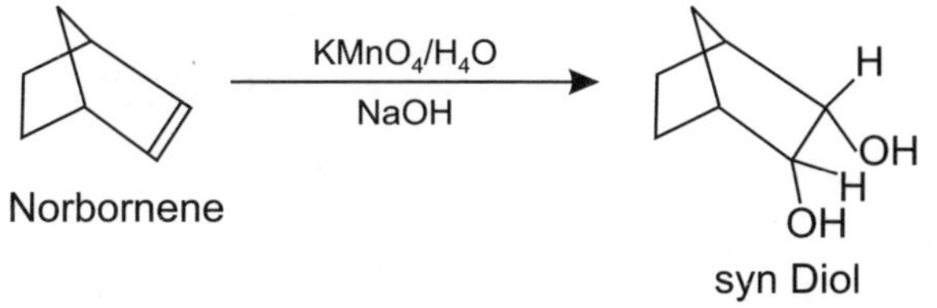
Norbornene
$KMnO_4/H_4O$
NaOH
H
OH
H
OH
syn Diol

The mechanism of syn hydroxylation with alkaline potassium permanganate is given below:

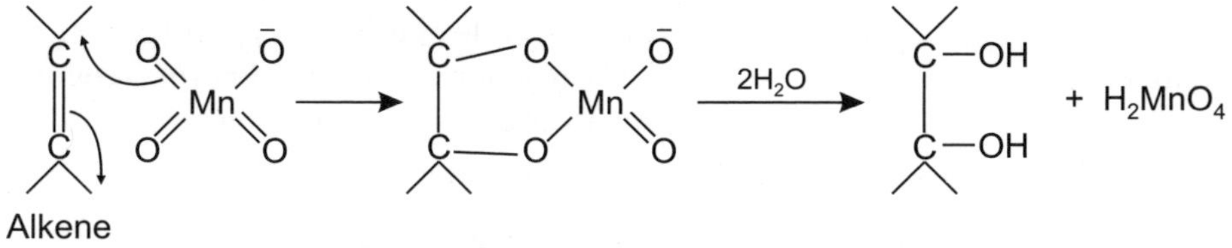
Mn
Mn
$2H_2O$
C—OH
C—OH
+ H_2MnO_4
Alkene

Syn hydroxylations are carried out by using dilute aqueous solutions (1–2%) of potassium permanganate at low temperatures (0–5°C). In this oxidation, yield of the dlios are generally low due to difficulties encountered in the isolation of water-soluble diols[50,51]. For more examples of oxidation by potassium permanaganate see potassium permanganate (section 1.110). In view of this, the alternative hydroxylation with osmium tetroxide is preferred.

Syn hydroxylation with osmium tetroxide.

Osmium tetroxide is toxic and very expensive so, usually catalytic amount of osmium tetroxide is used in presence of an oxidising agent, such as hydrogen peroxide, the mechanism of oxidation with osmium tetroxide is given below:

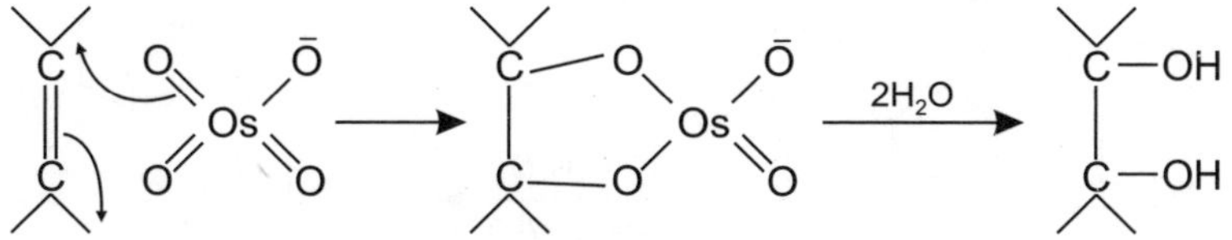
Os
Os
$2H_2O$
C—OH
C—OH

For more examples of oxidation of alkenes by osmium tetroxide, *see* section 1.88.

Syn hydroxylation of olefins has also been found to take place in the presence of enzymes (*see* section 3) and also by the Woodwards modification of the Prevost reaction involving oxidation of alkenes with iodine and silver carboxylate, (for details *see*, iodine and silver carboxylate section 1.59).

Anti Hydroxylation of Alkenes

Anti hydroxylation can be achieved by the reaction of alkenes with peracids to give the epoxides (for details *see* section 1.97) followed by hydrolytically cleaving by back side nucleophilic attack (inversion).

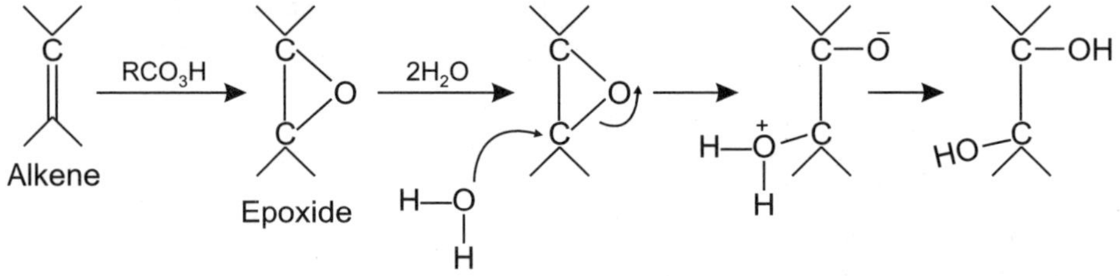

RCO_3H $2H_2O$

Alkene Epoxide

Alkenes can also be hydroxylated is the anti mode by hydrogen peroxide in the presence of vanadium pentoxide, molybdenum trioxide, selenium dioxide and tungsten trioxide[52,53]. Thus, cyclohexene is converted into DL-trans-1,2-cyclohexanediol[52].

Antihydroxylation of alkene can also be affected by the oxidation of alkenes with silver acetate or silver benzoate iodine followed by the hydrolysis of the intermediate ester under dry conditions. The reaction is known as **Prevost reaction**. For details of the process see iodine and silver carboxylate, section 1.59).

For more details *see* chiral oxidants, section 1.26).

2.1.2.7 Oxidation at Allylic Positions of Alkenes

Allytic positions of alkenes can be oxidised to give unsaturated hydroperoxides, alcohols, esters and carbonyl compounds. For this, singlet oxygen (*see* singlet oxygen section 1.125) generated by the irradiation of solution of oxygen in the presence of sensitizer is used. Thus, alkenes react as shown below[54,55].

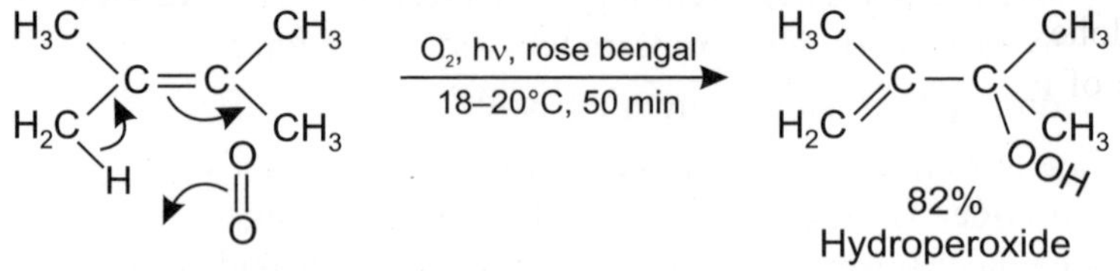

O_2, hν, rose bengal
18–20°C, 50 min

82%
Hydroperoxide

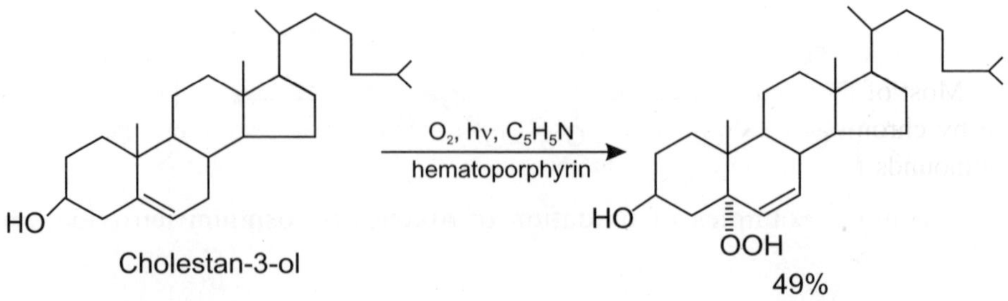

O_2, hν, C_5H_5N
hematoporphyrin

Cholestan-3-ol

49%

The hydroperoxides can be converted into alcohols[56] or carbonyl[57] compounds

OOH OH

O_2, hν / RT, 4 hr → NaOH / RT, 26 hr →

Cyclohexene 81% 82%

CHO

$AcOH, HClO_4$ / RT, 36 hr →

39%

Allylic esters are obtained by the reaction of alkenes with tert-butyl peroxyacetate[58] or peroxybenzoate[58] in presence of cuprous bromide

$t\text{-}BuO_2COPh$, CuBr / 80–82°C, 4 hr → $-OCOC_6H_5$

Cyclohexene 71–80%

The α, β-unsaturated ketones are obtained by the oxidation of allylic positions of alkenes and cycloalkenes. Thus, *trans*-5-decene is converted into *trans*-5-decen-4-one in 81% yield by oxidation with pyridine-α-seleninic anhydride prepared *in situ* from pyridine-α-diselenide and iodoxybenzene[59].

$E-C_4H_9CH=CHC_4H_9$ —(2-pyridyl-Se)$_2$, $PhIO_2$, C_6H_6 / 80°C, 81% → $(E)-C_3H_7COCH=CHC_4H_9$

trans-5-Decene

Cyclohexene can be oxidised into the allylic position by irradiation in the presence of mercuric bromide to give 2-cyclohexenone in 80% yield[60]

$HgBr_2$, t-BuOH / hν(254 nm), RT →

Cyclohexene 80%
2-cyclohexenone

Most of the allylic oxidations of methylene to carbonyl group can be carried out by chromium oxide or its complex with pyridine and perfomed on steroidal compounds (*see* chromium trioxide, section 1.31–1.33).

2.1.3 Oxidation of Dienes

Conjugated diene (and compounds that behave like conjugated dienes in Diels-Alder reaction) reacts with singlet oxygen (generated photochemically) to form cyclic peroxides. Thus, a dilute solution of isoprene in dichloromethane and methanol with methylene blue or rose bengal as sensitizer on irradiation in presence of oxygen with a 500-W iodine lamp yield 50% yield of 4-methyl-1,2-dioxa-4-cyclohexene[61].

H_3C Isoprene + O=O —(CH_2Cl_2, MeOH; hν, O_2; rose bengal)→ H_3C 50% 4-Methyl-1,2-dioxa-4-cyclohexene

In a similar way cyclic dienes are converted into bicyclic endoperoxides[61]

—(O_2, hν, iso PrOH; rose bengal, 6 hr)→ 35% endoperoxide

Using this procedure, naturally occurring ascaridole is synthesised from α-terpinene[62]

α-Terpinene —(O_2, hν, iso PrOH; Chlorophyll or eosin)→ 40% Ascaridole

Cyclopentadiene on oxidation with singlet oxygen and subsequent reduction with thiourea (one step reaction) yield unsaturated *cis* diols[63].

Cyclopenta diene —(O_2, hν, MeOH, $CS(NH_2)_2$; 0°C, 450 W high pressure Hg lamp with pyrex filter)→ [Unsaturated endoperoxide] → 60% Cyclopentene 3,5-*cis* diol (OH, OH)

Anthracene and its homolugues which behave like conjugated dienes in the Diels-Alder reaction also react-with siniglet oxygen to form 9,10-endoperoxides[64]

C_6H_5 ... O_2, microwave discharge, crystalline cellulose → C_6H_5 ... 79%

9,10-Diphenyl anthracene

9,10-Diphenyl anthracene 9,10-endoperoxide

The dienes can be hydroxylated at one or both double bonds depending on the amount of oxidant used. Thus, cyclo opentadiene on treatment with excess of hydrogen peroxide and osmium tetroxide as catalyst in tert-butyl alcohol gives[65] a mixture of 3,5–cyclopentanediol and 1,2,3,4-cyclopentanetetrol in 21 and 61% yields, respectively.

Cyclopentadiene — H_2O_2 (excess), OsO_4 → HO, OH: 21% 3,5-Cyclopentane diol + HO, OH, HO, OH: 61% 1,2,3,4-Cyclopentane tetrol

Oxidation of cyclopentadiene dimer with potassium permanganate and triethylbenzyl ammonium chloride (as PTC) gives 83% yield to exo, *cis*-diol; in this case hydroxylation of the double bond in the cyclohexene ring take place[66].

Cyclopentadiene dimer — (1) $KMnO_4$, $PhCH_2NEt_3Cl$, CH_2Cl_2, 0–3°C, 1.5 hr (2) 3% NaOH, RT, 8 hr → HO, HO: 83% exo, *cis*-diol

Ozonolysis of cyclic dienes give dialdehyde or 1,4-butane diol depending on the reaction conditions[67,68].

1,3-Cyclohexadiene — (1) 1.5 mol O_3, CH_2Cl_2, –78°C, 10 min (2) Me_2S, 0°C, overnight → CHO, CHO: 2-Hexenedial

1,3-Cyclohexadiene — (1) O_3 (excess), CH_2Cl_2, –78°C (2) $LiAlH_4$, Et_2O, –10°C to reflux, 15 min → CH_2OH, CH_2OH: 70% 1,4-Butanediol

2.1.4 Oxidation of Alkynes

Oxidation of diarylacetylenes (internal alkyne) with thallium nitrate in glyme solution containing perchloric acid gives α-diketones[69].

$$\underset{\text{Diphenylacetylene}}{C_6H_5C{\equiv}CC_6H_5} \xrightarrow[\text{glyme, reflux, 3 hr}]{Tl(NO_3)_3,\ 70\%\ HClO_4} \underset{\substack{85\% \\ \alpha\text{-Diketone}}}{C_6H_5COCOC_6H_5}$$

However, terminal alkynes on oxidation with potassium permanganate[80] give carboxylic acids with one less carbon atom.

$$C_8H_{17}C{\equiv}CH \xrightarrow[\substack{\text{AcOH, Aliquat 336,} \\ \text{cooling (ice), 5 hr}}]{KMnO_4,\ H_2O,\ C_5H_{12}} \underset{70\%}{C_8H_{17}CO_2H}$$

Terminal alkynes undergo oxidative coupling[71] to give diacetylenes by cupric salts.

$$\underset{\text{Phenyl acetylene}}{C_6H_5C{\equiv}CH} \xrightarrow[\text{40 min}]{O_2,\ CuCl,\ 34\text{–}40°C} \underset{\substack{70\% \\ \text{Diphenyl diacetylene}}}{C_6H_5C{\equiv}C{-}C{\equiv}CC_6H_5}$$

Long-chain terminal diacetylenes also undergo oxidative coupling to give[72] macrocyclic diacetlylenes and polyacetylenes.

$$HC{\equiv}C(CH_2)_{10}C{\equiv}CH \xrightarrow[\substack{\text{MeOH, } Et_2O \\ \text{heat, 72 hr}}]{Cu(OAc)_2,\ C_5H_5N} \underset{17\%}{\text{cyclo-}[(CH_2)_{10}{-}C{\equiv}C{-}C{\equiv}C]} + \underset{30\%}{\text{cyclo-}[(CH_2)_{10}{-}C{\equiv}C{-}C{\equiv}C{-}(CH_2)_{10}{-}C{\equiv}C{-}C{\equiv}C]}$$

Dialkyl alkynes on oxidation with thalium nitrate in presence of perchloric acid yield acyloins[69].

$$C_2H_5C{\equiv}CC_2H_5 \xrightarrow[\text{MeOH, RT, 30 min}]{Tl(NO_3)_3,\ HClO_4} \underset{70\%}{C_2H_5CH(OCH_3)COC_2H_5}$$

Oxidative rearrangement to esters has been observed[72] in alkylarylacetylenes

$$C_6H_5C{\equiv}CCH_3 \xrightarrow[\text{reflux, 2 hr}]{Tl(NO_3)_3,\ MeOH} \underset{80\%}{C_6H_5CH(CH_3)CO_2CH_3}$$

For more detail, about the oxidation of alkynes with thallium nitrate *see* thallium nitrate section 1.147.

References

1. A. Kleinzeller and Z. Fencl, *Chem. Listy,* 1952, **46**, 300; *Chem. Abstr.,* 1953, **47**, 4290.
2. D. G. H. Ballard, A. Courtis, I. M. Shively and S.C. Taylor, *J. Chem. Soc. Chem.,* Commun., 1983,954.
3. S.V. Ley and A. J. Redgrave, *Synlett*, 1990,393.
4. S.V. Ley and F. Sternfeld, *Tetrahedron Lett.,* 1988, **29**, 5305.
5. S.V.Ley, F. Sternfeld and S. Taylor, *Tetrahedron Lett.,* 1987, **28**, 225.
6. D. T. Gibson, J. R. Koch, C. L. Schuld and R. E. Kallia, *Biochemistry*, 1968,**7**,3795.
7. D. T. Gibson, M. Hensley, H. Yoshioka and T. J. Mabry, *Biochemistry*, 1970,**9**,1626.
8. T. Hudlicky and J. D. Price, *Synlett*, 1990,159.
9. T. Hudlicky, L. Luna and J. D. Price, *Tetrahedron Lett.,* 1989, **30**, 4053.
10. D. Sam and H. E. Simmons. *J. Am. Chem. Soc.,* 1972, **94**, 4024.
11. R. A. Pacaud and C. F. H. Allen, *Org., Synthesis,* Collective Volume 1943, **2**, 336.
12. L. Friedman, *Organic Synthesis,* 1963, **43**, 80.
13. W. F. Tuley and C. S. Marvel, *Org. Synth.,* Collective Volume, 1955, **3**, 822.
14. S. V. Lieberman, R. Connor, *Org. Synth.* Collective Volume, 1943, **2**, 441.
15. A. L. Etard, *Compd. Rend.,* 1880,**90**,1534; *Ann. Chem., Phys.,* 1881, **22**, 218;
16. Law and Perkin, *J. Am. Chem. Soc.,* 1944, **64**, 25.
17. W. R. Boon, S. *Chem. Soc.,* 1940, 5230.
18. M. Perviswamy and M. V. Bhatt. *Synthesis,* 1977, 330.
19. L. I. Smith and I. M. Webster, *J. Am. Chem. Soc.,* 1937, **59**, 662.
20. D. Pletcher and S. J. D. Tail, *J. Chem. Soc.,* Perkin Trans. 2, 1979,788.
21. R. P. Linstead and P Levine, *J. Am. Chem. Soc.,* 1942, **64**, 2022.
22. A. A. Levine and A. G. Cole, *J. Am.* Chem. Soc., 1932, **54**, 338.
23. P. S. Bailey and R. E. Erickson, *Org. Synth.* Collective Volume, 1973, **5**, 489.
24. J. H. Gardner and C. A. Naylor, Jr., *Org. Synth.* Collective Volume, 1943, **2**, 523.
25. D. Burn, V. Petrow and G. O. Weston, *Tetradron Lett.,* 1960,14.
26. M. Kulka, J. Am. *Chem. Soc.,* 1946, **68**, 2472.
27. A. Wohl, *Chem. Ber.,* 1899, **32**, 3486.
28. A. E. Tschitschibabin, *Chem., Ber.* 1923, **56**, 1879.

28a. R. U. Lemieux and E. Van Rudloff, Can *J. Chem.,* 1955, **33**, 1701; E. Von Rudloff, Can. *J. Chem.,* 1955, **33**, 1714; S.W. Pelletier, K. N. Iyer and C. W. J. Chang, *J. Org., Chem.,* 1970, **35**, 3535.

29. Bhushan, Rathore and Chandraelkaran, *Synthesis,* 1984, 431.
30. Sala and Sergent, J. Chem., *Chem. Commun.,* 1978, 253.

30 a. H. Wieland, *Ber.,* 1912, **45**, 484; P. Barbrer and R. Locquin, Compt., *Rend.,* 1913, **156**, 1443; C. Djerassi, *Chem. Rev.,* 1946, **38**, 526.

30 b. F. Freeman. R.H. DuBois and N. J. Yamachika, *Tetrahedron,* 1969, **25**, 3441; F. Freeman and N. J. Yamachika, *Tetrahedron Lett.,* 1969,3615; F. Freeman and N. J. Yamachika, *J. Am. Chem. Soc.,* 1970, **92**, 3730.

30 c. F. Frceman, R. H. Bu Bois and T. G. Mclaughlin, *Org. Synth.* Collective Volume, 1988, **6**, 1028.

31. L. A. Paquette and J. H. Barrett, *Org. Synth.* Collective Volume, 1973, **5**, 467.
32. A. Pasetti and D. Sianesi, Gazz, *Chim. Ital.,* 1968, **98**, 265.
33. R. W. White and W. D. Emmons. *Tetrahedron*, 1961, **17**, 31.
34. F. Camps, J. Coll, A. Messeguer and M. A. Pericias, *Tetrahedron Lett.,* 1981, **22**, 3895.
35. G. B. Payne, P. H. Deming and P. H. Williams. *J. Org., Chem.,* 1961, **26**, 659.
36. G. B. Payne and P. H. Williams, *J. Org. Chem.,* 1959, **24**, 54.
37. E. Weitz and A. Schefter, Chem., Ber., 1921, **54**, 2327.
38. R. W. Murray and R. Jeyaraman, *J. Org. Chem.,* 1985,**50**,2847.
39. D. J. Reif and H. O. House, *Org. Syn.* Collective Volume, 1963,**4**,860.
40. K. Gollnick and G. O. Schenek in 1,4-Cycloaddition reactions, J. Mammer (Ed.) *Academic Press, New York,* 1967 p. 255.
41. T. Kasturi and K. B. Sharpless, *J. Am. Chem., Soc.,* 1980, **102**, 5974.
42. B. E. Rossiler, T. Kasturi and K. B. Sharpless, *J. Am. Chem., Soc.,* 1981, **103**, 464.
43. A. G. Schultz and R. H. Schlessinger, *Tetrahedron Lett.,* 1970, 2721.
44. G. Rio and J. Berthelot, *Bull. Soc. Chim.,* Fr., 1971, 3555.
45. P. D. Bartlett and A. P. Schaap, *J. Am. Chem. Soc.,* 1970, **92**, 3223.
46. P. D. Bartlett, G. D. Mendenhall and D. L. Durham, *J. Org. Chem.,* 1980, **45**, 4269.
47. J. Tsujc, M. Nagashima and H. Nemoto, *Org. Synth.,* 1984, **62**, 9.
48. M. Roussel and H. Mimoun, *J. Qrg. Chem.,* 1980, **45**, 5387.
49. H. R. Rogers, J. X. McDermott and G. M. Whitesides, *J. Org. Chem.,* 1975, **40**, 3577.
50. P. D. Bartlett and A. Bavely, *J. Am. Chem., Soc.,* 1938, **60**, 2416.
51. K. B. Weberg and K. A. Saegebarth, *J. Am. Chem. Soc.,* 1957, **79**, 2822.
52. M. Mugdan and D. P. Young., *J. Chem., Soc.,* 1949, 2988.
53. G. B. Payne and P. H. Williams. *J. Org. Chem.,* 1959, **24**, 54.
56. E. H. Farmer and A. Sundralingam, *J. Chem. Soc.,* 1942, 121.
57. J. R. Scheffer and M. D. Ouchi, *Tetrahedron Lett.,* 1970, 223.
58. M. S. Kharasch, G. Sosnovsky and N. C. Yang, *J. Am. Chem. Soc.,* 1959, **81**, 5819
59. D. H. R. Barton and D. Crich, *Tetrahedron,* 1985, **41**, 4359.
60. N. Friedman, M. Gorodelsky and Y. Mazur, *Chem., Commun.,* 1971, 874.
61. K. Kondo and M. Matsumoto, *Chem. Commun.,* 1972, 1332.
62. G.O. Schenck, K. G. Kinkel and H.–J. Mertens, Justus Liebigs *Ann. Chem.,* 1953, **584**, 125.
63. C. Kaneko, A. Sugimoto and S. Tanaka, *Synthesis,* 1974, 876
64. J. R. Scheffer and M.D. Ouchi, *Tetrahedron Lett.,* 1970, 223.
65. N. A. Milas and L.S. Maloney, *J. Am. Chem., Soc.,* 1940, **62**, 1841.
66. T. Ogino and K. Mochizuki, *Chem., Lett.,* 1979, 443.
67. T. Hudlicky, H. Luna, G. Barbieri and L. D. Kwart, *J. Am. Chem. Soc.,* 1988, **110**, 4735.
68. G. B. Payne and P. H. Williams, *J. Org., Chem.,* 1959, **24**, 54
69. A. Mckillop, O. H. Oldenziel, B. P. Swann, E. C. Tayler and R. L. Robey, *J. Am. Chem. Soc.,* 1973, **95**, 1296.
70. A. P. Krapcho, J. R. Larson and J. M. Eldridge, *J. Org. Chem.,* 1977, **42**, 7749.
71. A. S. Hay., 1960, **25**, 1275.
72. E. Eglinton and A. R. Galbraith. *J. Chem. Soc.,* 1959,889.

2.2 OXIDATION OF ALCOHOLS

The oxidation of different types of alcohols depends on the nature of the oxidising agent and the reaction conditions.

2.2.1 Oxidation of Primary Alcohols

Primary alcohols can be oxidised to aldehydes, carboxylic acids and esters.

Oxidation of Primary Alcohols to Aldhydes

Following procedures are used for the oxidation of primary alcohols to aldehydes.

1. Catalytic dehydrogenation

Catalytic dehydrogenation of primary alcohols to aldehydes can be achieved by passing vapours of alcohols at 270–350°C over a catalyst supported on asbestos, silica gel, pumice etc. Thus, ethyl alcohol is converted[1] into acetaldehyde (38% yield) by passing its vapours at 270°C over mixtures of oxides of copper, cobalt and chromium on asbestos

$$\underset{\text{Ethylalcohol}}{CH_3CH_2OH} \xrightarrow[\text{on asbeseon}]{\text{270°C, oxides of Cu, Co, Cr}} \underset{\substack{88\% \\ \text{Acetaldehyde}}}{CH_3CHO}$$

An industrial dehydrogenation[2] is the conversion of methallyl alcohol to methacrolein .

$$\underset{\text{Methallyl alcohol}}{CH_2{=}CH(CH_3)CH_2OH} \xrightarrow{\text{Cu, 350°C, air}} \underset{\substack{95\%\text{ yield} \\ \text{Methacrolein}}}{CH_2{=}C(CH_3)CHO}$$

The above procedure is applicable[3] to saturated and unsaturated alcohols. Unsaturated primary alcohols, which cannot be vapourised can be dehydrogenated[4] by passing a current of air on alcohol through a solution of alcohol in heptane or ethylacetate in presence of platinum or better platinum dioxide

$$CH_3CH{=}C(CH_3)CH_2OH \xrightarrow[\text{60°C, 2 hr}]{O_2/PtO_2,\ C_7H_{16}} \underset{77\%}{CH_3CH{=}C(CH_3)CHO}$$

2. Oxidation with argentic oxide

Alcohols (aliphatic, aromatic as well as unsaturated) can be oxidised[5] in the liquid phase with argentic oxide in nitric acid or acetic acid from –10 to 60°C.

3. Oxidation with ceric ammonium nitrate

Benzylic alcohols are oxidised[6] with catalytic amount of ceric ammonium nitrate in presence of sodium bromate (as a reoxidant) to give the corresponding aldehyde.

$$\underset{\text{Benzylalcohol}}{C_6H_5CH_2OH} \xrightarrow[NaBrO_3,\ CH_3CN]{(NH_4)_2Ce(NO_3)_6} \underset{\substack{90\% \\ \text{Benzaldehyde}}}{C_6H_5CHO}$$

4. Oxidation with lead tetraacetate

Saturated and unsaturated primary alcohols are oxidised[7] to aldehydes

$$CH_2{=}CHCH_2OH \xrightarrow[\text{RT, 10 hr}]{Pb(OAc)_4,\ C_5H_5N} CH_2{=}CHCHO \quad 35\%$$

$$C_6H_5CH_2OH \xrightarrow[\text{RT, few hours}]{Pb(OAc)_4,\ C_5H_5N} C_6H_5CHO \quad 85\%$$

5. Oxidation with dinitrogen tetroxide

Dinitrogen tetroxide (obtained either commercialy or by heating lead nitrate) oxidises primary alcohols in chlorform at 0°C to aldehydes in excellent yields[8]

$$Cl{-}C_6H_4{-}CH_2OH \xrightarrow[\text{0°C, 15 min RT overnight}]{CHCl_3,\ N_2O_4} Cl{-}C_6H_4{-}CHO$$

p-Chloro benzylalcohol → 98% *p*-Chloro benzaldehyde

6. Oxidation with chromium (VI) reagents

A number of reagents have been used.

***(a)* Sodium or potassium dichromate:** It is used in presence of dilute sulphuric acid

$$3RCH_2OH + K_2Cr_2O_7 + 4H_2SO_4 \longrightarrow 3RCHO + Cr_2(SO_4)_3 + K_2SO_4$$

In this oxidation, the aldehydes are isolated by continuous distillation in order to prevent over oxidation.

'Neutral' sodium dichromate does not cause much overoxidation.

The oxidation of primary alcohol[10] with a solution of potassium dichromate in acetic acid in the presence of Adogen 464 (an PTC) and benzene give aldehyde at 55°C. Similar results are obtanied by using a chloroform solution of tetrabutylammonium chromate[11] at 60°C. The tetrabutylammanium chromate can easily by prepared in situ from sodium dichromate and tetrabutyl ammonium bisulphate in 3 M sulphuric acid in methylene chloride. Under such conditions, the oxidation is complete within minutes at room temerature.

***(b)* Pyridinium dichromate, $(C_6H_5NH)_2\ Cr_2O_7$:** It is prepared by the reaction of pyridine with a solution of chromium trioxide in minimum amount of water. It oxidises decanol in methylene chloride at 25°C to give 99% yield of decanal[12].

$$\underset{\text{Decanol}}{CH_3(CH_2)_8CH_2OH} \xrightarrow[CH_2Cl_2,\ 25°C]{(C_6H_5NH_2)_2Cr_2O_7} \underset{\text{99\% Decanal}}{CH_3(CH_2)_8CHO}$$

(*c*) **Jones reagent:** A solution of chromium trioxide in dilute sulphuric acid used in aqueous acetone is called **Jones reagent**[13]. It is used for the oxidation of primary alcohols to aldehydes.

(*d*) **Chromium trioxide adsorbed an silica gel:** With this reagent[14] all types of primary alcohols (saturated, unsaturated and aromatic) can be oxidised at room temperure to aldehydes

$$C_9H_{19}CH_2OH \xrightarrow[\text{RT, 5 min}]{10\%\ CrO_3/SiO_2,\ Et_2O} \underset{86\%}{C_9H_{19}CHO}$$

(*e*) **Collins reagent:** It is a complex of chromium troxide with two molecules of pyridine and is prepared[5] by adding in small portions, chromium trioxide to pyridine. The mixture is stirred and cooled in ice. It oxidises primary alcohols to aldehydes[16].

$$C_7H_{15}CH_2OH \xrightarrow[\text{RT, 15 min}]{CrO_3.2C_5H_5N,\ CH_2Cl_2} \underset{90\%}{C_7H_{15}CHO}$$

(*f*) **Pyridinium Chlorochromate, $C_5H_5NHCrO_3Cl$ (PCC):** It is prepared[17] by adding pyridine to the solution of chromium trioxide in 6 M hydrochloric acid and is superior to collins reagent. It oxidises even unsaturated primary alcohols to aldehydes[18].

$$C_5H_{11}C{\equiv}CCH_2OH \xrightarrow[\text{25°C, 1-2 hr}]{C_5H_5NH,\ CrO_3Cl} \underset{70\%}{C_5H_{11}C{\equiv}CCHO}$$

Since PCC is slightly acidic, it cannot be used to oxidise acid sensitive compounds. In such cases, the complex (PCC) is formed in the presence of alumina and then the reagent is stirred with alcohol at room temperature for 2 hr.; the formed aldehyde is isolated by filtration and evaporation of the filtrate[19].

(*g*) **Chromium trioxide–3,5-Dimethyl pyrazole complex:** It is obtained by adding 1 equivalent of 3,5-dimethyl pyrazole to a suspension of chromium trioxide in dichloromethane. It oxidises allylic and blenzylic alcohols to aldehydes[18].

$$RC{\equiv}CCH_2OH \xrightarrow[\text{RT, 30 min}]{C_5H_8N_2,\ CrO_3,\ CH_2Cl_2} \underset{70\%}{RC{\equiv}CCHO}$$

$C_5H_8N_2$ = (3,5-dimethylpyrazole: ring with CH_3, N, N–H, H_3C)

(*h*) **Poly (vinylpyridinium) chlorochromate:** It is prepared by the tretment of cross-linked poly (vinylpyridine) with chromium trioxide and hydrochloric acid. It is used for for the oxidation[20] of primary alcohols to aldehydes.

$$C_5H_{12}CH_2OH \xrightarrow[CH_2Cl_2,\ RT,\ 3\ day]{\text{Poly (vinylpyridinium) chlorochromate}} C_5H_{12}CHO \quad 91\%$$

(i) **Chromyl Chloride, CrO_2Cl_2:** It is prepared[21] from chromium trioxide or sodium dichromate, hydrochloric acid and sulphuric acid. It is adsorbed on alumina-silica gel from its solutions in dichlormethane, and this oxidises aliphatic and aromatic alcohols at room temperature in 77–100% yields[22].

(j) **Ditert-Butyl Chromate:** It is prepared *in situ* from chromyl chloride in tert-butyl alcohol at –70°C and is used for the oxidation[23] of primary alcohols to aldehydes in good yield

It is also prepared from chromium trioxide and tert-butyl alcohol. It oxidises[25] primary aliphatic and aromatic alcohols to the corresponding aldehydes at 1–2°C.

7. Oxidation with Hypochlorites

Potassium hypochlorite (contaning 24–35% available chroine) oxidises benzylic alcohols to aldehydes[24].

The tert-butyl hypochlorite oxidises[24] primary benzylic alcohols to aldehydes.

$$C_6H_5CH_2OH \xrightarrow[30\text{–}35°C,\ 20°C,\ 1\ hr]{t\text{-}BuOCl,\ C_5H_5N,\ CCl_4} C_6H_5CHO \quad 71\%$$

8. Oxidation with N-Chlorosuccinimide (NCS)

It oxidises[24] primary alcohols to aldehydes

$$C_6H_5CH_2OH \xrightarrow[boil,\ 15\ min]{C_5H_5N,\ NCS,\ C_6H_6} PhCHO \quad 89\%$$

9. Oxidation with Manganese dioxide, MnO_2

Manganese dioxide is commenly used as an oxidising agent[15,25–27] for the oxidation of simple and unsaturated primary alcohols to aldehydes.

$$\underset{\text{Cinnamyl alcohol}}{C_6H_5CH{=}CHCH_2OH} \xrightarrow[RT,\ 30\ min]{MnO_2,\ CCl_4} \underset{\text{Cinnamyl aldehyde}}{C_6H_5CH{=}CHCHO}$$

$$CH_2{=}CHCH_2OH \xrightarrow[RT,\ 19\ hr]{MnO_2,\ \text{Petroleum ether}} CH_2{=}CHCHO \quad 99\%$$

$$(CH_3)_2C{=}CHCH_2CH_2C(CH_3){=}CHCH_2OH \xrightarrow[RT,\ 96\ hr]{MnO_2,\ C_6H_{16}} (CH_3)_2C{=}CHCH_2CH_2C(CH_3){=}CHCHO \quad 63\%$$

$$C_6H_5CH_2OH \xrightarrow[\text{24 hr. chloroform}]{MnO_2,\ RT} C_6H_5CHO \quad 89\%$$

During oxidation with manganese dioxide, the configuration of the double bond is conserved. This is amply clear by the oxidation of Retinol (Vitamin A) to retinal

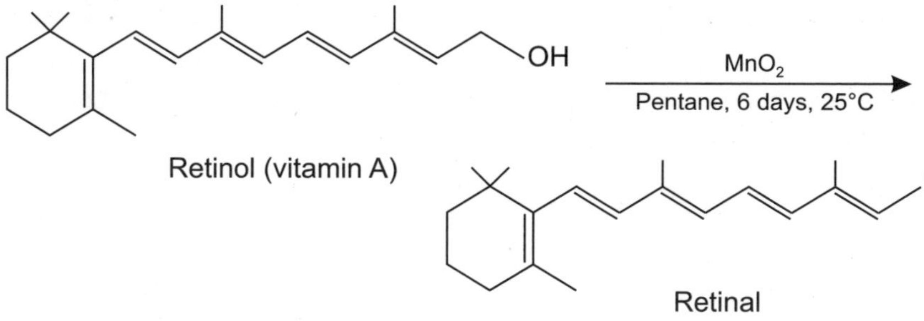

OH — MnO_2 / Pentane, 6 days, 25°C → CHO

Retinol (vitamin A)

Retinal

Mechanism

The oxidation with manganese dioxide is beleived[28,29] to proceed via a radical intermediate. Thus, the oxidation of benzyl alcohol involves adsorption on manganese dioxide, followed by the formation of a coordination complex (A). The coordination in this way allows transfer of electron, which generates a radical (B). This process is accompanied by reduction of Mn(IV) to Mn (III). This is followed by a second electron transfer, which generates the carbonyl product adsorbed on $Mn(OH)_2$(C). Finally, the product is adsorbed with loss of water to complete the oxidation.

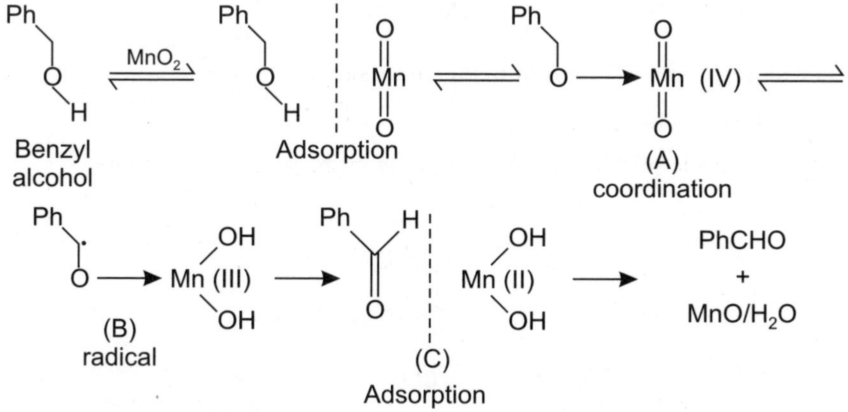

Ph O H — MnO_2 ⇌ Ph O H ¦ O=Mn=O ⇌ Ph O → Mn (IV) (O=Mn=O) ⇌

Benzyl alcohol — Adsorption — (A) coordination

Ph O → Mn (III) (OH)(OH) → Ph H O ¦ Mn (II) (OH)(OH) → PhCHO + MnO/H_2O

(B) radical — (C) Adsorption

10. Oxidation with Permanganates and Manganates

Manganates and permanganates are rarely used to convert alcohols into aldehydes because permanganate oxidises the formed aldehydes to acids. However, if potassium permanganate is adsorbed on molecular sieve, then the resultant oxidant oxidises benzyl alcohol and cinnamyl alcohol to the aldehydes[30] in 80 and 94%, yields respectively.

$$C_6H_5CH_2OH \xrightarrow[C_6H_6,\ 70°C,\ 2\ hr]{KMnO_4,\ molecular\ sieve\ 4Å} \underset{80\%}{C_6H_5CHO}$$

$$C_6H_5CH{=}CHCH_2OH \xrightarrow[C_6H_6,\ 70°C,\ 2\ hr]{KMnO_4,\ molecular\ sieve\ 4Å} \underset{90\%}{C_6H_5CH{=}CH{-}CHO}$$

Bis (2,2′-bipyridyl) Copper (II) permanganate (BBCP) oxidises[31] benzylic and allylic alcohols to aldehydes.

$$C_6H_5CH_2OH \xrightarrow[RT,\ 15\ min]{BBCP,\ CH_2Cl_2} \underset{95\%}{C_6H_5CHO}$$

$$PhCH{=}CHCH_2OH \xrightarrow[RT,\ 10\ min]{BBCP,\ CH_2Cl_2} \underset{91\%}{PhCH{=}CHCHO}$$

$$BBCP = (\text{2,2'-bipyridyl})_2\ Cu(MnO_4)_2$$

Similar results are obtained by using bis (2,2′-bipyridyl) silver permanganate[32].

For preparation of BBCP *see* bis (2,2′-bipyridyl) copper (II) permanganate, section 1.10.)

Barium manganate, $BaMnO_4$, oxidises benzylic alcohols to the aldehydes[33].

$$C_6H_5CH_2OH \xrightarrow[reflux,\ 9\ min]{BaMnO_4,\ C_6H_6} \underset{90\%}{C_6H_5CHO}$$

For more details *see* barium manganate, section 1.7).

11. Tetrachloro-o-benzoquinone

It oxidises[34] allylic primary alcohols into the corresponding aldehydes

$$\underset{\text{Cinnanyl alcohol}}{C_6H_5CH{=}CHCH_2OH} \xrightarrow[CCl_4]{\text{tetrachloro-o-benzoquinone}} \underset{\text{Cinnamaldehyde}}{C_6H_5CH{=}CHCHO}$$

20°C 15 hr	100%
35°C 8 hr	85%

Similar results are obtained by using tetrachloro-p-benzoquinone and 2,3-dichloro-5, 6 dicyano p-benzoquinone[34]

12. Dimethyl Sulfoxide (DMSO)

It is a very mild oxidant for the conversion of primary alcohols into aldehydes. The oxidation is performed in presence of an activator and a base. For details about the reaction mechanism and examples of oxidation see dimethyl sulfoxide, section 1.49

13. Benzenselenic Anhydride $(C_6H_5SeO)_2O$

The oxidation of primary alcohols to aldehydes can be accomplished by benzeneselenic anhydride (prepared *in situ* from diphenyl diselenide and tert-butyl hydroperoxide[35]

$$CH_3(CH_2)_8CH_2OH \xrightarrow[C_6H_6,\text{ reflux, 5 hr}]{Ph_2Se_2,\ t\text{-}BuOOH} CH_3(CH_2)_8CHO$$

14. Dess-Martin periodinane

It is used for the oxidation of alcohols under mild conditions and in high yields[36].

For more details *see* Dess-Martin periodinane section 1.43.

Oxidation of Primary Alcohols to Carboxylic Acids

For the oxidation of primary alcohols to carboxylic acids, normally the same oxidants that are used to oxidise alcohols to aldehydes are used in appropriate amounts, in different solvents, at higher tempreature or for longer reaction times. Following are given the procedures used for oxidation of primary alcohols to carboxylic acids.

1. Catalytic Oxidation

Primary alcohols are oxidised to carboxylic acids with air or oxygen with platinum-on-charcoal or better with platinum dioxide at 60°C for 2 hr[37].

$$C_{11}H_{23}CH_2OH \xrightarrow[60°C,\text{ 2 hr}]{O_2/PtO_2,\ C_7H_{16}} \underset{96\%}{C_{11}H_{23}CO_2H}$$

However, if the reaction is done for 15 minutes, only the aldehyde is obtained in 77% yield.

2. Potassium Hydroxide

At higher temperature, potassium hydroxide acts as an oxidant. For details *see* potassium hydroxide, section 1.106

3. Nitric Acid

The oxidation of primary alcohols to acids can be accomplished with nitric acid for details *see* nitric acid, section 1.85

4. Chromium Trioxide

A solution of chromium trioxide in water and sulphuric acid or acetic acid is used to oxidise primary alcohols to carboxylic acids[38]

$$RCH_2OH \xrightarrow[5°C,\text{ RT, overnight}]{4CrO_3,\ AcOH,\ H_2O} \underset{90\%}{RCOOH}$$

5. Pyridinium Dichromate

It converts primary alcohols in dichloromethane into aldehydes (*see* section 1.115). However, in dimethylformamide at 25°C, carboxylic acids are formed. Thus, cyclohexylmethanol gives cyclohexane carboxylic acid in 84% yield[12].

6. Manganese Dioxide

Normaly manganese dioxide oxidises primary alcohols to aldehydes (*see* section 1.75). However, under proper conditions, carboxylic acids are obtained[39].

$$\text{(HO)(CH}_2\text{OH)(CH}_2\text{OH)(H}_3\text{C)pyridinium chloride} \xrightarrow[\text{RT, 3 hr}]{\text{MnO}_2\text{, KOH, EtOH}} \text{(HO)(CO}_2\text{H)(CH}_2\text{OH)(H}_3\text{C)pyridine} \quad 82\%$$

For more details about application of manganese dioxode *see* section 1.75.

7. Permanganate Oxidation

Saturated and benzylic alcohols can be oxidised to carboxylic acid by permanganate. However, in unsaturated alcohols, there may be cleavage of the double bond. Thus, 2-benzoyl aminobutanol on oxidation gives α-benzoylaminobutyric acid in 67–72% yield.[40]

$$\underset{\text{2-Benzoyl aminobutanol}}{H_2NCH_2CH_2\underset{\substack{|\\ COC_6H_5}}{CH}—CH_2OH} \xrightarrow[40°C]{KMnO_4\text{, aq. NaOH}} \underset{\substack{67–72\%\\ \alpha\text{-Benzoylamino butyric acid}}}{H_2NCH_2CH_2\underset{\substack{|\\ COC_6H_5}}{CH}—CO_2H}$$

Sodium permanganate monohydrate can be used in non-aqueous media. This oxidant can oxidise even allylic alcohols[41].

$$CH_3(CH_2)_6CH_2OH \xrightarrow[69°C\text{, 5 hr}]{NaMnO_4.H_2O\text{, }C_6H_{14}} \underset{67\%}{CH_3(CH_2)_6CO_2H}$$

Cupric permanganate octa hydrate [$Cu(Mn_4)_2$. $8H_2O$] can oxidise saturated and benzylic alcohols at room temperatures or by refluxing in dichloromethane. Thus, decanol and benzyl alcohol on oxidation yield decanoic acid and benzoic acid, repectively at room temperature after 24 hr. in 81 and 84% yields, respecively[42].

Oxidations in homogeneous conditions can be accomplished by using tetrabutyl ammonium permanganate, which is soluble in dichloromethane, chloroform, acetone and pyridine; the catalyst is prepared by treating an aquous solution of potassium permanganate with tetrabutylammonium bromide and filtering the precipitated reagent[40,43].

The permanganate oxidations can also be carried out in a two-phase system by preparing the quaternary ammonium permanganate in situ by using only a catalytic amount of quaternary salt[44].

$$CH_3(CH_2)_6CH_2OH \xrightarrow[\text{Bu}_4\text{NBr (catalyst)}\ \text{RT, 3 hr}]{\text{KMnO}_4,\ \text{H}_2\text{O},\ \text{C}_6\text{H}_6} CH_3(CH_2)_6CO_2H\ \ 47\%$$

In place of quaternary ammonium salt, catalytic amount of crown ethers, for example, dicyclohexano-18-crown-6 can be used[44].

8. Nickel Peroxide, NiO_2

It is prepared by adding an alkaline solution of 6% sodium hypochlorite to a solution of nickel sulphate hydrate at 20°C. Oxidation of primary alcohols to carboxylic acids takes place in aqueous sodium hydroxide. This reagent can oxidise saturated, unsaturated and aromatic primary alcohols[45].

$$CH{\equiv}CCH_2OH \xrightarrow[\text{5°C, 0.5 hr.}]{\text{NiO}_2,\ \text{NaOH},\ \text{H}_2\text{O}} CH{\equiv}CCO_2H\ \ 50\%$$

9. Sodium and Potassium Ruthenate

Both these reagents oxidise alkylic and benzylic alcohols to carboxylic acids at room temperature. Thus, sodium ruthenate in 1 M sodium hydroxide oxidises cinnamyl alcohol into cinnamic acid in 70% yield[46].

$$C_6H_5CH{=}CHCH_2OH \xrightarrow[\text{1MNaOH}\ \text{10°C, 1 hr}]{\text{Sod. ruthenate}} C_6H_5CH{=}CHCOOH\ \ 70\%$$

If potassium ruthenate is used, only catalytic amount is required in presence of a persulphate, which reoxidises the reduced ruthenium salt[47].

Oxidation of Primary Alcohols to Esters

In oxidations under acidic conditions, esterification of the primary alcohols under acidic conditions may take place. Thus, ethanol on oxidation with peroxymonosulphate (oxone) in the presence of conc. sulphuric acid at 70°C gives quantitative yield of ethyl acetate[48].

In a similar way, butyl alcohol, on oxidation with sodium dichromate and sulphuric acid at 25–30°C gives 41–47% yield of butyl butyrate[49].

$$\underset{\text{Butyl alcohol}}{C_4H_9OH} \xrightarrow[< 20°C–35°C]{\text{Na}_2\text{Cr}_2\text{O}_7,\ \text{H}_2\text{SO}_4,\ \text{H}_2\text{O}} \underset{\text{41–47\%, Butyl butyrate}}{C_3H_7COOC_4H_9}$$

2.2.2 Oxidation of Secondary Alcohols

Secondary alcohols can be oxidised to ketones, α-hydroxyperoxides or to carboxylic acids.

Oxidation of Secondary Alcohols to Ketones

Most of the oxidants used for the oxidation of primary alcohols to aldehydes can be used for oxidation of secondary alchohols to ketones. Since, ketons are much less sensitive to overoxidation than aldehydes, more rigrous reaction conditions, *e.g.*, using excess of the oxidant, higher temperature and longer reaction times can be used.

Following methods are used for the oxidation of secondary alcohols to ketones:

1. Catayltic Dehydrogenation

Vapours of volatile alcohols are passed over copper[50] at 300°C or over copper chromite[51] at 275–325°C. For higher boiling alcohols, dehydrogenation can be affected in liquid phase by refluxing secondary alcohols with copper chromite in xylene[152] for 4–5 hr. Alternatively, high boiling secondary alcohols can be heated in paraffin oil with Raney nickel[53] and a catalytic amount of potassium hydroxide at 150–180°C.

Raney Ni, KOH (cat)
paraffin oil, 150–180°C, 6 hr
95%

Dehydrogenation can also be accomplished[54] in 90–100% yield by passing vapour of alcohol in a current of air over copper or silver catalyst at 350–400°C.

2. Catalytic Oxidation

Catalytic oxidations are affected by passing oxygen through solutions of alcohols in presence of metal catalysts. Thus, alcohols in ethylene carbonate are oxidised[55] by oxygen at room temperature in presence of sodium acetate and palladium chloride in 62–98% yield. Alternatively, oxygen is passed at room temperature under irradiation through a solution of catalytic ammounts of chloroplatinic acid and cuprous chloride; by this procedure secondary alcohols give up 98% ketones[56].

Oxidation of secondary alcohols in vapour phase[3] at 250–300°C or in liquid phase[57] at room temperature can be affected by oxides of copper and silver.

3. Photooxidation

Photooxidation of secondary alcohols to ketones can be affected by singlet oxygen via the formation of hydroperoxides[58].

$$\text{>—OH} \xrightarrow[\text{Sensitiser}]{O_2\ h\nu} \text{>C(O—OH)—OH} \longrightarrow \text{>=O} + H_2O_2$$

4. Ceric Ammonium Nitrate

Ceric ammonium nitrate oxidises[59] saturated and unsaturated secondary alcohols to ketones. The ceric salt is used only in catalytic amount with sodium bromate (as an reoxidant).

$$C_3C_7CHC_8H_{17}(OH) \xrightarrow[\text{MeCN, } H_2O, \text{ 80°C, 0.8 hr}]{0.2\ (NH_4)_2Ce(NO_3)_6,\ 2.0\ NaBrO_3} C_3C_7COC_8H_{17}\ \ 94\%$$

5. Oxidation with Chromium Reagents

(*a*) Chromium trioxide: It is the most common reagent for the oxidation[60] of secondary alcohols to ketone. The oxidation takes place in presence of sulphuric acid.

$$R_2CHOH + 2CrO_3 + 3H_2SO_4 \longrightarrow 3\ R_2C{=}O + Cr_2(SO_4)_3 + 6H_2O$$

The mechanism of the reaction is as given below:

$$R_2C(H){-}OH + CrO_3 \longrightarrow \underset{\text{ester}}{R_2C(H){-}O{-}Cr(OH)(=O)_2} \longrightarrow R_2C{=}O + Cr(OH)_2{=}O$$

Thus, p-substituted 1–phenylethanols are oxidised to the ketones.

$$CH_3{-}C_6H_4{-}CH(OH){-}CH_3 \xrightarrow{CrO_3,\ HClO_4,\ AcOH} CH_3O{-}C_6H_4{-}COCH_3$$

(*b*) Jones reagent (CrO_3, H_2SO_4, H_2O, acetone): It is used for the oxidation of secondary alcohols to ketones[61].

$$CH_3CH(OH)C{\equiv}CC_4H_9 \xrightarrow[5\text{–}10°C,\ 2.5\ hr]{CrO_3,\ H_2O,\ H_2SO_4,\ Me_2CO} CH_3COC{\equiv}CC_4H_9\ \ 74\%$$

(*c*) Tetrabutylammonium Chromate: Chromium trioxide is converted into tetrabutyl ammonium chromate by the addition of catalytic amount of tetrabutylammonium chloride (a PTC) in dichloromethane[62]. It is an excellent reagent for the oxidation of secondary alcohols to ketones.

(*d*) Sodium dichromate: Sodium or potassium dichromate in dilute sulphuric acid is a good reagent[63] for the oxidation of secondary alcohols to ketones.

$$\underset{\text{Menthol}}{\text{Menthol}} \xrightarrow{Na_2CN_2O_7/H_2SO_4} \underset{\text{Menthone}}{\text{Menthone}}$$

Menthol Menthone

(*e*) **Pyridinium chromate:** Pyridinium chromate[64] (prepared by adding 2 mol of pyridine to 1 mol of chromium trioxide) adsorbed on silica oxidises secondary alcohols to ketones.

(*f*) **Pyridinium dichromate:** Pyridinium dichromate[12] (prepared from chromium trioxide in minimum amount of water and pyridine) oxidise allylic secondary alcohols more rapidly than their saturated analogues.

$$CH_2{=}CHCH(OH)C_5H_{11} \xrightarrow[25°C,\ 9\ hr]{1.5(C_2H_5NH)_2Cr_2O_7,\ CH_2Cl_2} CH_2{=}CHCOC_5H_{11}\ \ 80\%$$

(*g*) **Collins reagent:** It is complex of chromium trioxide with two molecules of pyridine [$CrO_3 \cdot 2C_5H_5N$] and is useful for the oxidation of acid-sensitive secondary alcohols to ketones[15].

(*h*) **Chromium Trioxide –3,5–dimethylpyrazole complex:** The 1 : 1 complex[18] generated *in situ*) oxidises secondary alcohols to ketones in 10 min. at room temperatures giving ketones in good yield (78–100%).

RR′CHOH
N
O
O
Cr—OH
N
N
H
Cr
CRR′
O
NH
+
O=CRR′
Cr:
N
+
Cr
OH

(*i*) **Pyridinium chlorochromate (PCC):** PCC (for preparation, *see* section 1.114) oxidises[17] secondary alcohols to ketones in dichloromethane at room temperature within 1–2 hr. in good yields.

This reagent (PCC) can also be adsorbed on silica gel[19] or formed with cross-linked poly (vinylpyridine) resin[20]; these adsorbed resins are good for the oxidation of secondary alcohols.

(*j*) **Tetrabutylammonium chlorochromate:** It is prepared[65] by adding a solution of chromium trioxide in concentrated hydrochloric acid to a solution of tetrabutylammonium hydrogen sulphate in hydrochloric acid (5N). The secondary alcohols on refluxing with the reagent in chloroform affords ketone in good yield.

6. Oxidation with Halogens

Chlorine in carbon tetrachloride oxidises 5α-cholestane–3β, 19 diol in chloroform below 30°C giving 5α-cholestane –17–ol–3-one after 15 min in 77% yield[66].

Oxidation with chlorine (or bromine) in presence of hexamethylphosphoric triamide (HMPA) also give similar results.

The addition of a solution of chlorine in chloroform to a solution of secondary alcohol in a mixture of HMPA, dichlormethane and a solution of sodium dihydrogen posphate in water at 0°C give ketones in good yield[67].

$$\text{Cyclohexanol (H, OH)} \xrightarrow[\text{CH}_2\text{Cl}_2,\ \text{H}_2\text{O}]{\text{Cl}_2,\ \text{CHCl}_3,\ \text{HMPA}} \text{Cyclohexanone (=O)}\quad 81\%$$

7. Oxidation with Hypochlorites

Sodium hypochlorite in acetic acid oxidises secondary alcohols to ketones in few minutes[68].

Calcium hypochlorite in presence of an ion exchanger (1RA 900) oxidises secondary alcohol (*e.g.*, cyclohexanol to ketone cyclohexanone) at room temps in good yields (60–98%)[69]. Even tert-butyl hypochlorite in carbon tetrachloride in presence of pyridine oxidises cyclohexonal to cyclohexanone[24].

With bromine in dichloromethane in presence of hexamethylphosphoric triamide (HMPA) and aqueous sodium bicarbonate the oxidation of secondary alcohol is selective[67].

$$\underset{\text{OH}}{\text{CH}_3\text{CHC}_6\text{H}_{13}} \xrightarrow[\text{H}_2\text{O, NaHCO}_3,\ 0\text{–}5^\circ\text{C}]{\text{Br}_2,\ \text{CH}_2\text{Cl, HMPA}} \underset{95\%}{\text{CH}_3\text{COC}_6\text{H}_{13}}$$

8. Oxidation with Sodium Bromite and Sodium Bromate

Sodium bromite oxidises[70] secondary alcohols in preference to primary alcohol. Menthol on treating with sodium bromite in aqueous acetic acid gives 86% yield of menthone at room temperature after 5.5 hr.

Such an oxidation is also possible by sodium bromate[71] in presence of catalytic amount of ceric ammonium nitrate or ceric sulphate.

9. Oxidation with Managanese Dioxide

As with primary alcohols (*see* section 2.2.1), secondary alcohols are also oxidised to ketones. The oxidation is accelerated by removal to the formed water by azeotropic distillation[72] with benzene.

$$(\text{CH}_3\text{O})_2\text{C}_6\text{H}_3\text{–}\underset{\text{OH}}{\text{CH}}\text{–CH}_3 \xrightarrow[\text{RT, 15 hr.}]{\text{excess MnO}_2,\ \text{Me}_2\text{CO}} (\text{CH}_3\text{O})_2\text{C}_6\text{H}_3\text{–COCH}_3$$

Oxidation with manganese dioxide can also be used for the preparation of unsaturated and acetylenic ketones[73].

$$CH_3CH{=}CH\underset{\displaystyle OH}{\underset{|}{C}}HC{\equiv}COEt \xrightarrow[\text{RT, 4.2 hr}]{\text{excess } MnO_2,\ CH_2Cl_2} CH_3CH{=}CHCOC{\equiv}COEt$$

40%

10. Oxidation with Permanganates and Manganates

Potassium per manganate in water solution is generally not used for the oxidation of secondary alcohol; this is due to further oxidation of the formed ketones to carboxylic acids[74]. However, oxidation of secondary alcohols to ketones can be carried out in organic solvents with solid or supported permanganates. Thus, if potassium permanganate is adsorbed on molecular sieve, then the resultant oxidant oxidises secondary alcohols to ketones in 90% yield[30]. Still better results are obtained with potassium permanganate in toluene in presence of neutral alumina[75].

It is found that copper sulphate pentahydrate in presence of potassium permanganate also affects the oxidation in good yields[40].

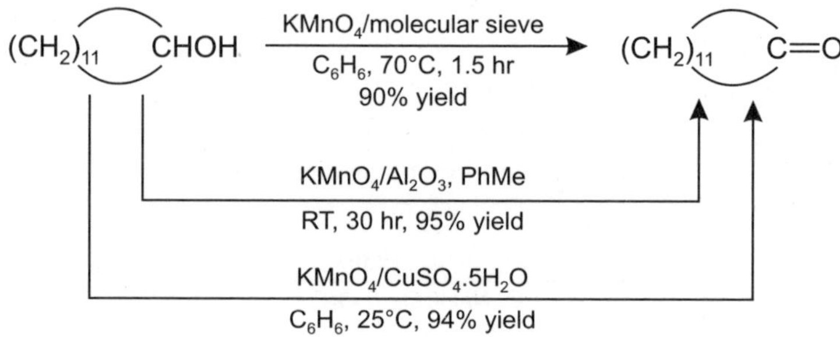

Similar results are obtained with copper permanganate octahydrate[42]

$$CH_3(CH_2)_7\underset{\displaystyle OH}{\underset{|}{C}}HCH_3 \xrightarrow[CH_2Cl_2,\ 5\ min]{Cu(MnO_4)_2.8H_2O} CH_2(CH_2)_7COCH_3$$

2-Decanol — 93% 2-Decanone

As already stated potassium permanganate may be dissoleved in organic solvents after complexing with crown ethers. Potassium permanganate can also be made soluble in organic solvents by its conversion into tetrabutylammonium permanganate.

Barium manganate (prepared[76] from potassium manganate and barium chloride) is used for quantitative oxidation of benzhydrol to benzophenone.

$$C_6H_5\underset{\displaystyle OH}{\underset{|}{C}}HC_6H_5 \xrightarrow[\text{reflux, 1 hr}]{BaMnO_4,\ C_6H_6} C_6H_5COC_6H_5$$

Benzhydrol — 100% Benzophenone

Sodium permanganate monohydrate in dichloromethane or toluene oxidises[41] secondary alcohols.

$$\text{2-Cyclohexen-1-ol} \xrightarrow[\text{hexane, reflux, 24 hr}]{Na_2MnO_4.H_2O} \text{2-Cyclohexen 1-one (47\%)}$$

2-Cyclohexen
-1-ol

47%
2-Cyclohexen
1-one

$$\text{Cyclohexanol} \xrightarrow[\text{hexane, reflux, 1.5 hr}]{Na_2MnO_4.H_2O} \text{cyclohexanone (100\%)}$$

Cyclohexanol

100%
cyclohexanone

11. Oxidation with Potassium Ferrate, K_2FeO_4

In aqueous solution at room temperature, potassium ferrate oxidises[27] secondary alcohols to ketones.

$$C_6H_5CH(OH)CH_3 \xrightarrow[\text{RT, 18 min}]{K_2FeO_4,\ H_2O} C_6H_5COCH_3$$

1 phenylethanol

90%
Acetophenone

Double bonds are unaffected by this reagent.

12. Oxidation with Ruthenium Salts

Ruthenium tetroxide oxidises[78] cyclohexanol to cyclohexanone in 93% yield in carbon tetrachloride at room temperature. Instead of costly ruthenium tetroxide, catalytic amounts of ruthenium trichloride may be used in the presence of sodium hypocholrite as a reoxidant[79].

Sodium ruthenate[80] (which is prepared from ruthenium dioxide and sodium peroxidate in sodium hydroxide solution) and potassium ruthenate[81] (which is prepared from ruthenium trichloride and potassium persulphate solution) also oxidise secondary alcohols to ketones at room temperature.

13. Oppenauer Oxidation[82]

The secondary alcohols are oxidised to ketones by refluxing with aluminium isopropoxide in acetone and benzene or toluene. In this process, acetone acts as a hydrogen acceptor and is converted into isopropyl alcohol. For ditails about Oppenauer oxidations and its meechanism *see* Oppenaur oxidation, section 1.2

14. 2,3–Dichloro–5,6–dicyano–p–benzoquinone (DDQ)

It is useful for the oxidation of allylic alcohols in steroids, either alone or as catalyst with periodic acid as a reoxidant[84].

DDQ, C_6H_6
RT, 5–15 hr.

70%

15. N–Chlorosuccinimide (NCS)

It oxidises[24], cyclohexanol into cyclohexanone by heating for 5–10 min on a steam bath.

A mixture of N-Chlorosuccinimide and dimethyl sulphide oxidises secondary alcohols to ketones under very mild conditions. Thus, using this procedure 4–tert–butylcyclohexanol is oxidised to the correspoding ketone[85]

$NCS–(CH_3)_2S$, 0–25°C
Et_3N–toluene

$C_4H_9(t)$
4-tert-Butyl cyclohexanol

$C_4H_9(t)$
4-tert-Butyl cyclohexanone (90–93%)

16. Dimethyl Sulfoxide (DMSO)

As in the case of primary alcohols, dimethyl sulfoxide oxidises secondary alcohols to ketones under mild conditions. For more details about oxidation with DMSO and the mechanism of the reaction *see* dimethyl sulfoxide section 1.49.

17. Dess-Martin periodinane

As in the case of primary alcohols, Dess-martin periodinane is also used for the oxidation of secondary alcohols to ketones.

For more details, see Dess-Martin periodinane section 1.43.

Oxidation of Secondary Alcohols to α-hydroxy Peroxides

Photooxidation of secondary alcohols with oxygen in presence of benzophenone give α-hydroxy peroxides[86]

$$R{-}\underset{\displaystyle OH}{\underset{|}{CH}}{-}R' \xrightarrow[12°C,\ 17\text{–}57\ hr]{O_2,\ h\nu,\ Ph_2CO} R{-}\overset{\displaystyle OOH}{\overset{|}{\underset{\displaystyle OH}{\underset{|}{C}}}}{-}R'$$

$R = CH_3$

$R' = CH_3, C_2H_5, C_3H_7$

16–25%

α-Hydroxyperoxide

The α-hydroxyperoxides show the presence of active oxygen and give 2,4-dinitrophenyl hydrazones of the corresponding ketones. These peroxides tend to explode and so must be handeled with care.

Oxidation of Secondary Alcohols to Carboxylic Acids

Oxidation of secondary alcohols, particularly the cyclic alcohols with powerful origising agents like nitric acid or potassium permanganate give carboxylic acids via the formation of the intermediate ketones. Some examples are given below.[74,87,88]

Cyclohexanone (cyclohexyl–OH) $\xrightarrow[50\text{–}60°C,\ >1\ hr;\ yield\ 58\text{–}60\%]{50\%\ HNO_3/NH_4VO_3}$ $HO_2C(CH_2)_4CO_2H$ Adipic acid

Cyclohexanone $\xrightarrow[15\text{–}38,\ yield\ 70\%]{KMnO_4,\ Na_2CO_3,\ H_2O}$ Adipic acid

2-Methyl cyclohexanol (CH_3, –OH) $\xrightarrow[30°C,\ 1\ hr,\ RT]{CrO_3,\ H_2SO_4,\ H_2O}$ $CH_3CO(CH_2)_4CO_2H$

46–55%

δ-Acetylvaleric acid

2.2.3 Oxidation of Tertiary Alcohols

Tertiary alcohols resist oxidation. However, they can be dehydrated in acidic media to alkenes, which can be degraded to carboxylic acids[89] (**Barbier-Loguin and Wieland degration of carboxylic acids**).

$$Me_2CHCH_2CO_2Et \xrightarrow[Et_2O]{MeMgI} \underset{\text{tert-alconol}}{Me_2CHCH_2CMe_2OH}$$

$$\xrightarrow[reflux]{[O]\ CrO_3,\ 10\%\ H_2SO_4} Me_2CHCOOH$$

It is found that some tertiary alcohols are converted into tetiary hydroperoxides on reaction with hydrogen peroxide in sulphuric acid[90,91]. An example is the reaction of dimethylphenyl-carbinol with 87% hydrogen peroxide and sulphric acid at <0°C gives 94% yield of the corresponding hydroperoxide[90].

In case of acetylenic alcohols, in which tertiary hydroxyl group adjacent to the triple bond also give the corresponding hydroperoxide[91].

$$Me_2C(OH)C\equiv CH \xrightarrow[-4 \text{ to } 0°C,\ 5.5\ hr]{50\%\ H_2O_2,\ H_2SO_4} Me_2C(OOH)C\equiv CH \quad (98\%)$$

A rare example[92] of the oxidation of tertiary alcohol is the degradation of 3-ethyl-3pentanol to 3-pentanone and iodoethane by acetylhypiodite in acetic acid and chlorobenzene (irradiation 1 hr. 20–25°C)

$$CH_3CH_2C(OH)(CH_2CH_3)-CH_2CH_3 \xrightarrow[\text{Chlorobenzene, } \nu,\ 1\ hr,\ 20\text{–}25°C]{CH_3COOI/HOAc} CH_3CH_2COCH_2CH_3 + ICH_2C_2H_5$$

3-Ethyl-3-pentanol → 3-Pentanone (90%) + Iodoethane (84%)

Another interesting example of oxidation of tertiary alcohols is the oxidation of trans–9,10–dihydroxy–1,4,5,8,9,10–hexahydronaphthalene to 3,8–cyclodecadiene–1,6–done and its tetramethyl acetal using lead tetracetate.[93]

trans-9,10-Dihydroxy-1,4,5,8,9,10-hexahydro naphthalene $\xrightarrow[\text{MeOH, <30°C, 15 min, RT, 4 hr}]{Pb(OAc)_4,\ CCl_3CO_2H}$ 3,8-Cyclodecadiene--1,6-dione + tetramethyl acetal (CH_3O, OCH_3, CH_3O, OCH_3) 7.4%

References

1. J. M. Church and H. K. Joshi. *Ind. Eng. Chem.*, 1951, **43**, 1804.
2. J. M. Church and L. Lynn, *Ind. Eng., Chem.*, 1950, **42**, 768.
3. M. Y. Sheikh and G. Eadon, *Tetrahedron Lett*, 1972, 257.
4. K. Heyns and L. Blazejewicz, *Tetrahedron*, 1960, **9**, 67.
5. L. Syper, *Tetrahedron Lett.*, 1967, 4193.
6. W. S. Trahanovsky and W. S. Trahanovsky, *J. Org. Chem.*, 1967, **32**, 2349.
7. R. E. Partch, *Tetrahedron Lett.*, 1964, 3071.
8. B. O. Field and J. Grundy, *J. Chem., Soc.*, 1955, 1110.
9. D. G. Lee and U. A. Spitzer, *J. Org. Chem.*, 1970, **35**, 3589.
10. R. O. Hutchins, N. R. Natale and W. J. Cook, *Tetrahedron Lett.*, 1977, 4167.
11. S. Cacchi, *Synthesis*, 1979, 356.

12. E. Corey and G. Schmidt, *Tetrahedron Lett.,* 1979, 399.
13. J. C. Sauer, *Org., Synth.* Collective Volume, 1963, **4**, 813.
14. E. Santaniello, F. Ponti and A. Manzocchi, *Synthesis,* 1978, 534.
15. J. R. Holum, *J. Org. Chem.,* 1961, **26**, 4814.
16. R. Ratcliffe and R. Rodehorst, *J.Org. Chem.,* 1970, **35**, 4000.
17. E. Corey and J. W. Suggs, *Tetrahedron Lett.,* 1975, 2647.
18. E. Corey and G. W. J. Fleet, *Tetrahedron Lett.,* 1973,4499.
19. Y. –S. Cheng, W. –L. Liu and S. –H. Chen, *Synthesis*, 1980,223.
20. J. M. J. Frechet, J. Wärnock and M. J. Farrall, *J. Org. Chem,* 1978,**43**,2618.
21. J. Rocek and A. E. Radkarsky, *J. Am. Chem. Soc.,* 1973,**95**,7123.
22. J. San Filippo, Jr., and C. –I. Chern, *J. Org., Chem.,* 1977,**42**,2182.
23. K. B. Sharpless and K. Alashi, *J. Am. Chem., Soc.,* 1975,**97**,5927.
24. C.A. Grob and H. J. Schmid, *Helv. Chim. Acta.* 1953,**36**,1763.
25. J. Attenburrow, A.F.B. Cameron, J. H. Chapman, R. M. Evans, B. A. Hems, A.B.A. Jansen and T. Walkor, *J. Chem. Soc.,* 1952,1094.
26. R. J. Highet and W. C. Wildman, *J. Am. Chem. Soc.,* 1955,**77**,4399.
27. R. J. Gritter and T. J. Wallace, *J. Org. Chem.,* 1959,**24**,1051
28. J. M. Goldman, *J. Org. Chem,* 1969,**34**,3289.
29. H. B. Henbest and M. J. W., Strattord, Chem. Ind (London), 1961,1170.
30. S. L. Regen and C. Koteel, *J. Am. Chem. Soc.,* 1977,**99**,3837.
31. H. Firouzabadi, M. Naderi, A. Sardarian and B. Vessal, *Synth. Commun.,* 1983,**13**,611
32. H. Firouzabadi, B. Vessal and M. Naderi, *Tetrahedron Lett.,* 1982,**23**,1847.
33. H. Firouzabadi and E. Ghaderi, *Tetrahedron Lett.,* 1978,839.
34. E. A. Braude, R. P. Linstead and K. R. Wooldridge, *J. Chem. Soc.,* 1956,3070.
35. I. Kuwajima, M. Shimizu and H. Urable, *J. Org. Chem.,* 1982,**47**,837.
36. D. B. Dess and J. C. Martin, *J. Org., Chem.,* 1983,**48**,4155.
37. Y. Hayakawa, M. Uchiyama and R. Noyori, *Tetrahedron Lett.,* 1986,**27**,4191.
38. F. L. M. Pattison, J. B. Stothers and R. G. Wooltord, 1956,**78**,2255.
39. H. Ahrens and W. Korytnyl, J. Heterocyl. Chem., 1967,**4**,625.
40. F. M. Menger and C. Lee, *J. Org., Chem.,* 1979,**44**,3446.
41. F. M. Menger and C. Lee., *Tetrahedron Lett.,* 1981,**22**,1655.
42. N. A. Noureldin and D. G. Lee, *J. Org. Chem.,* 1982,**47**,2790.
43. T. Sala and M. V. Sargent, Chem. Commun., 1978,253.
44. A. W. Harriott and D. Picker, *Tetrahedron Lett.,* 1974,1511.
45. K. Nakagawa, R. Konaka and T. Nakata, *J. Org. Chem.,* 1962,**27**,1597.
46. D. G. Lee, D. J. Hall and J. H. Cleland, Can. *J. Chem.,* 1972,**50**,3741.
47. M. Schröder and W. P. Griffith, Chem., Commun, 1979,58.
48. R. J. Kennedy and A. M. Stock, *Chem. Ber.,* 1900,**33**,533.
49. G. R. Rubertson, *Org. Synth.,* Collective Volume, 1932,**1**,138.
50. G. B., Neave, *J. Chem. Soc.,* 1912,**101**,513.
51. R. E. Dunbar and M. R. Arnold, *J. Org. Chem.,* 1945,**10**,501.
52. W. R. Nes, *J. Org. Chem.,* 1958,**23**,899.
53. K. Alder and H. Wirtz, *Justus Lebis Ann. Chem.*, 1956, **601**, 138.
54. C. Moureu and G. Mignonac, C. R. *Hebd. Seances Acad. Sci*, 1920, **170**, 258.

55. T. F. Blackburn and J. Schwartz., *Chem. Commun.,* 1977,154.
56. R. E. Comeron and A. B. Bocarsly, *J. Am. Chem. Soc.,* 1985,**107**,6117.
57. L. Syper, *Tetrahedron Lett.,* 1967,4193.
58.. H. H. Wasserman and J. E. Van Verth, *J. Am. Chem., Soc.,* 1974,**96**,585.
59. S. Kanemoto, H. Tomioka, K. Oshima and H. Nozaki, *Bull. Chem., Soc., Jpn.,* 1986,**59**,105.
60. O. Mancera, G. Rosenkranz and C. Djerassi, *J. Org., Chem.,* 1951,**16**,192.
61. K. Bowden, I. M. Heilbron, E. R. H. Jones and B. C. L. Weedon, *J. Chem. Soc.,* 1946,39.
62. T. Brunelet, C. Jouitteau and G. Gelbard, *J. Org., Chem.,* 1986,**51**,4016.
63. H. C. Broun and C. P. Garg, *J. Am. Chem. Soc.,* 1961,**83**,2952.
64. R. P. Singh, H. N. Subbaro and S. Dev, *Tetrahedron*, 1979,**35**,1789.
65. E. Santaniello, F. Milani and R. Casati, *Synthesis,* 1983,749.
66. J. Wicha and A. Zarecki, *Tetrahedron Lett.,* 1974,3059.
67. M. A. Neirabeyeh, J. –C. Ziegler and B. Gross, *Synthesis,* 1976,811.
68. R. V. Stevens, K. T. Chapman, H. N. Weller, *J. Org., Chem.,* 1980,**45**,2030.
69. M. Schneider, J. –V. Weber and P. Faller, *J. Org. Chem.,* 1982,**47**,364.
70. T. Kageyama and M. Okawara, *Synthesis,* 1983,815.
71. S. Kanemoto, H. Tomioka, K. Oshima and H. Nozaki, Bull. Chem., Soc. Jpn., 1986,**59**,105.
72. E. Adler and H. –D.Becker, Acta Chem., Seand, 1961,**15**,849.
73. G. Stork and M. Tomasz, *J. Am. Chem. Soc.,* 1964,**86**,475.
74. V. J. Braun and G. Lemke, *Chem. Ber.,* 1922,**55**,3526.
75. S. Quici and S. L. Regen, *J. Org. Chem.,* 1979,**44**,3436.
76. F. Firouzabadi and Z. Mostafavipoor, Bull. *Chem. Soc.* Jpn., 1983,**56**,914.
77. R. J. Audette, J. W. Quail and P. J. Smith., *Tetrahdron Lett.,* 1971,279.
78. L. M. Berkowitz and P. M. Rylander, *J. Am. Chem. Soc.,* 1958,**80**,6682.
79. S. Wolfe, S. K. Hasan and J. R. Campbell, *Chem. Commun.,* 1970,1420.
80. D. G. Lee, D. T. Hall and J. H. Cleland, Can. *J. Chem.,* 1972,**50**,3741.
81. M. Schröder and W. P. Griffith., *Chem., Commun.,* 1979,58.
82. R. V. Uppeanauer, Rec. Trav. Chim., 1937,**56**,137; C. Dgerassi. Org., React., 1951,**6**,207.
83. D. Burn, V. Petrow and G. O. Weston, *Tetrahedron Lett.,* 1960. 14.
84. S. Cacchi, *Synthesis*, 1978,848.
85. E. J. Corey and C. U. Kim., *J. Am. Chem. Soc.,* 1972,**94**,7586.
86. G. O. Schenck, H. –D. Becker, K.–H. Schulte–Eliter and C. H. Kranch, *Chem. Ber.,* 1963,**96**,509.
87. C. Djerassi, Org. React. 1951,**6**,207.
88. J. R. Schaeffer and A. D. Snoddy, *Org., Synth.* Collective Volume, 1963,**4**,19.
89. P. Babier and R. Locquin, C. R. Hebd. Seances Acad. Sci., 1963,**156**,1443.
90. H. Ross and R. Rüttel, *Chem., Ber.,* 1956,**89**,2641.
91. N. A. Milas and O. L. Mageli, *J. Am. Chem. Soc.,* 1952,**74**,1471.
92. T. R. Beebe, B. A. Barnes, K. A. Bender, A. D. Halbert, R. D. Milley, M. L. Ramsay and M. W. Ridenour, *J. Org. Chem.,* 1975,**40**,1992.
93. A. G. Brook, *J. Chem. Soc.,* 1952,5040.

2.3 OXIDATION OF PHENOLS

Phenols undergo different types of oxidative reactions. Some of these are dehydrogenative coupling, oxidation to free radicals, hydroxylation in the aromatic ring and oxidation to quinones.

2.3.1 Dehydrogenative Coupling

Phenols undergo oxidative coupling in ortho or para positions. Thus, β-naphthol on heating with ferric chloride in ether give α, α′–bi–β–naphthol[1] and 2,4-bis (tert-butyl)–5–methyl phenol on heating with chromic acid yield the correspoding dimer.[2]

OH

$FeCl_3$-ether

reflux

β-Naphthol

OH HO

60%

α,α′-bi-β-Naphthol

OH

$K_2Cr_2O_7$, AcOH

H_2SO_4, H_2O, 50–55°C

OH OH

2,4-bis (tert-butyl) -5-methylphenol

Dimer

2.3.2 Oxidation of Phenols to Phenoxy Radicals

Oxidation of sterically hindered phenols with potassium ferricyanide give the corresponding phenoxy raidcals[3,4]

R R–OH R

$K_3Fe(CN)_6$, KOH, H_2O, C_6H_6

RT, 1-2 hr

R R–Ȯ R

R = $C(CH_3)_3$ or C_6H_5

R = $C(CH_3)_3$, yield 99–100%

R = C_6H_5, yield 81–91%

2.3.3 Hydroxylations of Phenols in the Aromatic Ring

Phenols undergo oxidation[5] with potassium persulphate in alkaline solution. This reaction is known as **Elbs persalphate oxidation**. The hydroxyl group normally goes to the para position to form a hydroquinone derivative. However, if the para position is blocked, the hydroxyl group goes to the ortho position to form a catechol derivative. Some examples of persulphate oxidation are given below[6].

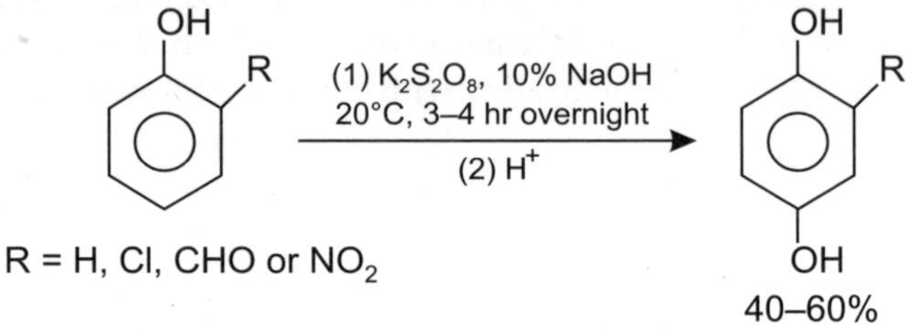

(1) $K_2S_2O_8$, 10% NaOH 20°C, 3–4 hr overnight; (2) H^+

R = H, Cl, CHO or NO_2

40–60%

The mechanism[7,8] of the reaction is as shown below. The initially formed persulaphate (1) is hydrolysed to the hydroquinone(2).

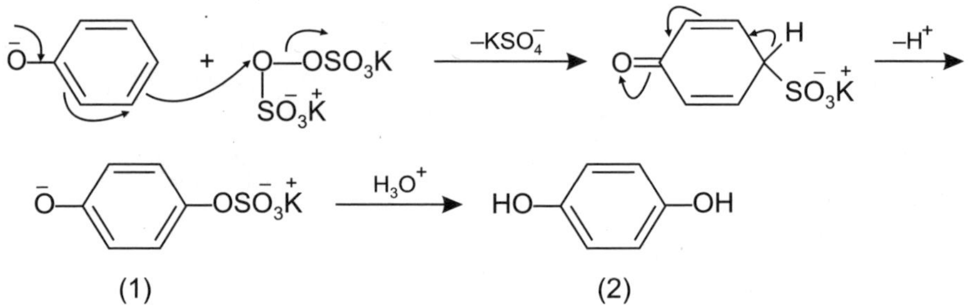

$-KSO_4^-$; $-H^+$; H_3O^+

(1) (2)

2.3.4 Oxidation of Phenols to Quinones

Phenols can be converted into ortho-and para-quniones by fermy salt [Potassium nitroso disulfonate, • $ON(SO_3K)_2$]. Some other oxidising agents like chromium trioxide, bromine, DDQ, meruric oxide and mercuric trifluoroacetate can also be used. Some examples[9, 10] are given below.

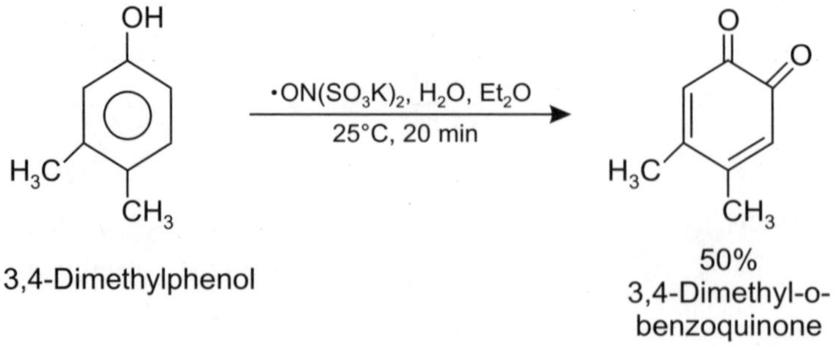

$\cdot ON(SO_3K)_2$, H_2O, Et_2O 25°C, 20 min

3,4-Dimethylphenol

50%
3,4-Dimethyl-o-benzoquinone

$$\text{2,3,6-Trimethyl phenol} \xrightarrow[\text{12°C, 4 hr}]{\cdot ON(SO_3K)_2,\ H_2O,\ C_7H_{16}} \text{2,3,6-Trimethyl } p\text{-benzoquinone}$$

Oxidation to *p*-benzoquniones occurs even if the para position is occupied by a substituent and even if an ortho position is free. It is found that the substituent in the para position is either eliminated or modified. Some typical examples are given below.

$$\text{2,4,6-Tribormo } m\text{-cresol} \xrightarrow[\text{70–75°C, 10 min}]{CrO_3/HOAc} \text{2,6-Dibromo-} m\text{-toluquinone (70\%)}$$

$$\text{2-Bromo-4-Hydroxy--5-methoxybenzyl alcohol} \xrightarrow[\text{pH 6, r.t., 1 hr}]{\text{Fremy salt}} \text{2-Bromo-5-methoxy } p\text{-benzoquinone (84\%)}$$

$$\text{4-Benzyl-2,6-dibromo-phenol} \xrightarrow[\text{100°C, 4 hr}]{Br_2,\ CCl_4,\ \text{sealed tube}} \text{2,4-Dibromophenyl chinomethide (80-85\%)}$$

CH_3O — OH → HgO, MOH, RT; yield, 88% → O= =O

2,5-Di-tert-butyl 4-methoxyphenol

2,5-di-tert-butyl p-benzoquinone

$Hg(OCOCF_3)_2$, MeOH; polyvinyl pyridine, RT, yield 79%

In some cases, the oxidation of qunions is accompanied by doubling of the molecule[15].

OCH_3, OH, OCH_3 → DDQ, MeOH; RT, 1 min → CH_3O, O=, CH_3O, OCH_3, =O, OCH_3 + O=, OCH_3, =O, OCH_3

62% 13%

Naphthols can be oxidised to quinones by ceric ammonium nitrate[16].

CH_3O, OH, CH_3, CH_3O → $(NH_4)Ce(NO_3)_6$; MeCN, RT, yield 45% → CH_3O, O, CH_3, CH_3O, O

CH_3O, OH, CH_3, CH_3O, Cl → $(NH_4)Ce(NO_3)_6$; MeCN, RT, yield 82%

The oxidiations to quinones are easier in case two hydroxyl groups are present in *ortho* or *para* position[17].

OH
Hydroquinone
$Na_2Cr_2O_7$, H_2SO_4, H_2O
20-30°C, 30-45 min
76-81%
p-Benzoquinone
$KBrO_3/H_2SO_4$
35-45°C, yield 70%

The well-known DDQ (2,3-dichloro-5,6-dicyano-p-benzoquinone) can be prepared by the oxidation[18,19] of 2,3-dichloro-5,6-dicyanohydroquinone

2,3-Dichloro- -5,6-dicyano hydroquinone
PbO_2, 5% HCl, EtOH, C_6H_6
RT, 5 min, 83% yield
2,3-Dichloro- 5,6-dicyano- *p*-benzoquinone
$N_2O_4(HNO_3 + As_2O_3)$, CCl_4
RT, 10 min, 93% yield

Lead tetracetate in acetic acid converts *p*, *p*′-dihydroxybiphenyls into diphenoquinones[20,21]

$x = y = H$
or
$x = Cl$, $y = H$
p,*p*′-Dihydroxy-biphenyls
$Pb(OAc)_4$, AcOH
30°C, RT, overnight
70-75%
Di-phenoquinones

Ortho quinions are obtained by oxidation of o-dihydroxy compounds with sodium iodate[22]. An example is given below:

OH OH NH — $NaIO_3$, H_2O, aqueous EtOH, RT, 20 min → O O N H 81%

2.3.5 Oxidative Ring Opening of Phenols

Organic peroxyacids oxidise[23] phenols with ring opening. One such example is given below:

OH β-Naphthol — AcO_2H, AcOH, <40°C, 15 hr → COOH CH=CH CO_2H 70% o-Carboxyl cinnamic acid

References

1. H. Waldev, *Chem., Ber.*, 1882, **15**, 2166.
2. H. E. Albert, *J. Am., Chem., Soc.*, 1954, **76**, 4983.
3. C. D. Cook and R. C. Woodworth, *J. Am. Chem. Soc.*, 1953, **75**, 6242.
4. K. Blmroth, A. Berndt, H. Perst and C. Reichardt, *Org. Synth.* Collective Volume, 1973, **5**, 1130.
5. E. Elbs, J. Prakt. Chem, 1893,**48**,179; S. M. Sethna, Chem., Rev., 1951, **91**, 49.
6. K. Kondo, A. Negishi and M. Fukuyama, *Tetrahedron Lett.*, 1969,2461.
7. W. Baker, CN. Brown, *J. Chem. Soc.*, 1948,2303.
8. E. J. Bergmann, *J. Am. Chem. Soc.*, 1958, **80**, 3717.
9. H.–J. Teuber, *Org. Synth.* Collective Volume, 1988, **6**, 480.
10. P. A. Wehrli and F. Pigott., *Org. Synth.* Collective Volume, 1988,**6**,1010.
11. L. I. Smith and D. J. Byers, *J. Am. Chem. Soc.*, 1941, **63**, 612.
12. J. M. Saa′, J. Morey and A. Costa, *Tetrahedron Lett.*, 1986,**27**,5125.
13. T. Zincke and W. Walter, Justus Liebgs, *Ann. Chem.*, 1904, **334**, 367.
14. A. Mckillop and D. W. Young, *Synth.* Commun., 1977,**7**,467.
15. H. D. Becker, *J. Org. Chem.*, 1965,**30**,982.
16. H. Laatsch, Justus Liebigs *Ann. Chem.*, 1986,**1655**,1669.
17. H. T. Clarke and W. W. Hartman, *Org. Synth.* Collective Volume, 1932,**1**,543.
18. P. W. D. Mitchell, *Can. J. Chem.*, 1963,41,550.
19. A. G. Brook, *J. Chem. Soc.*, 1952,5040.

20. B. R. Brown and A. R. Todd, *J. Chem. Soc.,* 1954, 1280.
21. K.–H. König, W. Schulze and G. Möller, *Chem. Ber.,* 1960, **93**, 554.
22. B. M. Trost, T. N. Salzmann and K. Hiroi, *J. Am. Chem. Soc.,* 1976, **98**, 4887.
23. F. P. Greenspan, *Ind. Eng. Chem.,* 1947, **39**, 847.

2.4 OXIDATION OF ETHERS

Ethers are believed to be unreactive and are comparatively inert substances. However, under certain conditions they are oxidised.

2.4.1 Autooxidation

Most aliphatic ethers, in air are slowly converted into unstable peroxides. Though, the peroxides are present in low concentration, but are very dangerous and may cause violent explosion during distillation.

$$(C_2H_5)_2O + [O] \xrightarrow{\text{air}} (C_2H_5)_2\ddot{O}: \longrightarrow O$$

Diethyl ether — Peroxide

In order to prevent the formation of peroxide, it is advisable to store ether in coloured bottles.

2.4.2 Formation of Hydroperoxides

Ethers having α-hydrogens, on irradiation with UV light in presence benzophenone (a sensitizer) are converted into α-hydroperoxides. Thus, when oxygen is bubbled through tetrahydrofuran at 12°C for 28 hr with simultaneous irradiation, the product obtained is the correspoding α-hydroperoxide

$$\text{Tetrahydrofuran} \xrightarrow[\text{28 hr}]{O_2,\ \nu,\ 12°C} \text{Tetrahydrofuran } \alpha\text{-hydroperoxide (–O—OH)}\ (86\%)$$

Tetrahydrofuran

86%
Tetrahydrofuran
α-hydroperoxide

In a similar way, other ethers like dioxane, methyl butyl ether, diisopropyl ether, dibutyl ether and diisoamyl ether, give the corresponding α-hydroperoxides in low yields (6–15%).

2.4.3 Conversion of Ethers into Esters

Ethers on oxidation with ozone[2], benzyltriethylammonium permanganate[3] and ruthenium tetroxide[4] give esters.

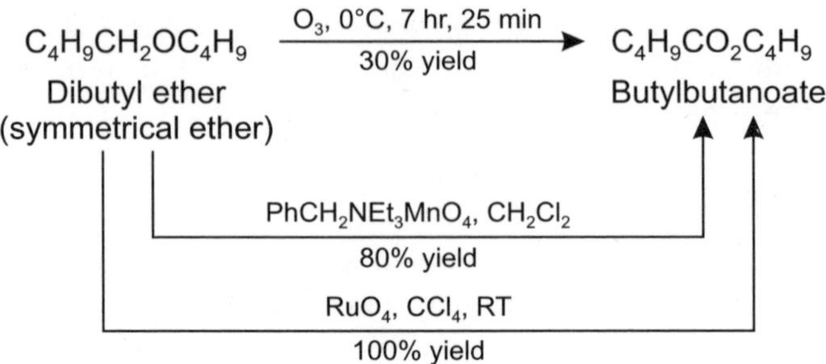

In unsymmetrical ethers like benzyl methyl ether and benzyl butyl ether, the oxidation gives methyl benzoate[5] and butyl benzoate[3], respectively.

$$\underset{\text{Benzyl methyl ether}}{C_6H_5CH_2OCH_3} \xrightarrow[20°C,\ 7\ min]{RuCl_3.H_2O,\ NaIO_3,\ MeCN,\ CCl_4,\ H_2O} \underset{\substack{89\% \\ \text{Methylbenzoate}}}{C_6H_5CO_2CH_3}$$

$$\underset{\text{Benzyl butyl ether}}{C_6H_5CH_2OC_4H_9} \xrightarrow{C_6H_5CH_2NEt_3MnO_4,\ CH_2Cl_2} \underset{\substack{\text{Butyl benzoate} \\ 99\%}}{C_6H_5CO_2C_4H_9}$$

It is formed that in case of unsymmetrical ethers, the products obtained differ when different oxidising agents are used. Thus, as already stated, benzyl methyl ether on oxidation with ruthenium tetroxide gives methyl benzoate in 87% yield. However, hexadecyl methyl ether on oxidation with chromic acid in acetic acid gives[6] palmitic and hexadeyl formate and hexadecyl ethyl ether gives palmitic and and hexadecyl acetate.

$$\underset{\text{Hexadecylmethyl ether}}{C_{16}H_{33}OCH_3 \text{ or } C_{15}H_{31}CH_2OCH_3} \xrightarrow[20°C,\ 1\ min]{CrO_3,\ AcOH, CH_2Cl_2} \underset{\text{Palmitic acid}}{C_{15}H_{31}CO_2H} + \underset{\text{Hexadeceyl formate}}{HCOOC_{16}H_{33}}$$

$$\underset{\text{Hexadecyl ethyl ether}}{C_{15}H_{31}CH_2OC_2H_5} \xrightarrow[20°C,\ 1\ hr]{CrO_3,\ AcOH,\ CH_2Cl_2} \underset{\substack{55\% \\ \text{Palmitic acid}}}{C_{15}H_{31}CO_2H} + \underset{\text{Hexadecyl acetate}}{C_{16}H_{33}OCOCH_3}$$

In case of secondary benzyl ethers, cleavage to the ketone predominates over the formation of alkyl benzoates.

$$C_6H_{13}\underset{\substack{| \\ CH_3}}{C}HOCH_2C_6H_5 \xrightarrow[0°C,\ 12\ hr]{4CrO_3,\ H_2SO_4,\ Me_2CO} \underset{79\%}{C_6H_{13}COCH_3} + \underset{21\%}{C_6H_5CO_2CH_2C_6H_{13}}$$

2.4.4 Oxidation of Cyclic Ethers to Lactones

Cyclic ethers, for example, tetrahydro furan on oxidation with zincdichromate[7], Zinc permanganate[8] supported on silica gel and ruthenium tetroxide[4] gives lactones.

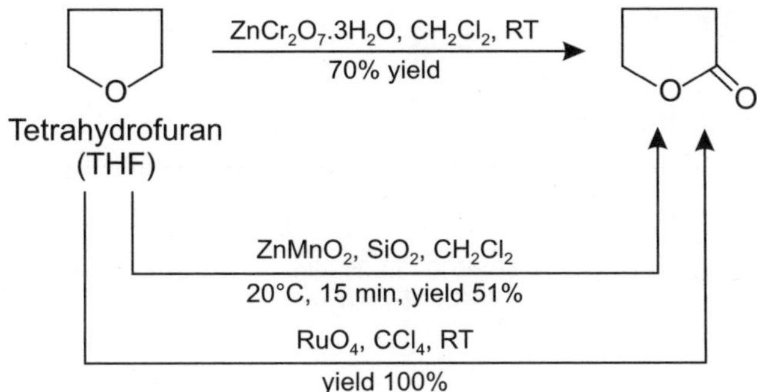

Three membered cyclic ethers, (epoxides) on oxidative ring opening give anti hydroxylation products of alkenes (for more details *see* 2.1.2.6).

References

1. R. Criegee, *Chem. Ber.,* 1944, **77**, 22.
2. F. G. Fisher, Justers Liebigs *Ann. Chem.,* 1929, **476**, 233.
3. H.–J. Schmidt and H. J. Schäfer, *Angew. Chem. Int. Ed. Engl.,* 1979, **18**, 68.
4. L. M. Berkowitz and P. N. Rylander, *J. Am. Chem. Soc.,* 1958, **80**, 6682.
5. P. H. J. Carlsen, T. Katsuki, V. S. Martin and K. B. Sharpless, *J. Org. Chem.,* 1981, **46**, 3936.
6. I. T. Harrison and S. Harrison, *Chem. Commun.,* 1966, 752.
7. H. Firouzabadi, A. R. Saddrian, H. Moosavipour and G. M. Afsari, *Synthesis,* 1986, 285.
8. S. Wolfe and C. F. Ingold, *J. Am. Chem. Soc.,* 1983, **105**, 7755.

2.5 OXIDATION OF ALDEHYDES

Aldehydes are easily oxidsed to carboxylic acids. Following are given some of the important methods used:

1. Auto Oxidation

Aldehydes on contact with air are spontaneously oxidised to carboxylic acids. For example, benzaldehyde if left in open bottle, is slowly converted into benzoic acid at room temperature. The oxidation is acclerated if aldehyde is exposed to oxygen, especially under irradiation with ultraviolet light.

2. Catalytic Oxidation

Oxygen in the presence of catalyst like copper acetate and nickel acetate converts aldehydes into carboxylic acid. Thus, oxygen in presence of catalyst oxidise methacrolein into methacrylic acid in 99% yield at room temperature[1].

$$CH_2{=}C(CH_3)CHO \xrightarrow[h\nu]{O_2,\ \text{copper acetate, Nickel acetate}} CH_2{=}C(CH_3){-}CO_2H$$

Methacrolein → 99% Methacrylic acid

Furtural on treatment with oxygen in aqueous alkaline solution in presence of cuprous oxide and silver oxide at 50–55°C is converted into furoic acid[2].

$$\text{Furtural (2-furyl–CHO)} \xrightarrow[\text{50-55°C, 20-25 min; 40°C, 15-30 min}]{O_2,\ NaOH,\ H_2O,\ Cu_2O,\ Ag_2O} \text{2-furyl–}CO_2H$$

Furtural → 86-90% Furoic acid

3. Oxidation with Hydrogen Peroxide

Aldehydes are oxidised to carboxylic acids by hydrogen peroxide and its derivatives, in presence of catalysts. Thus, acrolein is oxidised to acrylic acid in 90% yield by treatment with 12% hydrogen peroxide solution in tert. amyl alcohol at 40°C for 3 hr. in presence of selenium dioxide[3]. In a simlar way, methacrolein is converted[3] into methacrylic acid by 90% hydrogen peroxide in presence of selenium dioxide at 60°C.

$$CH_2{=}C(R)CHO \xrightarrow[\text{40-60°C, 1 hr}]{H_2O_2/SeO_2} CH_2{=}C(R){-}CO_2H$$

Acrolein R = H; Methacrolein R = CH_3 → Acrylic acid, R = H; 90%; Methacrylic acid, R = CH_3, 75%

4. Oxidation with Oxone

Oxone (potassium peroxymonosulfate, $2KHSO_5\ KHSO_4.\ K_2SO_4$) oxidises benzaldehyde in chloroform and dilute sulphuric acid at room temperature within 2 days in 70% yield[4].

5. Oxidation with Peroxysulphuric Acid

Peroxysulphuric acid (H_2SO_5, prepared by adding 90% hydrogen peroxide to concentrated sulphuric acid at <15°C and keeping the mixture for 2 hr. at room temperature) oxidises aldehydes to carboxylic acid. Thus, methacrolein is oxidised to methacrylic acid in 29% yield. In this case, if methanol is used as a solvent, methyl methacrylate is obtained in 91–97% yield[5].

6. Oxidation with Silver (I) Oxide (Argentous Oxide, Ag_2O)

Silver oxide is a mild oxidising agent and is a reagent of choice for the oxidation of aldehydes to carboxylic acids. The reagent (silver oxide is generally prepared *in situ* from silver nitrate in alkaline medium[6]. Some examples of oxidation[6,7] by silver oxide are given below:

3-Cyclohexene--1-carboxaldehyde → ($AgNO_3$, KOH, EtOH; Cooling with ice, 1 hr; RT, 2-3 hr) → 3-Cyclohexene-1-carboxylic acid, 62.5%

3-Thienaldehyde → ($AgNO_3$, NaOH, H_2O; Cooling with ice, 5 min) → 3-Thienoic acid, 97%

Furfural → (Ag_2O, NaOH) → 2-Furoic acid

The mechanism of the oxidation of aldehydes to carboxylic by silver oxide is given below:

R(H)C=O + Ag—O—Ag → R(H)C(O⁻)—O—Ag → Ag—Ag + R—C(=O)OH

6a. Silver (II) Oxide (Argentic oxide, AgO) is very powerful oxidising agent. The reagent is prepared[8] by the oxidation of silver nitrate with potassium peroxydisulfate ($K_2S_2O_8$) in alcoholic solvent. Thus, cyclohexene–1–carboxaldehyde on oxidation with silver (II) oxide gives 3–cyclohexene–1–carboxylic acid[9]

Cyclohexene -1-carboxaldehyde → (4AgO, THF, H_2O (9:1); 25°C, 14 hr) → 3-Cyclohexene-1--carboxylic acid, 97%

Cyclohexene -1-carboxaldehyde → (AgO, NaCN; 97% yield) → 3-Cyclohexene-1--carboxylic acid

It was found[10] that a mixture of silver (II) oxide and sodium cyanide also affected the above oxidation.

7. Oxidation with Fuming Nitric Acid

Some aldehydes like β-chloropropionaldehyde are oxidised[11] to the corresponding carboxylic acids by fuming nitric acd.

$$\underset{\beta\text{-Chloropropionaldehyde}}{ClCH_2CH_2CHO} \xrightarrow[30\text{-}35°C,\ 25\ min]{\text{Fuming } HNO_3} \underset{\substack{60\text{-}65\% \\ \beta\text{-Chloroproponic acid}}}{ClCH_2CH_2CO_2H}$$

8. Oxidation with Potassium Dichromate

Oxidation of furfural with potassium dichromate and dilute sulphuric acid (heating 30–45 min, 100°C) gives 75% yield of furoic acid[12].

9. Oxidation with Sodium and Potassium Hydroxide

Vanillin on heating for 5 min with a mixture of sodium and potassium hydroxide at 180–195°C gives vanillic acid[13] in 89–95% yield.

$$\underset{\text{Vanillin}}{4\text{-HO-}3\text{-}CH_3O\text{-}C_6H_3\text{-CHO}} \xrightarrow[180\text{-}195°C,\ 5\ min;\ (2)\ H^+]{(1)\ NaOH/KOH} \underset{\substack{89\text{-}95\% \\ \text{Vanillic acid}}}{4\text{-HO-}3\text{-}CH_3O\text{-}C_6H_3\text{-}CO_2H}$$

10. Oxidation with Sodium Chlorite

Sodium Chlorite is a suitable reagent for the oxidation of aldehydes to carboxylic acids. However, in this case, the byproducts of the reaction, NaOCl and ClO_2 being themselves oxidants decrease the yield. In view of this, certain scavengers like hydrogen peroxide[14], resorcinol[15] are added to the solution of the substrate and sodium chlorite in water, aqueous acetonitrile or tert-butyl alcohol. Thus, sodiumchlorite oxidises cinnamaldehyde to cinnamic acid without affecting the double bond.

$$\underset{\text{Cinnamaldehyde}}{C_6H_5CH{=}CHCHO} \xrightarrow[\text{resorcinol, acetate buffer, RT, 45 min}]{NaClO_2,\ t\text{-BuOH},\ H_2O} \underset{\substack{72\% \\ \text{Cinnamic acid}}}{C_6H_5CH{=}CHCO_2H}$$

Alternative route (Cinnamaldehyde → Cinnamic acid): $NaClO_2$, 35% H_2O_2, MeCN, H_2O, 10°C, 1 hr, 95% yield

11. Oxidation with Manganese Dioxide

Since manganese dioxide is used to oxidise toluene into benzaldehyde, it is rather surprising that manganese dioxide can convert[16] benzaldehyde into benzoic acid in 75% yield by refluxing with manganese dioxide in petroleum ether for 24 hr.

It is reported that manganese dioxide can oxidise unsaturated aldehydes via the formation of cyanohydrin.[16a]

$$\underset{\text{Cinnamaldehyde}}{C_6H_5CH{=}CH{-}CHO} \xrightarrow[\text{MeOH}]{\text{NaCN}} \underset{\text{Cyanohydrin}}{\left[C_6H_5CH{=}CH{-}\overset{OH}{\overset{|}{C}}HCN\right]} \xrightarrow{MnO_2/MeOH}$$

$$[C_6H_5CH{=}CHCOCN] \xrightarrow[\text{2 hr}]{25°C} \underset{\substack{>90\% \\ \text{Cinnamic acid}}}{C_6H_5CH{=}CHCOOH}$$

12. Oxidation with Potassium Permanganate

It is commonly used to oxidise both aliphatic[17] and aromatic[18] aldehydes to the corresponding carboxylic acids. However, much better yields of carboxylic acids are obtained by oxidation of aldehydes with tetrabutylammonium permanganate[19] or benzyltriethylammonium permanganate[20]

Piperonal $\xrightarrow[\text{yield, 99\%}]{Bu_4N\ MnO_2,\ C_5H_5N,\ RT}$ Piperonylic acid

Piperonal $\xrightarrow[CH_2Cl_2,\ 1.25\ hr.\ 80\%\ yield]{PhCH_2NEt_3MnO_4,\ AcOH}$ Piperonylic acid

13. Oxidation with Nickel Peroxide (NiO_2)

Nickel peroxide, an unusual oxidant can convert aromatic aldehydes into the corresponding carboxylic acid in 100% yield[21]. Thus, piperonal on heating with nickel peroxide in sodium hydroxide and water at 60°C for 3 hr give 100% yield of pipronylic acid.

14. Oxidation with Ruthenium Salts

Pure ruthenium tetroxide[22] gives very poor yields in the oxidation of aldehydes to carboxylic acid. However, potassium ruthenate (prepared *in situ* from ruthenium trichloride and potassium persulphate in water and used in catalytic amounts) gives 99% yield of *m*-nitrobenzoic acid (from *m*-nitrobenzaldehyde) at room temperature after 2 hr.

15. Oxidation with Iodosobenzene

Iodoso benzene in presence of tris (triphenyl phosphine) ruthenium dichloride converts benzaldehyde into benzoic acid (96% yield) at room temperature[23].

16. Cannizzaro Reaction

Aromatic aldehydes that do not have α-hydrogen atom (s) on treatment with concentrated aqueous alkali undergo self-oxidation and reduction to give alcohol and the salt at the corresponding carboxylic acid.

$$C_6H_5CHO + C_6H_5CHO \xrightarrow{KOH} C_6H_5CH_2OH + C_6H_5CO\bar{O}\overset{+}{K}$$

The reaction is known as Cannizzaro reaction[24]. The yields of the acids can be 50% at a maximum.

The mechanism involves the formation of an anion (1) by the interaction of the aldehyde with hydroxide ion. The anion (1) then transfers a hydride ion intermolecularly to a carbonyl atom of another molecule, forming the carboxylic acid and the alkoxide ion (2). The shift of proton from (2) to (1) gives the final product.

$$RCHO + :\bar{O}H \rightleftharpoons R-\underset{OH}{\overset{O^-}{C}}-H \quad (1)$$

$$(1) + RCHO \rightleftharpoons RCOOH + R-\underset{H}{\overset{O^-}{C}}-H \quad (2)$$

$$\rightleftharpoons R-\overset{O}{\overset{\|}{C}}-O^- \text{ (Acid salt)} + R-\underset{H}{\overset{OH}{C}}-H \text{ (Alcohol)}$$

17. Oxidation with Dimethyl Sulfoxide

The aldehyde is first converted into aldehyde-bisulphite adduct, which in turn is oxidised by dimethyl sulfoxide[35]

$$Ph-C_6H_4-CH(CH_3)CHO \xrightarrow{NaHSO_3} Ph-C_6H_4-CH(CH_3)CH(OH)SO_3Na \xrightarrow[RT,\ 24\ hr]{DMSO,\ Ac_2O}$$

$$[Ph-C_6H_4-CH(CH_3)COSO_3Na] \xrightarrow[H_2O,\ H^+]{K_2CO_3} Ph-C_6H_4-CH(CH_3)CO_2H$$

18. Oxidation of Aldehydes to Peroxy Acids

Aldehydes can be oxidised to peroxy acids by passing oxygen under irradation through the solution of the aldehyde in acetone[26] or by passing a current of ozonised oxygen through a solution of aldehyde in ethyl acetate at 20–22°C for 1–2 hr[27].

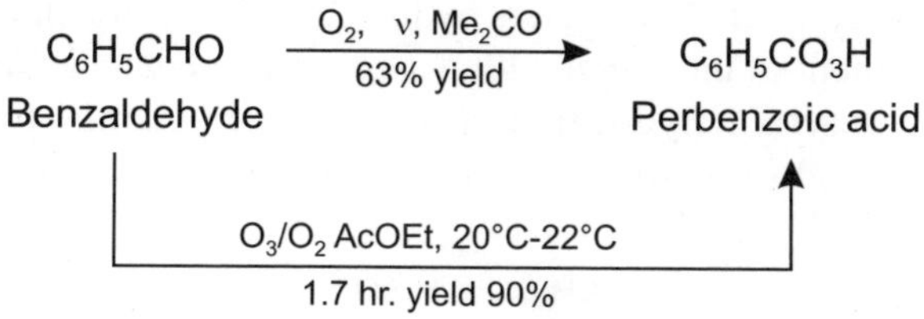

$$\underset{\text{Benzaldehyde}}{C_6H_5CHO} \xrightarrow[\text{63\% yield}]{O_2,\ \nu,\ Me_2CO} \underset{\text{Perbenzoic acid}}{C_6H_5CO_3H}$$

O_3/O_2 AcOEt, 20°C-22°C

1.7 hr. yield 90%

19. Dakin Reaction

The replacement of the aldehyde (or acetyl group) in phenolic aldehydes (or ketones) by a hydroxyl group on reaction with alkaline hydrogen peroxide is known as Dakin reaction[28]. Thus, salicylaldehyde is converted into catechol[29].

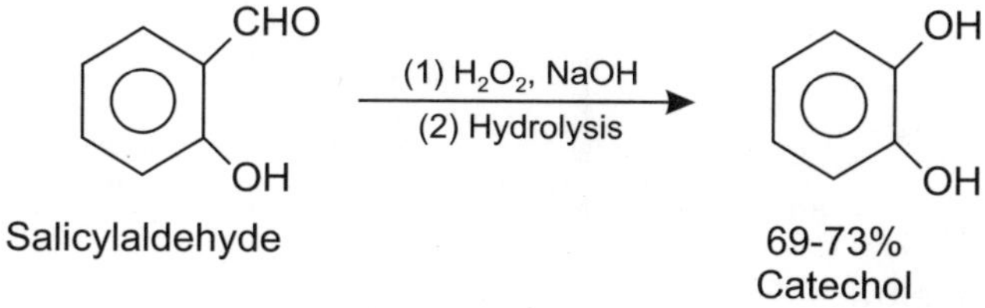

CHO OH (1) H_2O_2, NaOH (2) Hydrolysis OH OH

Salicylaldehyde

69-73%
Catechol

Reaction mechanism: It is beleived that the carbonyl carbon is attacked by the hydroperoxide anion. The resulting intermediate (1) undergoes migration of aryl group with subsequent elimination of hydroxide ion of give formic ester (2), which on hydrolysis gives catechol.

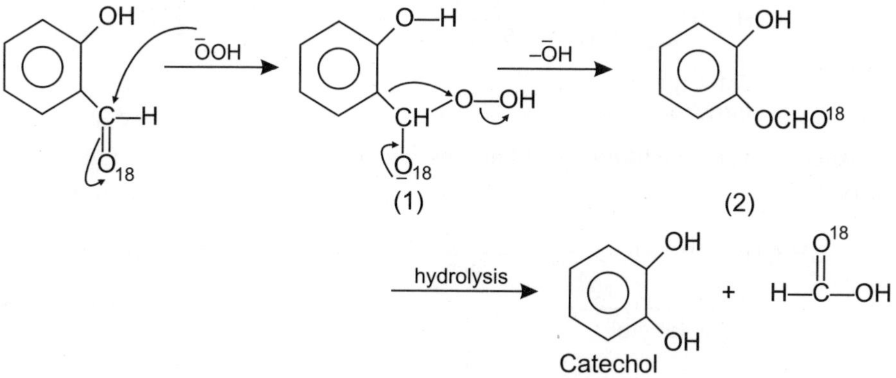

OH C—H O_{18} $\bar{O}OH$ O—H CH O—OH O_{18} (1) $-\bar{O}H$ OH $OCHO^{18}$ (2)

hydrolysis OH OH Catechol + O^{18} H—C—OH

The above mechanism[30] has been confirmed by taking the o-hydroxybenzaldyde in which oxygen CHO group is labelled by ^{18}O as shown above.

In Dakin's reaction, in place of hydrogen peroxide, other peroxyacids like peroxyacetic acid, peroxybenzoic acid and *m*-chloroperoxybenzoic acid can also be used.

20. Oxidation of Aldehydes to Esters

Aldehydes can be converted into esters by the conversion of aldehyde into acetals followed by treatment of the acetal with peroxyacetic acid and sulphuric acid as a catalyst. Thus, butyraldehyde diethyl acetal on treatemnt for 30 min with peroxyacetic acid and sulphuric acid furnishes ethyl butyrate is 60% yield[31].

$$CH_3CH_2CH_2CH(OC_2H_5)_2 \xrightarrow{CH_3CO_3H/H_2SO_4} CH_3CH_2CH_2C(=O)OC_2H_5$$

Butyraldehyde diethylacetal → 60% Ethyl butyrate

Heptanal dimethyl acetal on reaction with ozone in dichloromethane at –78°C for 15 hr or at room temperature for 1.5 hr. yield, methyl heptanoate[32]

$$C_6H_{13}CH(OCH_3)_2 \xrightarrow[-78^\circ C,\ 15\ min]{O_3,\ CH_2Cl_2} [\text{intermediate}] \longrightarrow C_6H_{13}C(=O)OCH_3$$

Heptanal dimethyl acetal → 90% Methyl heptanoate

Cyclic acetals react much faster. Thus, tetrahydropyranyl ethers are cleaved to hydroxy esters[32]

$$C_6H_{13}CH\langle O\text{-}CH_2\text{-}CH_2\text{-}O\rangle \xrightarrow[-78^\circ C,\ 10\ min]{O_3,\ CH_2Cl_2} C_6H_{13}CO_2CH_2CH_2OH$$

Tetrahydropyrany deriv. of heptanal → 100%

Acetals from unsaturated aldehydes can be transformed into unsaturated esters[33].

$$CH_3CH{=}CHCH(OC_4H_9)_2 \xrightarrow[40^\circ C,\ 1\ hr]{AcO_2H,\ AcOEt} CH_3CH{=}CHCO_2C_4H_9$$

Crotonaldehyde dibutylacetal → 73% Butyl crotanate

References

1. J. M. Church and L. Lynn, *Ind. Eng. Chem.*, 1950, **42**, 768.
2. R. J. Harrison and M. Moyle, *Org. Synth.* Collective Volume, 1963, **4**, 493.
3. C. W. Smith and R. T. Holm, *J. Org. Chem.*, 1957, **22**, 746.

4. R. J. Kennedy and A. M. stock, *J. Org. Chem.,* 1960, **25**, 1901.
5. A. Nashihara and I. Kubola, *J. Org. Chem.,* 1968,**33**,2525.
6. H. Filsselann, Chem. Ber., 1942,**75**,881.
7. E. J. Corey, N. W. Gilman and B. E. Ganem, *J. Am. Chem., Soc.,* 1968,**90**,5616.
8. R. N. Hammer and J. Klemberg, *Inorg. Synth.,* 1953,**4**,12; L. Fieser and L. Fieser, Reagents for Organic Synthesis, Wiley, New York, Vol 2, page 369.
9. E. Campaigne and W. M. Lesuer, *Org. Synth.* Collective Volume, 1963,**4**,919.
10. H. Rapport and H. N. Reist, *J. Am. Chem. Soc.,* 1955,**77**,490.
11. C. Moureu and R. Chaux, *Org. Synth.* Collective Volume, 1932, **I**, 166.
12. C. D. Hurd, J. W. Garrel and E. N. Osbone, *J. Am. Chem. Soc.,* 1933,**55**,1082.
13. I. A. Pearl, *Org. Synth.* Collective Volume, 1963,**4**,974.
14. E. Dalcanale and F. Montanari, *J. Org. Chem.,* 1986,**51**,567.
15. B. O. Lindgren and T. Nilsson, Acta Chem. Seand., 1973,**27**,888
16. M. Z. Barakat, M. F. Abdel-Wahals and M. M. El–Sadr, *J. Chem. Soc.,* 1956,4685.
16 a. E. J. Corey, N. W. Gilman and B. E. Ganem, *J. Am. Chem. Soc.,* 1968,**90**,5616.
17. R. Ruhoff, *Org. Synth.* Collective Volume, 1943, **2**, 315.
18. R. L. Shriner and E. C. Kleiderer, *Org. Synth.* Collective Volume, 1943, **2**, 538.
19. T. Sala and M. W. Sargent, Chem., Commun., 1978, 253.
20. D. Scholz, Monatsh. *Chem.,* 1979, **110**, 1471.
21. K. Nakawara, S. Mineo and S.Kawamura, Chem. Pharm. Bull., 1978,**26**,229.
22. L. M. Berkowitz and P. N. Rylander, *J. Am. Chem., Soc.,* 1958,**80**,6682.
23. P. Müller and J. Godoy, *Tetrahedron Lett.,* 1981,**22**,2361.
24. S. Cannizzaro, Ann., 1853,**88**,129; K. List and H. Limpricht, 1854,**90**,180.
25. P. G. M. Wuts and C. L. Bergh, *Tetrahedron Lett.,* 1986, **27**, 3995.
26. W. P. Jorissen and P. A. A. Van Der Beek, Rec. Trav. Chem. Pays –Bas, 1926,**45**,245.
27. C.R. Dick and R. F. Hanna, *J. Org. Chem.,* 1964,**29**,1218.
28. H. D. Dakin, OS, 1941, **1**, 149; J. E. Lettler, *Chem. Rev.,* 1949,**45**,385.
29. H. D. Dakin, *Org. Synth.* Collective Volume, 1932,**1**,149.
30. C. A. Bunton in Peroxide Reaction Mechanisim, Edited by J. O. Edwards, Interscience.
31. D. L. Heywood and B. Phillips, *J. Org. Chem.,* 1960,**25**,1699.
32. P. Deslongchamps and C. Moreau. *Can. J. Chem.,* 1971,**49**,2465.
33. D. L. Heywood and B. Phillips, *J. Org., Chem.,* 1960, **25**,1699.

2.6. OXIDATION OF KETONES

Ketones on oxidation give different products depending on the oxidising agent and the reaction conditions. Following are given some of the important oxidations:

Oxidation of Ketones to Carboxylic Acids

We have known that aldehydes are more easily oxidised than ketones. Aldehydes can be oxidised by mild oxidising agents (like silver oxide) or strong oxidising agents like potassium permanganatge

$$R{-}\overset{O}{\overset{\|}{C}}H \xrightarrow{KMnO_4,\ \bar{O}H} R{-}\overset{O}{\overset{\|}{C}}\bar{O} \xrightarrow{H_3O^+} R{-}\overset{O}{\overset{\|}{C}}{-}OH$$

As seen in the oxidation of aldehydes, they loose the hydrogen that is attached to the carbonyl carbon atom. Since ketones lack this hydrogen, they are more resistant to oxidation. However, oxidation of ketones with chromic acid or permanganate under vigorous reaction conditions vigorous enough to bring about reaction usually leads to rupture of the carbon chain adjacent to the carbonyl group resulting in the formation of carboxylic acids. These oxidations are of little value in synthesis. However, ketones can be oxidised under different conditions.

Willgerodt Reaction

Treatment of ketones with a mixture of sulphur and ammonium sulphide (prepared by saturating a concentrated aquous solution of ammonia with hydrogen sulphide[1,2]; the reaction mixture is heated to 160–222°C (using a autoclave or a sealed tube). The reaction products, amide of ω-arylcarboxylic acids together with smaller amounts of the corresponding ammonium salt of the carboxylic acid, which on hydrolysis with concentrated hydrochloric acid gives the coresponding acid ($-COCH_3$ of the ketone is converted into $-CH_2COOH$)

$$\underset{\text{Ketone}}{ArCOCH_3} \xrightarrow[H_2O,\ 160\text{-}220°C]{(NH_4)_2S,\ S} ArCH_2CONH_2 + ArCH_2C\bar{O}_2\overset{+}{N}H_4 \xrightarrow{\text{Hydrolysis, HCl, }\Delta} ArCH_2COOH$$

The above reaction is known as Willgerodt reaction[3]. An improvement of the above reaction is known as **Kindler variation of willgerodt reaction**[3,4]. This consists in heating the ketone with approximately equal amount of sulphur and a dry amine (*e.g.*, morpholine) instaed of ammonium poly sulphide. The main product is a thioamide, which on subsequent hydrolysis with acids or alkalies yields the carboxylic acids.

$$\underset{\text{Ketone}}{ArCOCH_3} + S + \underset{\text{Morpholine}}{HN\langle\ \rangle O} \xrightarrow{\Delta} ArCH_2CSN\langle\ \rangle O \xrightarrow[\Delta]{HCl} ArCH_2COOH + \underset{\text{Morpholine}}{HN\langle\ \rangle O} + H_2S$$

Using Willgerodt reaction (in the modified form) a large number of carboxylic acids have been obtained as given below.

Starting Ketone	***Product formed***	***Reference***
$C_6H_5COCH_2CH_3$	$C_6H_5CH_2COOH$	5
Methyl 2-napthyl ketone (naphthalene–$COCH_3$)	2-Naphthyl acetic acid (naphthalene–CH_2CO_2H), 90%	

Starting Ketone	*Product formed*	**Reference**
CH_3OC– (acenaphthene) 1-Acetyl acenaphthene	HO_2CH_2– (acenaphthene) 57% Acenaphthyl-1-acetic acid	6
$COCH_3$ (pyridine) 3-Pyridylketone	CH_2CO_2H (pyridine) 74% 3-Pyridylacetic acid	7

Oxidation with Potassium Hydroxide and Carbon Tetrachloride

Ketones having α and α′ hydrogens are oxidised[8] with powdered potassium hydroxide, carbon tetrachloride, water and tert. butyl alcohol.

$$(C_6H_5)_2CHCOCH_3 \xrightarrow[25\text{-}80°C,\ 10\ min,\ 1\ hr]{CCl_4,\ KOH,\ t\text{-}BuOH,\ H_2O} \underset{70\%}{Ph_2CHCH_2CO_2H}$$

In case, ketones have α-hydrogen, but not α′-hydrogen, the product obtained is the same as that obtained by oxidation with hypohalites[8].

Oxidation with Thallium Trinitrate

Anyl methyl ketones are oxidised with thallium trinitrate in methanolic solution in presence of perchloric acid to give the corresponding methyl arylacetates in 61–94% yeild.

$$ArCOCH_3 \rightleftharpoons ArC(OH){=}CH_2 \xrightarrow[HClO_4]{Tl(NO_3)_2,\ MeOH} Ar{-}C(OH)(OCH_3){-}CH_2{-}Tl(ONO_2){-}ONO_2$$

$$\xrightarrow{-HNO_3,\ -TlNO_3} \underset{\substack{61\text{-}94\% \\ \text{Methyl arylacetate}}}{ArCH_2CO_2CH_3}$$

Oxidation with Potassium Permanganate

Some ketones like acetophenone, *p*-methyl acetophenone and 3,4–dimethyl acetophenone on oxidation with potassium permanganate give the corresponding benzoic acids. However, 2,4–dimethylacetophenone on shaking with 1% aqueous potassium permanganate at room temperature gives [10,11] 66–72% of 2, 4–dimethylphenylglyoxylic acid. Another ketone, 2,5–dimethyl propiophenone on treatment with cold aqueous potassium permanganate yields a β-Keto acid (2,5-dimethylbenzoyl acetic acid)[12].

Haloform Reaction

Halogens (mostly iodine) and alkali oxidise methyl ketones to give the corresponding carboxylic acid salt and haloform (iodoform if iodine is used).

$$RCOCH_3 + 4NaOH + 3I_2 \xrightarrow{KI} RCOONa + CHI_3 + 3NaI + 3H_2O$$

The reaction is known as haloform reaction[13] or Lieben iodoform reaction[14]

The reaction mechanism involves halogenation of the methyl group to a trihalogenated derivative, which is subsequently hydrolysed with alkali to give haloform and salt of the acid.

For mechanism of the reaction *see* sodium hypoiodite (section 1.134).

Some examples of oxidation of ketones are given below.

$$\underset{\text{Pinacolone}}{(CH_3)_3CCOCH_3} \xrightarrow[\text{(2) 0-10°C, 1 hr, RT 3 hn, (3) steam distn. (4) } H_2SO_4]{\text{(1) } Br_2\text{, NaOH, 0-10°C, 15-20 min}} \underset{\substack{\text{71-74\%} \\ \text{Pivalic acid}}}{(CH_3)_3CCO_2H}$$

$$\underset{\text{Mesityl oxide}}{(CH_3)_2C{=}CHCOCH_3} \xrightarrow[\text{reflux 3-4 hr, } H_2SO_4]{KOCl\text{, } H_2O\text{, dioxane}} \underset{\substack{\text{49-53\%} \\ \beta,\beta\text{-dimethyl acrylic acid}}}{(CH_3)_2C{=}CHCO_2H}$$

$$\underset{\substack{\text{Isopropenyl} \\ \text{methyl ketone}}}{CH_2{=}C(CH_3)COCH_3} \xrightarrow[<20°C\text{, > 1 hr, } H_2SO_4]{\text{3N NaOCl, 6N NaOH}} \underset{\substack{\text{41\%} \\ \text{Methacrylic acid}}}{CH_2{=}C(CH_3)CO_2H}$$

$$\text{5-}C_4H_9\text{-2-}COC_3H_7\text{-pyridine} \xrightarrow[\text{(2) } NaHSO_3\text{, } H_2SO_4]{\text{(1) } Br_2\text{, NaOH, } H_2\text{, RT, 15 hr}} \underset{\text{79\%}}{\text{5-}C_4H_9\text{-pyridine-2-}CO_2H}$$

Oxidation of Ketones to α-dicarbonyl compounds

Methyl and methylene groups adjacent to carbonyl group are easily oxidised to carbonyls to yield α-diketones. The oxidation is brought about by selenium dioxide. Some examples are given below.[19–21]

$$p\text{-}CH_3OC_6H_4COCH_2C_6H_5 \xrightarrow[\text{1.1 } SeO_2\text{, 70\% AcOH, 89°C, 12 hr}]{\text{1.1 } SeO_2\text{, ioxane, } H_2O\text{, reflux, 5-8 hr OR}} \underset{\text{83-100\%}}{p\text{-}CH_3OC_6H_4COCOC_6H_5}$$

Cyclohexanone $\xrightarrow[\text{Cooling, 3 hr, RT, 11 hr}]{H_2SeO_2,\ H_2O,\ \text{dioxane}}$ Cyclohexane 1,2-dione (60%)

For more examples *see* selenium dioxide section 1.121.

Oxidative Cleavage of Cyclic Ketones

The oxidation of cyclopentanone with oxygen[22] or with nitric acid[23] gives glutaric acid.

Cyclopentanone $\xrightarrow[\text{yield 100\%}]{O_2,\ \text{MeONa, HMPA, 80°C, 23 hr}}$ $CO_2H-CH_2-CH_2-CH_2-CO_2H$ Glutaric acid

Cyclopentanone $\xrightarrow[\text{yield 82\%}]{13\%\ HNO_3,\ 100°C}$ Glutaric acid

In a similar way, cyclohexanone can be oxidised to adipic acid[22,23]

Cyclohexanone $\xrightarrow[\text{yield 100\%}]{O_2,\ \text{MeONa, HMPA, 23.5°C, 20 hr}}$ $CO_2CH_2CO_2H$ / $CH_2CH_2CO_2H$ Adipic acid

Cyclohexanone $\xrightarrow[\text{yield about 100\%}]{33\%\ HNO_3,\ \text{warm, gradual addn.}}$ Adipic acid

Baeyer-Villiger Oxidation

The oxidation of ketones to esters by peracetic acid (or hydrogen peroxide) is knwon as Baeyer-Villiger oxidation.

$$R_2C{=}O \xrightarrow{C_6H_5COOOH} R-\overset{O}{\overset{\|}{C}}-OR + C_6H_5COOH$$

Ketone → Ester

For mechanism of the reaction *see* peroxy acids 1.97.

Some example of Balger, Vollger oxidations are given below:

p-O_2N—C_6H_4—CO—C_6H_5 —(CH_3CO_3H / H_2PO_4, CH_3CO_2H, 25°C)→ p-O_2N—$C_6H_5COOC_6H_5$

p-Nitrobenzophenone → 95% *p*-Nitrophenylbenzoate

Cyclopropyl methyl ketone (—$COCH_3$) —(CF_3CO_3H / CF_3CO_2H, Na_2HPO_4, CH_2Cl_2, reflux)→ Cyclopropylacetate (—$OCOCH_3$) 55%

The oxidation of cydic ketones with peracids serve as a useful preparative route to lactones[26,27].

Cyclohexanone —($C_6H_5CO_3H$ / $CHCl_3$, 25°C)→ lactone 71%

Cyclopentanone —(CF_3CO_3H / CF_3CO_2H, 10-15°C)→ lactone 81%

Oxidation of Oximes of Ketones

Hydrazones of ketones are converted into diazo compounds. The oxidation can be affected by mercuric oxide[28] or lead tetracetate[29]

R_2C=$NHNH_2$ —(HgO, KOH, Et_2O, EtOH / 70-90% yield)→ R_2C=N—N Diazo compound

R_2C=$NHNH_2$ —($Pb(OA)_4$, PhCN, 0-25°C / 76-79% yield)→ Diazo compound

Thus, acetone hydrazone is converted into 2–diazopropane[28], hexafluroacetone into 2–diazohexafluoropropane[29]. Similarly, 2–octanone hydrazone gives 4–diazooctane[30].

References

1. D. Crouse, M. M. Wheeler, M. Goemann, P. S. Tabai, S. K. Basu and D. M. S. Wheeler, *J. Org. Chem.,* 1981, **46**, 1814.
2. H. Hartmann and W. Bosshard, *Helv. Cheim. Acta,* 1941, **24**E, 28.
3. C. Willgerodt, *Ber.,* 1887, **20**, 24 67.
4. K. Kindler, *Ann.,* 1923, **43**1, 193.
5. E. Schwenk and D. Papa, *J. Org. Chem.,* 1946, **II**, 798.
6. I. L. Fieser and G. W. Kilmer, *J. Am. Chem., Soc.,* 1940, **62**, 1354.
7. H. Hartmann and W. Bosshard, *Helv. Chem. Acta,* 1941,**24**E,28.
8. C. Y. Meyers, A. M. Malte and W. S. Matthews, *J. Am. Chem. Soc.,* 1969, **91**, 7510.
9. A. Mckillop, B. P. Swann and E. C. Taylor, *J. Am. Chem. Soc.,* 1971, **93**, 4919.
10. A. Claus and R. Wollner, *Chem. Ber.,* 1885,**18**,1856.
11. A. Claus, Chem. *Ber.,* 1886, **19**, 230.
12. A. Claus and E. Flckert, *Chem., Ber.,* 1886,**19**,3182.
13. A. Lieben, Ann (suppl), 1887, **7**, 218; R. C. Fuson and B. A. Bull, *Chem. Rev.,* 1934,**15**,275.
14. D.A. Evans, V. N. Nelson, E. Vogel and T. R. Taber, *J. Am. Chem. Soc.,* 1981,**103**,3099.
15. L. T. Sandborn and E. W. Bousquet, *Org. Synth.* Collective Volume, 1932,**1**,526.
16. L. I. Smith, W. W. Prichard and L. J. Spillane, *Org. Synth.* Collective Volume, 1955,**3**,302.
17. T. While. *J. Chem. Soc.,* 1943,238.
18. E. Hardegger and E. Nickles, *Helv. Chim. Acta,* 1957,**40**,2428.
19. E. R. Bockstahler and D. L. Wright, *J. Am. Chem. Soc.,* 1949,**71**,3760.
20. E. J. Corey and J. P. Schalfer, *J. Am. Chem. Soc.,* 1960,**82**,910.
21. C. C. Hach, C. V. Banks and H. Diehl, *Org. Synth.* Collective Volume, 1963,**4**,229.
22. T. J. Wallace, H. Pobiner and A. Schriesheim, *J. Org. Chem.,* 1965,**30**,3768.
23. E. Boedtker, J. Pharm. Chem., 1932,**15**,225
24. W. Von. E. Doeving and L. Speers, *J. Am. Chem. Soc.,* 1950,**72**,5515
25. W. D. Emmons and G. B. Lucas, *J. Am. Chem. Soc.,* 1965,**77**,2287.
26. S. L. Friess, J. Am. Chem. Soc., 1949, **71**, 2571; J. L. Maleos and H. Menchacer, *J. Org. Chem.;* 1964,**29**,2026.
27. W. F. Sager and A. Duckworth, *J. Am. Chem. Soc.,* 1955,**77**,188.
28. S. D. Andrews, A. C. Day, P. Raymond and M. C. Whiting, *Org., Synth.* Collective Volume, 1970,**50**,27.
29. M. Middleton and D. M. Gale, *Org. Synth.* Collective Volume, 1988, **6**, 161.
30. A. C. Day, P. Raymond, R. M. Southam and M. C. Whiting, *J. Chem., Soc.,* 1966,467.

2.7 MISCELLANEOUS OXIDATIONS

2.7.1 Oxidation of Carboxylic Acids

The oxidation of carboxylic acids, with hydrogen peroxide gives peroxy acids[1,2] For details *see* hydrogen peroxide, section 1.56.

$$\underset{\text{Benzoic acid}}{C_6H_5CO_2H} + 70\%\ H_2O_2 \longrightarrow \underset{\text{Peroxybenzoic acid}}{C_6H_5CO_3H} + H_2O$$

The esters of carboxylic acids undergo oxidative coupling in the α-position with respect to ester group. Thus, ethyl acetate on treatment with diisopropylamide in tetrahydrofuan at –70°C followed by reaction with anhydrous ferric chloride in dimethylformamide yield diethyl succinate[3]

$$\underset{\text{Ethylacetate}}{CH_3CO_2Et} \xrightarrow[\text{(2) FeCl}_3\text{/DMF}]{\text{(1) Diisoproylamide/THF}} \underset{\substack{69\% \\ \text{Diethylsuccinate}}}{\begin{array}{c} CH_2CO_2Et \\ | \\ CH_2CO_2Et \end{array}}$$

The esters can also be hydroxylated in α-position under similar conditions[4]. Simlar results are obtained by oxidation with iodobenzenediacetate[5].

$$\underset{\text{Ethylphenyl acetate}}{C_6H_5CH_2CO_2Et} \xrightarrow[\text{(2) Mo}_2\text{O}_3]{\text{(1) Lithiumdiisoproylamide}} \underset{\substack{58\% \\ \text{Ethylmandalate}}}{\begin{array}{c} C_6H_5CHCO_2Et \\ | \\ OH \end{array}}$$

PhI(OAc)$_2$, 3MeONa

RT, 3 days, 70% yield

Hydroxylation of enolates in the α-position can occur enantioselectively with chiral oxidants, such as (+)–2R.8aS and (–)–2S.8aR camphorsulfonyloxyaziridines[6]

$$\underset{\text{Methyl phenyl acetate}}{C_6H_5CH_2COOCH_3} \xrightarrow[\text{(2) Oxidant (A)}]{\text{(1) LDA, THF, –78°C, 25-30 min}} \underset{\substack{(-)-R \\ 84\% \\ 54\%\ ee}}{C_6H_5CH(OH)COOCH_3}$$

A = (camphorsulfonyloxaziridine: N, O, SO$_2$)

The methylene group in α-position with respect to carboxylic group can be oxidised to carbonyl group to give α-keto esters; the oxidation is affected with air[7] in presence of cobalt benzoate at 110–115°C or with nitrogen sesquioxide[8]

$$\underset{\text{Methyl phenyl acetate}}{C_6H_5\text{—}CH_2\text{—}\overset{\displaystyle O}{\overset{\|}{C}}\text{—}OCH_3}\ \xrightarrow[\text{110-115°C}]{\text{air, }(C_6H_5COO)_2Co} \underset{\substack{86\% \\ \text{Phenyl glooxylate}}}{C_6H_5\text{—}\overset{\displaystyle O}{\overset{\|}{C}}\text{—}\overset{\displaystyle O}{\overset{\|}{C}}\text{—}OEt}$$

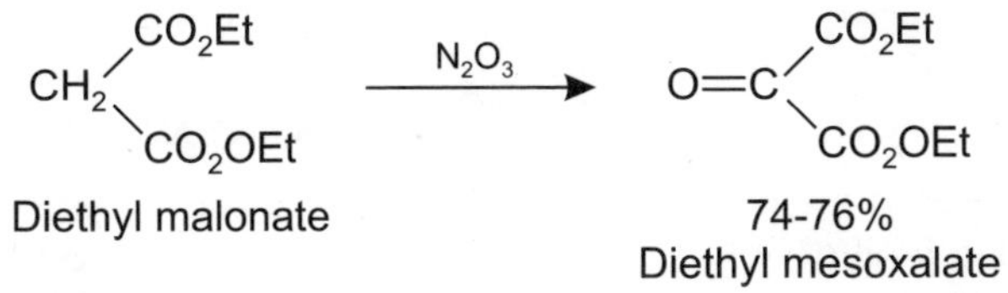

$$CH_2(CO_2Et)(CO_2OEt) \xrightarrow{N_2O_3} O{=}C(CO_2Et)(CO_2OEt)$$

Diethyl malonate → Diethyl mesoxalate (74-76%)

2.7.2 Oxidation of Amides

Oxidation of primary amides with lead tetracetate gives isocyanates[9].

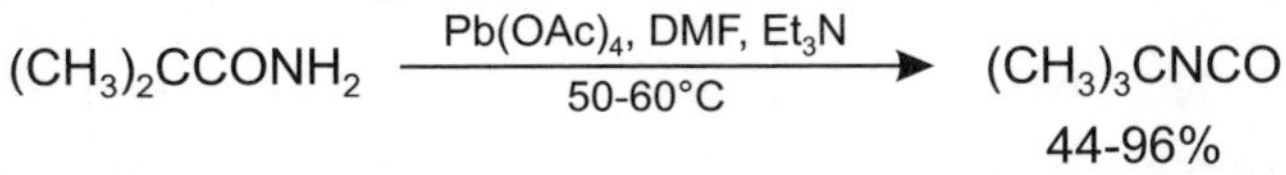

$$(CH_3)_2CCONH_2 \xrightarrow[50\text{-}60°C]{Pb(OAc)_4,\ DMF,\ Et_3N} (CH_3)_3CNCO$$

44-96%

2.7.3 Oxidation of Hydrazides

Hydrazides on oxidation with potassium ferricyanide in the presence of aqueous ammonia gives the corresponding aldehydes[11]. However, on oxidation with ceric ammonium nitrate, the corresponding carboxylic acids are obtained[11].

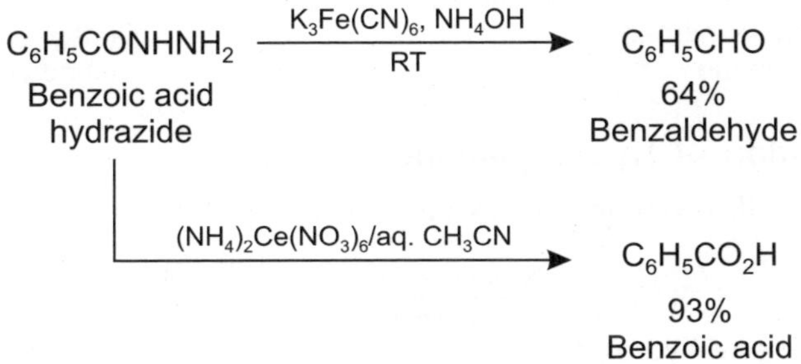

$$C_6H_5CONHNH_2 \xrightarrow[RT]{K_3Fe(CN)_6,\ NH_4OH} C_6H_5CHO$$

Benzoic acid hydrazide → Benzaldehyde (64%)

$$C_6H_5CONHNH_2 \xrightarrow{(NH_4)_2Ce(NO_3)_6/aq.\ CH_3CN} C_6H_5CO_2H$$

Benzoic acid hydrazide → Benzoic acid (93%)

2.7.4 Oxidation of Nitro Compounds

Primary and secondary nitroparffins in alkaline medium on oxidation with potassium permanganate gives aldehydes and ketones[12]. This is a variation of **Nef reaction**.

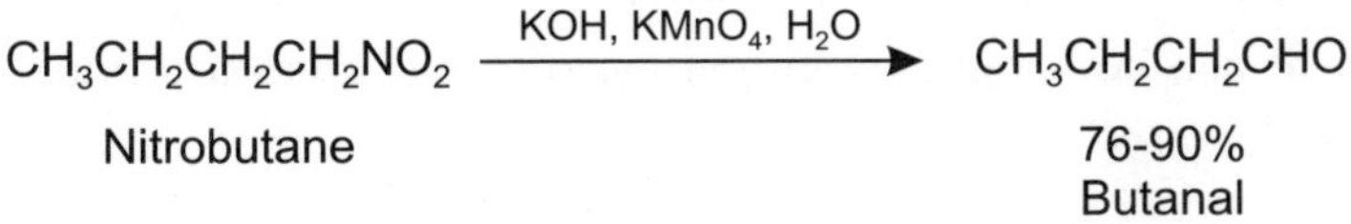

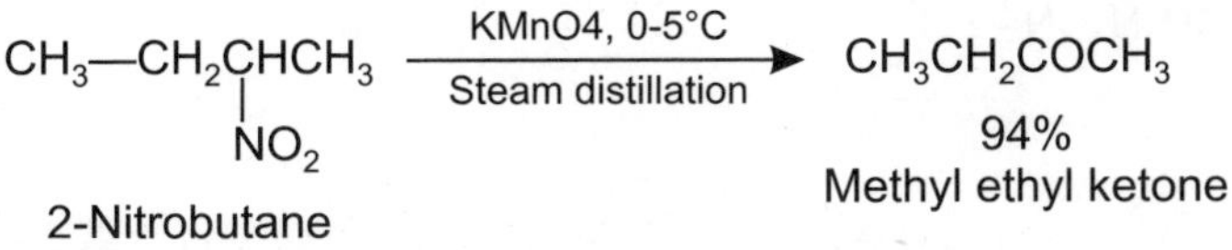

$$CH_3CH_2CH_2CH_2NO_2 \xrightarrow{KOH,\ KMnO_4,\ H_2O} CH_3CH_2CH_2CHO$$

Nitrobutane → Butanal (76-90%)

$$CH_3{-}CH_2CH(NO_2)CH_3 \xrightarrow[\text{Steam distillation}]{KMnO4,\ 0\text{-}5°C} CH_3CH_2COCH_3$$

2-Nitrobutane → Methyl ethyl ketone (94%)

2.7.5. Oxidation of Nitroso Compounds

The oxidation of nitroso compounds to the nitro compounds can be affected by hydrogen peroxide or potassium dichromate[13,14]. Under the above reaction conditions, the primary amino groups are not affected.

$$\text{2-Nitroso-5-nitrotoluene} \xrightarrow[65°C,\ 1.75\ hr]{K_2Cr_2O_7,\ H_2SO_4,\ H_2O} \text{2,5-Dinitrotoluene (55-66\%)}$$

2-Nitroso-5-nitro toluene → 55-66% 2,5-Dinitrotoluene

$$\text{3-Nitroso-2,6-diaminopyridine} \xrightarrow{H_2O_2,\ 35°C,\ 1.5\ hr} \text{3-Nitro-2,6-diaminopyridine (96\%)}$$

3-Nitroso-2,6-diamino pyridine → 96% 3-Nitro-2,6-diaminopyridine

Diethyl N-nitroso amine on oxidation with peroxytrifluoroacetic acid give the corresponding nitroamines[15].

$$(C_2H_5)_2NNO \xrightarrow[RT,\ 30\ min,\ 1\ hr\ reflux]{90\%\ H_2O_2,\ (CF_3CO)_2O,\ CH_2Cl_2} (C_2H_5)_2NNO_2$$

Diethyl-N-nitroso amine → Diethyl-N-nitro amine (76%)

2.7.6 Oxidation of Hydroxylamines

Oxidation of hydroxylamines gives nitrosocompounds[16,17].

$$C_6H_5NHOH \xrightarrow[0°C,\ 3\ min,\ yield\ 90\%]{K_2Cr_2O_7,\ H_2SO_4,\ H_2O} C_6H_5NO$$

Phenyl hydroxyl amine → Nitrosobenzene

$$C_6H_5NHOH \xrightarrow[few\ hrs,\ yield\ 89\%]{EtO_2CN\text{—}NCO_2Et,\ Et_2O,\ 0°C} C_6H_5NO$$

2.7.7 Oxidation of Azo Compounds

Oxidation of azo compounds with perbenzoic acid gives a mixture of azoxy compounds[18].

$$C_6H_5CH_2N{=}NC_6H_5 \xrightarrow[0°C,\ 19\ hr]{C_6H_5CO_3H,\ Et_2O} C_6H_5CH_2\text{—}N{=}\overset{+}{N}(O^-)\text{—}C_6H_5 + C_6H_5CH_2\text{—}\overset{+}{N}(O^-){=}N\text{—}C_6H_5$$

trans 47.5%
cis 3.0%

5%

2.7.8 Oxidation of Azides

Azides on oxidation by a sequence of reactions give[19] a mixture of the corresponding nitro compounds and aldehydes.

$$C_8H_{17}N_3 \xrightarrow[\text{RT, 5 min}]{Ph_3P,\ CH_2Cl_2} \left[C_8H_{17}N{=}PPh_3\right] \xrightarrow[-78°C,\ 10\ \text{min}]{O_3,\ CH_2Cl_2} C_8H_{17}NO_2\ 70\% + C_7H_{15}CHO\ 13\%$$

2.7.9 Oxidation of Hydrazo Compounds

Hydrdzo compounds on oxidation are dehydrogenated to azo compounds[20].

$$(CH_3)_2CHNHNHCH(CH_3)_2HCl \xrightarrow[\text{7 days heat}]{Cu} (CH_3)_2CHN{=}NCH(CH_3)_2$$

sym-Diisopropylhydrazine hydrochloride → 2,2′-Bisazoisopropane

Hydrazobenzene is oxidised by manganese dioxide[21], by diethyl azodicarboxylate[22] and by diphenyl picryl hydrazyl[23] into azo benzene.

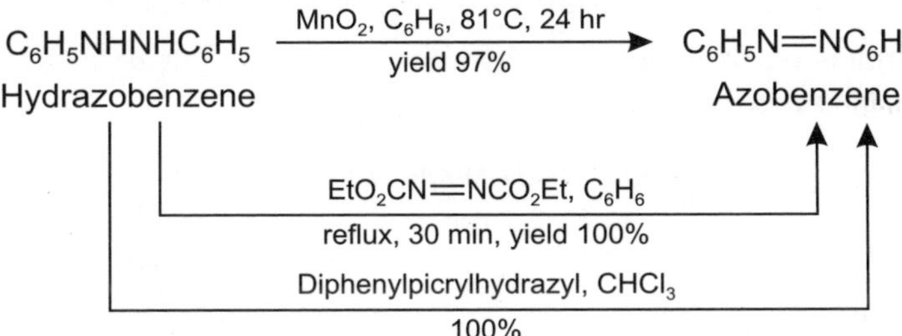

Diethyl and di-tert-butyl hydrazodicarboxylate are oxidised to the respective azo compounds by nitric acid[24] or by N-bromosucinimide[25]

$$EtO_2CNHNHCO_2Et \xrightarrow[5°C,\ 2\ \text{hr}]{\text{Fuming } HNO_3} EtO_2CN{=}NCO_2Et\ \ 70\text{-}80\%$$

Diethylhydrazodi-carboxylate → Diethylazodicarboxylate

$$RO_2CNHNHCO_2R \xrightarrow[CHCl_3]{\text{Diphenylpicrylhydrazyl}} RO_2CN{=}NCO_2R\ \ 90\text{-}95\%$$

Di-tert-butylhydrazo dicarboxylate, $R = (CH_3)_3C$ → Ditert. butylazo dicarboxylate

2.7.10 Oxidation of Primary Amines

Primary amines can be oxidised to azo, azoxy, nitroso, nitro, imines, aldehyde or ketones and nitriles depending on the structure of the primary amine and the oxidant used.

Oxidation of Primary Amines into Azo Compounds

Aromatic primary amines are dehydrogenated to azo compounds by mangansse dioxide[26], barium manganate[27] or oxygen and cobalt peroxide[28].

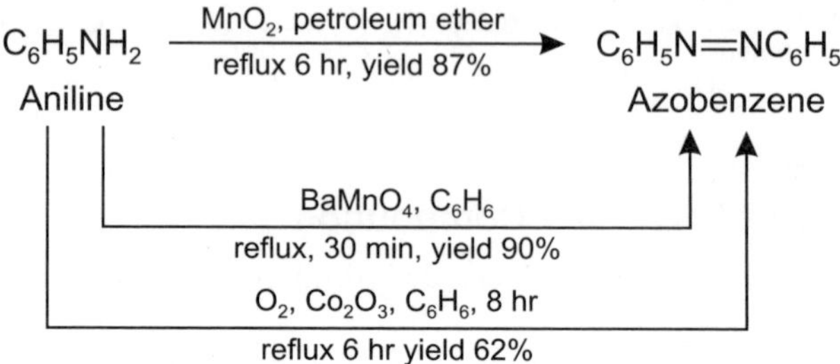

Oxidation of *p*-chloroaniline to *p–p′*-dichloro azobenzene is affected by relfuxing for 15 min with silver permanganate-pyridine complex[29] in benzene (yield 85%). Nickel peroxide in benzene converts *p*-nitroaniline into *p-p′*-dinitroazo benzene[30] on refluxing in benzene for 6 hr; yield 44%.

Oxidation of Primary Amines into Azoxy Compounds

Primary amines on oxidation with hydrogen peroxide[31] or peroxyacetic acid[32] yield azoxy benzene.

$$\underset{\text{Aniline}}{C_6H_5NH_2} \xrightarrow[20°C,\ 7\ min]{50\%\ H_2O_2,\ MeCN,\ MeOH,\ NaOH} \underset{\text{Azoxybenzene}}{C_6H_5N(\rightarrow O)=NC_6H_5}$$

Aniline → Azoxybenzene: 45% AcO_2H, H_2O; 40°C, RT, overnight, yield 60%

During oxidation with peroxacetic acid, 11% of nitrobenzene is also obtained.

Oxidation of Primary Amines into Nitroso Compounds

Oxidation of aromatic primary amines with aqueous permanganate and concentrated sulphuric acid give the corresponding nitroso compounds[33,34].

CH3, NH2, R′, R → (aq. $KMnO_4$ + conc. H_2SO_4) → CH3, NO, R′, R

3-Nitro-o-toluidine (R = NO_2; R′ = H)
5-Nitro-o-toluidine (R = H; R′ = NO_2)

3-Nitro-2-nitrosotoluene
(R = NO_2; R′ = H) yield 60%
5-Nitro-2-nitrosotoluene
(R = H; R′ = NO_2) yield 55-66%

2, 6-Dihaloanilines are converted into 2, 6–dihalonitrosobenzenes on oxidation with organic peoxy acids[35] and hydrogen peroxide[36] benzenes.

2,6-Dihaloanilines X = F or Cl —(BzO_2H, $CHCl_3$, RT yield 85% OR 30% H_2O_2, AcOH, RT, 48 hr yield 91%)→ 2,6-Dihalo nitrosobenzenes

Oxidation of Primary Amines into Nitro Compounds

Aliphatic primary amines can be oxidised to the correspodning nitro compounds. Thus, tert-butylamine is oxidised[37,38] to 2-methyl 2-nitropropane by aqueous potassium permanganate at 45°C for 8 hr.

$$(CH_3)_3C{-}NH_2 \xrightarrow{\text{aq. } KMnO_4,\ 45°C,\ 8\text{ hr}} CH_3{-}C(NO_2)(CH_3){-}CH_3$$

tert-Butytamine → 83% 2-Methyl-2-nitropropane

Under similar conditions 4–amino-2,2,4-trimethylbutane yields 4-nitro-2,2,4-trimethyl pentane is 69–82% yields[39].

$$H_2NCH(CH_3)CH_2C(CH_3)_2CH_3 \xrightarrow{\text{aq. } KMnO_4,\ 45°C,\ 8\text{ hr}} O_2NCH(CH_3)CH_2C(CH_3)_2CH_3$$

4-Amino-2,2,4-trimethyl butane → 69-82% 4-Nitro-2, 2,4- -trimethylbutane

Peroxyacetic acid[40] or dimethyldioxirane[41]. are also used to convert aliphatic primary amines to the corresponding nitro compounds.

$$CH_3CH_2CH(CH_3)NH_2 \xrightarrow{90\%\ H_2O_2,\ Ac_2O,\ H_2SO_4,\ (CH_2Cl_2)\ \text{reflux 2 hr, 65\% yield}} CH_3CH_2CH(CH_3)NO_2$$

2-Aminobutane → 2-Nitrobutane

Alternatively: $Me_2C(O_2)$ (dimethyldioxirane), Me_2CO, RT, 3 min; 87% yield

Primary aromatic amines on oxidation with hydrogen peroxide in presence of sulphuric acid or trifluoroacetic anhydride gives nitrobenzene[40,42,43]

$$C_6H_5NH_2 \xrightarrow[\text{reflux 80 min, yield 83\%}]{90\%\ H_2O_2,\ Ac_2O,\ H_2SO_4,\ CHCl_3} C_6H_5NO_2$$

Aniline → Nitrobenzene

$$C_6H_5NH_2 \xrightarrow[\text{reflux, 1.5 hr, yield 89\%}]{90\%\ H_2O_2,\ (CF_3CO)_2O,\ CH_2Cl_2} C_6H_5NO_2$$

Some other examples are given below.

2,6-Dichloro aniline $\xrightarrow{CF_3CO_3H}$ 2,6-Dichloro nitrobenzene (59-73%)

2-Amino-4-methyl pyridine $\xrightarrow[\text{10-25°C, 50 hr}]{30\%\ H_2O_2/\text{Fuming}\ H_2SO_4}$ 4-Methyl-2-nitro pyridine

Dehydrogenation of Primary Amines to Imines or Aldheydes or Ketones

The reaction of primary amines with a suitable oxidising agent results in dehydrogenation involving elimination of hydrogen from nitrogen of the amino group and from the carbon of the adjacent methyene group resulting in the formation of imines which in the presence of water gives aldehydes[27,46,47].

$$R\text{—}CH_2NH_2 \xrightarrow{-H_2} [RCH{=}NH] \xrightarrow{H_2O} RCHO + NH_3$$

Thus, the dehydrogenation of primary amine, benzylamine gives benzaldehyde

$$C_6H_5CH_2NH_2 \xrightarrow[\text{yield 95\%}]{BaMnO_4,\ C_6H_6,\ \text{reflux 1 hr}} C_6H_5CHO$$

Benzylamine → Benzaldehyde

$$C_6H_5CH_2NH_2 \xrightarrow[\text{50-75°C, steam distillation, yield 61\%}]{KMnO_4,\ ZnSO_4,\ t\text{-BuOH},\ H_2O} C_6H_5CHO$$

In case, the primary amino group is attached to a secondary carbon (as in the case of cyclohexylamine), dehydrogenation results in the formation of ketones[47].

$$\text{Cyclohexylamine} \xrightarrow[\text{70°C, distillation}]{KMnO_4,\ CaSO_4,\ H_2O,\ \text{t-BuOH}} \underset{75\%}{\text{Cyclohexanone}}$$

In certain cases, as in the case of benzhydrylamine, the product obtained by oxidation with potassium permanganate depends upon the cation of the buffer used in the reaction[48].

$$\underset{\text{Benzhydrylamine}}{(C_6H_5)CHNH_2} \xrightarrow[H_2O,\ \text{t-BuOH, 70°C, reflux 8 hr}]{KMnO_4,\ CaSO_4,\ 2H_2O} \underset{73\%}{(C_6H_5)_2C{=}N{-}N{=}C(C_6H_5)_2}$$

$$(C_6H_5)CHNH_2 \xrightarrow[\text{RT, 45 min}]{KMnO_4,\ ZnSO_4,\ H_2O,\ Me_2CO} \underset{74\%}{(C_6H_5)_2C{=}N\ CH(C_6H_5)_2}$$

Dehydrogenation of Primary Amines to Nitriles

Aliphatic primary amines having two hydrogen atoms on the adjacent carbon, give nitriles on dehydrogenation[49,50]

$$C_6H_{13}CH_2NH_2 \xrightarrow[\text{60°C, 30 min, yield 65\%}]{Pb(OAc)_4,\ C_6H_6,\ C_5H_5N} \underset{65\%}{C_6H_{13}CN}$$

$$C_6H_{13}CH_2NH_2 \xrightarrow[\text{yield 55\%}]{K_3Fe(CN)_6} C_6H_{13}CN$$

Some other examples of the dehydrogenation of primary amines to nitriles include:

- Conversion of hexylamine into capronitrile on refluxing with nickel perioxide in benzene for 1.5 hr[51] (73% yield).
- Conversion of octylamine into ocatanenitrile by treatment with sodium hypochlorite and tetrabutyl ammonium hydrogen sulphate in ethyl acetate at room temperature for 35 min[52] (yield 60%).
- Conversion of benzylamine into benzonitrile by reaction with AgO in water at 25–70° (yield 77%)[53].

2.7.11 Oxidation of Secondary and Tertiary Amines

Secondary amines on oxidation with hydrogen peroxide (6%) at room temperaure give the corresponding dialkylhydroxylamines[54].

$$\underset{\text{Diethylamine}}{Et_2NH} + 6\%\ H_2O_2 \xrightarrow{\text{RT}} \underset{\text{N, N-Diethyl hydroxylamine}}{Et_2N{-}OH}$$

Oxadation of tertiary amines with hydrogen peroxide and peroxyacetic acid give the corresponding oxide[55].

Bu
Bu — N
Bu
Tributyl-amine

$\xrightarrow[\text{0°C to r.t, 25 hr}]{H_2O_2\,(30–35\%)}$

Bu
Bu — $\overset{+}{N}$ — $\bar{O}$
Bu
958
Tributylamine oxide

Oxidations[56–58] of pyridine with hydrogen peroxide, peracetic acid or dimethyldioxirane gives the pyridine N-oxide

Pyridine $\xrightarrow[\text{NaOH, pH8, 25–30°C, 3 hr. 79\% yield}]{50\%\ H_2O_2,\ PhCN,\ MeOH}$ Pyridine oxide (N^+—O^-)

Pyridine → 40% AcO_2H, 85°C, 50–60 min, 78–83% yield → Pyridine oxide

Pyridine → Me_2C(dioxirane), H_2O, KOH, pH 7.5–8.0, RT, 2hr. yield 93% → Pyridine oxide

Pyridine oxide is an oxidant, which is capable of hydroxylating aromatic rings[59].

An intersting phenomenon is that the presence of oxygen on the nitrogen of the pyridine ring reverses the direction of electrophilic substituions of the pyridine ring. Normally, electrophilic attack on pyridine occurs in β-position, however, attack on pyridine oxide occur in α and *r* positions. Thus, after the introduction of the electrophile the pyridine oxide is converted into pyridine by mild reduction, such as treatement with salts of iron or titanium.

Like pyridine, pyridine homologues like α-picoline, 2,5–dimethyl pyridine and nicotinamide are converted into their N-oxides.

The best way to accomplish the oxidation of pyridine is to use *m*-chloroper benzoic acid. As expected, the electron rich pyridines are more easily oxidised than electron-deflicent one.

Pyridine + H—O—O—C(=O)—R $\xrightarrow{m\text{-CPBA}}$ Pyridine oxide (N^+—O^-) + RCOOH

R = CH_3—C_6H_4—

A new oxidant, methyl trioxorhenium (MTO) has recently been found to oxidise successfully both electron-deficient and electron rich pyridines. In this case, the active oxidant is most likely the peroxy complex (shown below, formed from MTO and H_2O_2) which adds oxygen to the pyridine nitrogen.

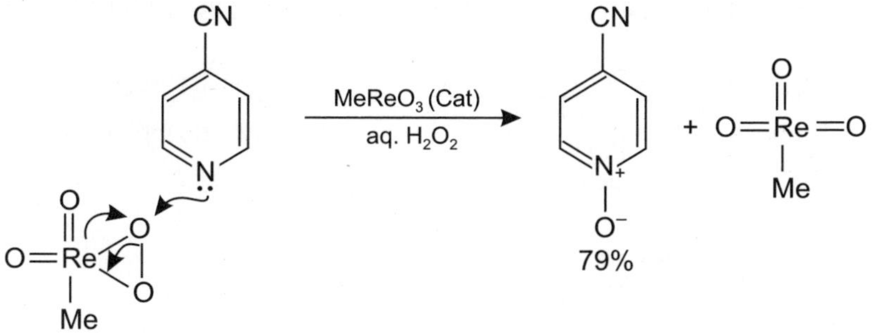

CN CN

$MeReO_3$ (Cat) / aq. H_2O_2 → + O=Re(=O)(=O)Me

79%

2.7.12 Oxidation of Aromatic Amines to Quinones

Aromatic amines can be oxidised to *p*-benzoquinones. Thus, aniline on oxidation with potassium dichromate and sulphuric acid at room temperature gives *p*-benzoquinone in 86% yield.[60] Similarly, 2,5-dimethylaniline at 80°C gives 55% yield of *p*-xyoquinone[61]. Oxidation of aromatic amines can also be affected by Fermy salt (potassium nitrosodisulfonate). The oxidation of amino group occurs even if it is acetylated. Some examples of oxidation by Fermy salt are given below[62,63].

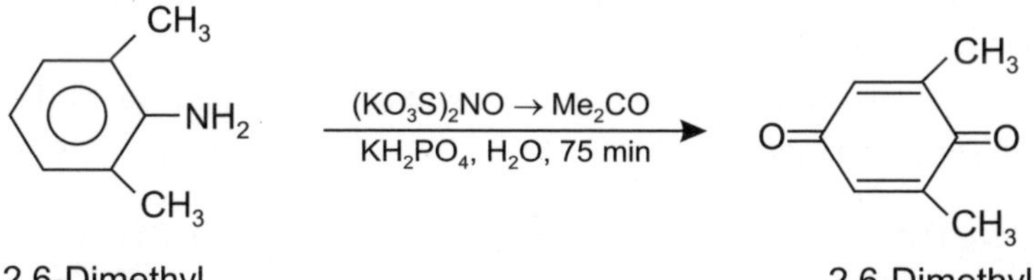

$(KO_3S)_2NO \rightarrow Me_2CO$ / KH_2PO_4, H_2O, 75 min →

2,6-Dimethyl aniline → 2,6-Dimethyl *p*-benzoquinone

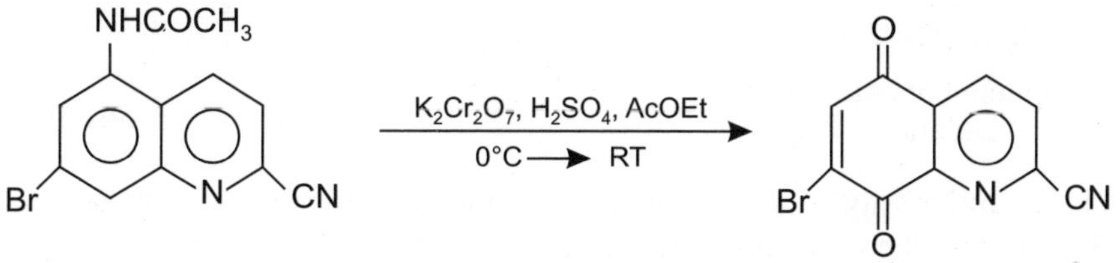

$K_2Cr_2O_7$, H_2SO_4, AcOEt / 0°C → RT

73%

2.7.13 Oxidation of Phosphorus Compounds

Tertiary phosphines, like tertiary amines are oxidised to the corresponding oxides. Thus, tributylphosphine on oxidation with ozone adsorbed in silica gel at –70°C give tributyl phosphine oxide in 92% yield[64].

$$(Bu)_3P \xrightarrow[CH_2Cl_2 - 70°C]{Ozone/Silicagel} (Bu)_3P{=}O$$

Tributyl phosphine → Tributyl phosphene oxide

Similarly triphenylphosphine is oxidised[27,64,65] to triphenylphosphine oxide

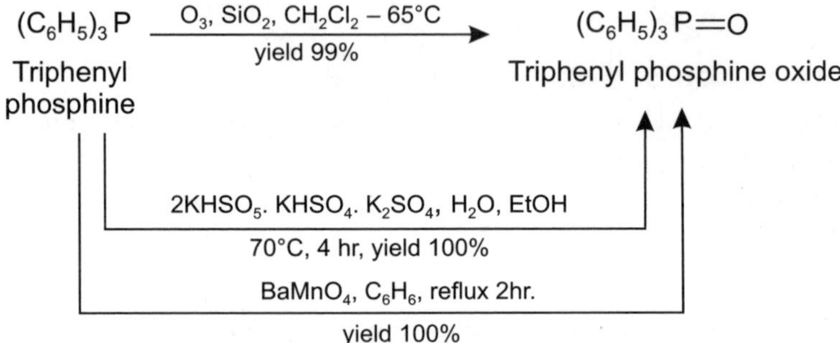

Similarly, trialkyl or triaryl phosphites are oxidised to the corresponding phosphate[64]

$$(C_6H_5O)_3P \xrightarrow{O_3,\ CH_2Cl_2Cl_2,\ 5\text{–}10°C} (C_6H_5O)_3P{=}O$$

Triphenyl phosphite → Triphenyl phosphate

The oxidation of phosphites to phosphates is important in the chemistry of nucleoside phosphites[66].

Trialkyl and triarylphosphine sulphides and selenides also on oxidation give trialkyl and triaryl phosphine oxides[67]

$$(C_6H_5)_3P{=}S \xrightarrow[-75°C]{O_3,\ CH_2Cl_2} (C_6H_5)_3P{=}O\ \ 93\%$$

Triphenyl phosphine sulphide → Triphenyl phosphine oxide

$$(C_6H_5)_3P{=}Se \xrightarrow[-75°C]{O_3,\ CH_2Cl_2} (C_6H_5)_3P{=}O$$

Triphenyl phosphine Selenide → Triphenyl phosphine oxide

In a similar way trialkyl or triaryl thiophosphates and selenophosphates on oxidation yield the corresponding phosphates[67].

$$(C_2H_5O)_3P{=}S \xrightarrow[-75°C]{O_3,\ CH_2Cl_2} (C_2H_5O)_3P{=}O\ \ 93\%$$

Triethylthio phosphate → Triethyl phosphate

$$(C_2H_5O)_3P{=}Se \xrightarrow[-75°C]{O_3,\ CH_2Cl_2} (C_2H_5O)_3P{=}O$$

Triethyl seleno phosphate → Triethyl phosphate

2.7.14 Oxidation of Arsenic Compounds

Tertiary arsines on oxidation with hydrogen peroxide give the corresponding arsine oxides[68].

$$\underset{\text{Triphenylarsine}}{(C_6H_5)_3As} \xrightarrow[\text{25–30°C, 20–30 min + 30 min}]{\text{30\% } H_2O_2, \text{ Me}_2CO} \underset{\substack{\text{84–87\%} \\ \text{Triphenylarsine oxide}}}{(C_6H_5)_3As{=}O}$$

2.7.15 Oxidation of Sulphur Compounds

Products obtanied by oxidation of sulphur compounds depend on the nature of the sulphur compound and the oxidising agent used.

Oxidation of Thiols

Thiols can be oxidised under mild conditions to give disulphides. Some examples are given below.

Reference

$$\underset{\text{Butanethiol}}{C_4H_9SH} \xrightarrow[\text{RT, 4 hr.}]{\text{air, } Al_2O_3.C_6H_6} \underset{\substack{\text{89\%} \\ \text{Dibutyldisulphide}}}{C_4H_9SSC_4H_9}$$ 69

70

$$CH_2{=}CH\ CH_2SH \xrightarrow[\text{reflux, 5–6 hr.}]{MnO_2,\ CHCl_3} CH_2{=}CH\ CH_2SSCH_2CH{=}CH_2$$

$$C_6H_5SH \xrightarrow[\text{RT. 6 hr.}]{\text{air, } Al_2O_3,\ C_6H_6} C_6H_5SSC_6H_5$$ 69

$$H_2N{-}C_6H_4{-}SNa \xrightarrow[\text{2hr.}]{\text{30\% } H_2O_2, \text{ 65–70°C}} H_2N{-}C_6H_4{-}S{-}S{-}C_6H_4{-}NH_2$$ 71

58–64%72

$$\underset{\text{HCl Salt}}{\text{2-pyridyl}{-}CH_2CH_2SH} \xrightarrow[\text{RT, 30 min}]{\text{30\% } H_2O_2, \text{ HCl}} \text{2-pyridyl}{-}CH_2CH_2SSCH_2CH_2{-}\text{2-pyridyl}$$

In the last example, oxidation with nitric acid gives pyridine α-ethyl sulfonic acid in 90% yield[73].

Oxidation of Sulphides

Sulphides (dialkyl, alkyaryl, diaryl and cyclic) on oxidation give the corresponding sulfoxides[74] with one equivalent of the oxidant) or sulfones[24] (with two equivalents of the oxidant).

$$C_4H_9SC_4H_9 \xrightarrow[0°C,\ RT]{1\ CF_3CO_3H} C_4H_9\overset{O}{\overset{\|}{S}}C_4H_9$$

Dibutylsulphide → 83% Dibutylsulfoxide

$$C_4H_9SC_4H_9 \xrightarrow[0°C,\ 30°C,\ few\ hrs]{2CF_3CO_3H} C_4H_9\underset{O}{\underset{\|}{\overset{O}{\overset{\|}{S}}}}C_4H_9$$

80% Dibutylsulfone

Certain oxidants are selective. Thus, managanese dioxide[75] (petroleum ether, rt, 85 hr) and chromic acid[75] (C_5H_5N, rt. overnight) yield only the corresponding sulfoxide even with excess of the reagent.

Organic sulphides are oxidised to sulfoxides with organic peroxy acids[76] (like *m*-$ClC_6H_4CO_3H$) at room temerature and to sulfones with hydrogen peroxide[76] at higher temperature. Under these conditions double bonds if present in the substrate (as in $C_6H_5SCH = CH_2$) resist epoxidation[75]. Sulfones are also obtained by the oxidation of sulphides with tetrabutylamonium persulphate[77].

Cyclic sulphides (contaiming sulphur in the ring) are oxidised to cyclic sulfoxides[78,79]

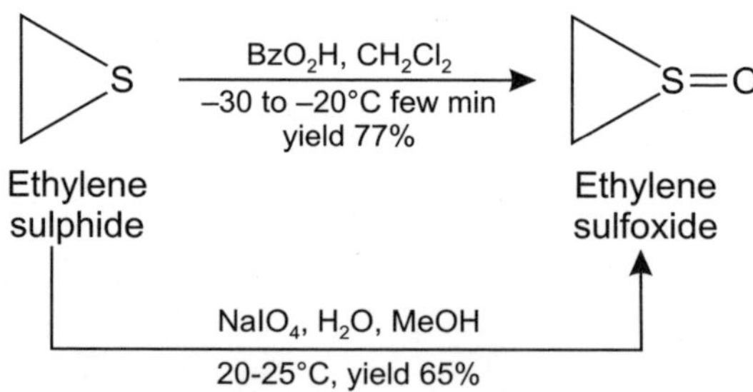

Tetrahydrothiophene (tetramethylene sulphide) gives the corresponding sulfoxide in 90% yield on refuxing with tetrabutylammonium periodate[80]; the sulfoxide is also obtained with one equivalent of hydrogen peroxide at room temperature. However, refluxing with two equivalents of hydrogen peroxide gives to corresponding sulfone[81].

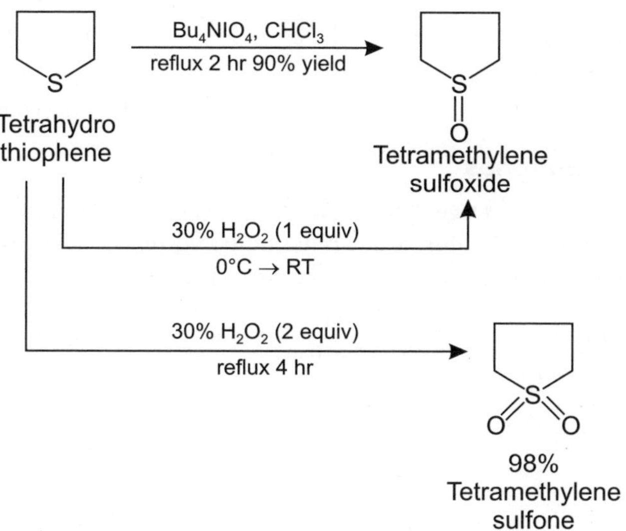

Oxidation of Thioacetals

2-Methyl-1,3–dithiane (the cyclic mercaptal of acetaldehyde) on oxidation[82] with sodium periodate or by *m*-chloroperoxybenzoic acid gives a mixture of diasteromeric *cis* and *trans* monosulfoxides.

$NaIO_4$, MeOH, H_2O

–7 to 0°C, –5, 1 hr 0°C overnight

80% *trans* and 7% *cis*

2-Methyl-1,3--dithiane

cis + *trans*

m-$ClC_6H_4CO_3H$

CH_2Cl_2, –15°C to –20°C, *trans* 49%, *cis* 4%

Oxidation of Sulfoxides

Sulfoxides on oxidation with air[83] in the presence of nobel metal salts or with sodium permanganate[84] give the corresponding sulfones.

$$(CH_3)_2S{=}O \xrightarrow[\text{heat, 16 hr}]{\text{air, RhCl}_3\text{, i-PrOH/H}_2\text{O (9:1)}} (CH_3)_2SO_2$$

Dimethyl Sulfoxide → 80% Dimethylsulfone

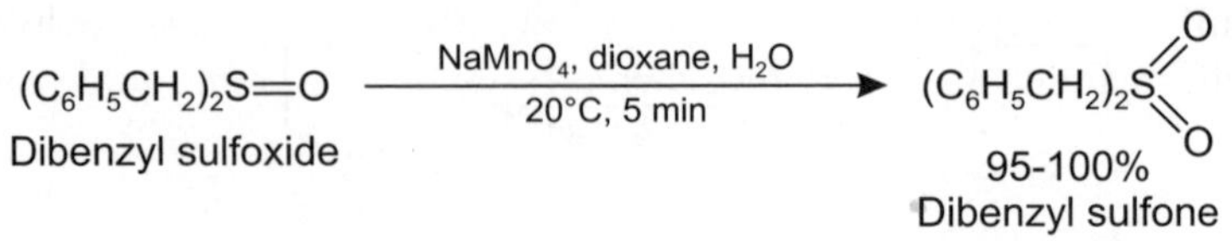

$$(C_6H_5CH_2)_2S{=}O \xrightarrow[20°C,\ 5\ min]{NaMnO_4,\ dioxane,\ H_2O} (C_6H_5CH_2)_2S(=O)_2$$

Dibenzyl sulfoxide → 95-100% Dibenzyl sulfone

Oxidation of Sulfones

Sulfones undergo a different type of reaction. Thus, methyl phenyl sulfone on reaction with carbon tetrachloride and potassium hydroxide yields phenyl trichloro methyl sulfone, which on hydrolysis give benzenesulfonic acid[85].

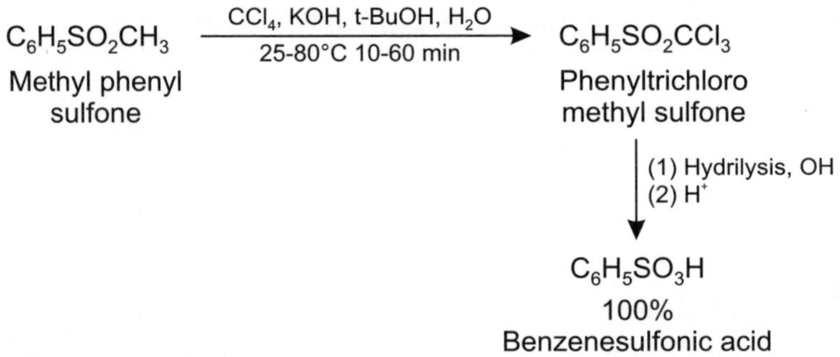

$$C_6H_5SO_2CH_3 \xrightarrow[25\text{-}80°C\ 10\text{-}60\ min]{CCl_4,\ KOH,\ t\text{-}BuOH,\ H_2O} C_6H_5SO_2CCl_3$$

Methyl phenyl sulfone → Phenyltrichloro methyl sulfone

$$C_6H_5SO_2CCl_3 \xrightarrow{(1)\ Hydrilysis,\ OH\quad (2)\ H^+} C_6H_5SO_3H$$

100% Benzenesulfonic acid

Under the above reactions dibenzyl sulfone is quantitatively converted into *trans*-stilbene

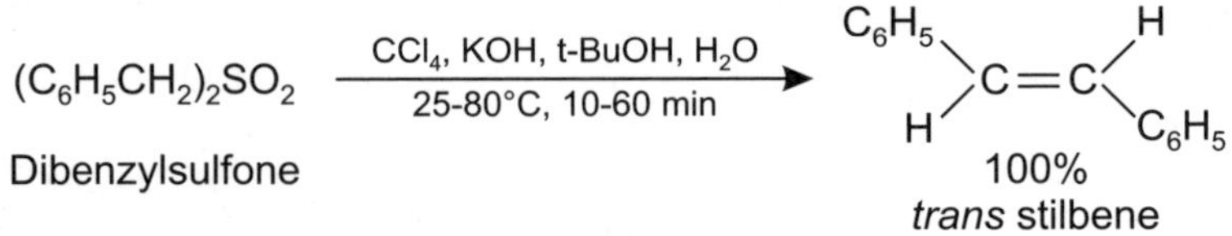

$$(C_6H_5CH_2)_2SO_2 \xrightarrow[25\text{-}80°C,\ 10\text{-}60\ min]{CCl_4,\ KOH,\ t\text{-}BuOH,\ H_2O} C_6H_5(H)C{=}C(H)C_6H_5$$

Dibenzylsulfone → 100% *trans* stilbene

Oxidation of Disulfides

Disulfides on oxidation with peroxy benzoic acid give[86] monosulfoxides (alkyl thiosulfinates)

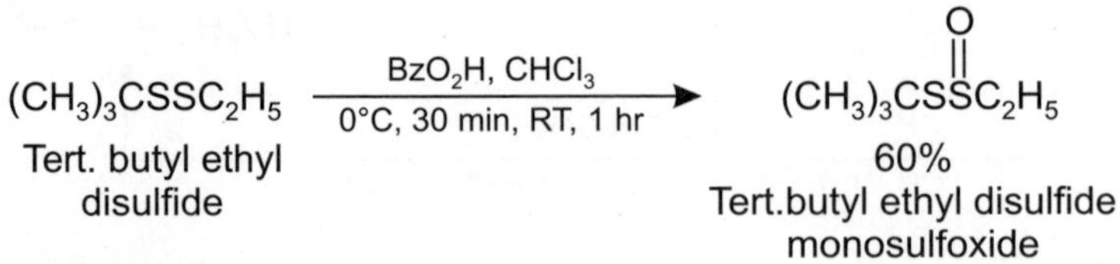

$$(CH_3)_3CSSC_2H_5 \xrightarrow[0°C,\ 30\ min,\ RT,\ 1\ hr]{BzO_2H,\ CHCl_3} (CH_3)_3CSS(=O)C_2H_5$$

Tert. butyl ethyl disulfide → 60% Tert.butyl ethyl disulfide monosulfoxide

Dimethyl disulfide on oxidation with chlorine and acetic anhydride gives methanesulfinyl chloride (as shown below).

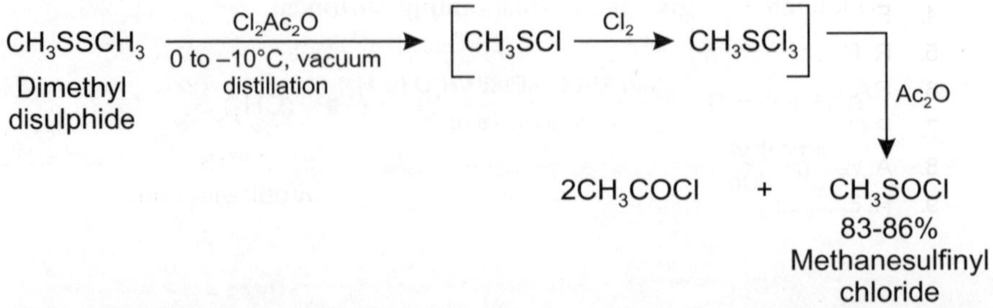

$$CH_3SSCH_3 \xrightarrow[0\ to\ -10°C,\ vacuum\ distillation]{Cl_2Ac_2O} \left[CH_3SCl \xrightarrow{Cl_2} CH_3SCl_3\right] \xrightarrow{Ac_2O} 2CH_3COCl + CH_3SOCl$$

Dimethyl disulphide → 83-86% Methanesulfinyl chloride

Disulfides on oxidation with oxygen[87] in a solution of potassium hydroxide in dimethylformamide at 23.5°C give the corresponding sulfonic acid in 90% yied.

$$C_4H_9SSC_4H_9 \xrightarrow[\text{23.5°C; (2) } H^+]{\text{(1) } O_2\text{, KOH, DMF}} C_4H_9SO_3H$$

Dibutyl disulfide → Butanesulfonic acid, 90%

Disulfides can be converted into sulfonyl chlorides by reaction with chlorine[88] in presence of hydrochloric acid and nitric acid

o,o′-Dinitrobenzene disulfide $\xrightarrow[\text{70°C, 1.5 hr}]{Cl_2\text{, HCl, }HNO_3\text{, }H_2O}$ o-Nitrobenzenesulfonyl chloride (84%)

Oxidation of Sulfinic Acids

Sulfinic acids on oxidation with sodium permanganate[89] and chlorine in water[90] give the corresponding sulfonic acids and sulfonyl chlorides, respectively

$$C_{12}H_{25}\text{-}C_6H_4\text{-}SO_2H \xrightarrow[\text{35-40°C, RT, overnight}]{NaMnO_4\text{, NaOH}} C_{12}H_{25}\text{-}C_6H_4\text{-}SO_3H$$

Sulfinic acid → Sulfonic acid, 70%

$$CH_3C_6H_4SO_2H \xrightarrow[\text{30°C}]{Cl_2\text{, KOH, }H_2O} CH_3C_6H_4SO_2Cl$$

Toluene o-sulfinic acid → o-Methyl benzene sulfonyl chloride

References

1. L. S. Silbert, E. Siegel and D. Swern, *Org. Synth.* Collective Volume, 1973, **5**, 904.
2. F. P. Greenspan, R. J., Gall and D. G. Mackellar, *J. Org. Chem.,* 1955, **20**, 215.
3. R. H. Frazier Jr., and R. L. Harlow, *J. Org. Chem.,* 1980, **45**, 5408.
4. E. Vedejs, D. A. Engler and J. E. Telschow, *J. Org. Chem.,* 1978, **43**, 188.
5. R. R. Coates and J. W. Cook, *J. Chem., Soc.,* 1942, 599.
6. F. A. Davis, M. S. Haque, T. G. Ulatowski and J. C. Towson, *J. Org., Chem.,* 1986, **51**, 2402.
7. P. G. Sergeev and A. M. Sladkov, Zh. Obshich. Khim., 1957, **27**, 819.
8. A. W. Dox, *Org. Synth,* Collective Volume, 1932, **1**, 266.
9. H. E. Baumgarten and A. Stakles, *J. Am. Chem. Soc.,* 1965, **87**, 114.

10. L. Kaly and O. Gross, *Chem., Ber.,* 1926, **59**, 727.
11. W. Rigby, *J. Chem., Soc.,* 1950,1907.
12. H. Shecter and F. T. Williams, Jr., *J. Org. Chem.,* 1962, **27**, 3699.
13. W. D. Langley. *Org. Synth.* Collective Volume, 1955, **3**, 334.
14. E. C. Taylor and A. Mckillop, *J. Org. Chem.,* 1965, **30**, 3153.
15. A. E. Tschitchibabin, *Chem., Ber.,* 1923,**56**,1879.
16. G. H. Coleman, C. M. Mecloskey and F. A. Stuart, *Org. Synth.* Collective Volume, 1955, **3**, 668.
17. E. C. Taylor and F. Yoneda, *Chem.* Commun., 1967,199.
18. P. A. Wehrli and F. Pigott, *Org. Synth.* Collective Volume, 1988,**6**,1010.
19. E. J. Corey, B. Samuelsson and F. A. Luzzio, J. Am. Chem. Soc., 1984,**106**,3682.
20. E. B. Hershberg and L. F. Fieser, *Org. Synth.* Collective Volume, 1943,**2**,423.
21. E. E. Pratt and T. P. McGovern, *J. Org. Chem.,* 1964,**29**,1540.
22. F. Yoneda, K. Suzuki and Y. Nitta, *J. Org. Chem.,* 1967,**32**,727.
23. E. A. Braude, A. G. Brook and R. P. Linstead, *J. Chem. Soc.,* 1954,3574.
24. J. C. Kauer, *Org. Synth.* Collective Volume, 1963,**4**,411.
25. L. A. Carpino and P. J. Crowley, *Org. Synth.* Collective Volume, 1973,**5**,160.
26. M. Z. Barakat, M. F. Abdel-Wahab and M. M. El–Sadr, *J. Chem., Soc.,* 1956, 4685.
27. M. Firouzabadi and Z. Mostafavipoor, *Bull. Chem. Soc.,* Jpn., 1983, **56**, 914.
28. J. S. Belew, C. Garza and J. W. Mathieson, Chem., Commun., 1970, 634.
29. H. Firouzabadi, B. Vessal and M. Naden, *Tetrahedron Lett.,* 1982,**23**,1847.
30. K. Nakagawa and T. Tsuji, *Chem., Pharm. Bull.,* 1963,**11**,296.
31. G. B. Payne, P. H. Deming and P. H. Williams, *J. Org. Chem.,* 1961,**26**,659.
32. F. P. Greenspan, Ind. Eng. Chem., 1947,**39**,847.
33. J. Meisenheimer and E. Hesse, *Chem. Ber.,* 1919,**52**,1161.
34. W. D. Langley, *Org. Synth.* Collective Volume, 1955,**3**,334.
35. R. R. Holmes and R. P. Bayer, *J. Am. Chem. Soc.,* 1960,**82**,3454.
36. L. Di Nunno; S. Florio and P. E. Todesco, *J. Chem. Soc. C.* 1970,1433.
37. A. Calder, A. R. Forrester and S. P. Hepburn, *Org. Synth*. Collective Volume, 1988,**6**,803.
38. N. Kornblum and R. J. Clutter, *J. Am. Chem. Soc.,* 1954,**76**,4494.
39. N. Kornblum and W. J. Jones, *Org. Synth.* Collective Volume, 1973,**5**,845.
40. W. D. Emmons, *J. Am. Chem. Soc.,* 1957,**79**,5528.
41. R. W. Murray, R. Jeyaraman and L. Mohan, *Tetrahedron Lett.,* 1986,**27**,2335.
42. W. D. Emmons, *J. Am. Chem. Soc.,* 1954,**76**,3470.
43. W. D. Emmons and A. F. Ferris, *J. Am. Chem. Soc.,* 1953,**75**,4623.
44. A. S. Pagano and W. D. Emmons, *Org. Synth.* Collective Volume, 1973,**5**,367.
45. R. H. Wiley and J. L. Hartman. *J. Am. Chem. Soc.,* 1951,**73**,494.
46. S. S. Rawalay and H. Shechter, *J. Org. Chem.,* 1967,**32**,3129.
47. K. H. Scheit and W. Kampe, *Angew. Chem.,* 1965,**77**,811.
48. H. Shechter and S. S. Rawalay, *J. Am. Chem. Soc.,* 1964,**86**,1706.
49. A. Stojiljkovic', V. Andrejevic' and M. L. Mihailovil, *Tetrahdron,* 1967,**23**,721.
50. G. I. Nikishin, E. I. Troyankii and V. A. Joffe, Izv. Akad. Nauk SSSR, 1982,2758; Chem. Abstr., 1983,142911.
51. K. Nakagawa and T. Tsuji; *Chem. Pharm. Bull.,* 1963,**11**,296.

52. G. A. Lee and H. H. Freedman, *Tetrahedron Lett.,* 1976,1641.
53. T. G. Clarke, N. A. Hampson, J. B. Lee, J. R. Morley and B. Scanlon, *Tetrahedron Lett,* 1968, 5685.
54. W. R. Dimstan and E. Goulding, *J. Chem. Soc.,* 1899,**75**,1004.
55. A. C. Cope and H. H. Lee, *J. Am. Chem. Soc.,* 1957,**79**,964.
56. G. B. Payne, P. H. Deming and P. H. Williams, *J. Org. Chem.,* 1961,**26**,659.
57. H. S. Mosher, L. Turner and A. Carlsmith., *Org. Synth.* Collective Volume, 1963,**4**,828.
58. R. W. Murray and R. Jeyaraman, *J. Org. Chem.,* 1985,**50**,2847.
59. G. Serra-Errante and P. G. Sammes. *Chem. Commun,* 1975,573.
60. K. Schniter, *Chem. Ber.,* 1887,**20**,2282.
61. F. Kehrmann and T. E. Stiller, *Chem. Ber.,* 1912,**45**,3346.
62. H.-J. Teuber and M. Hasselbach, *Chem. Ber.,* 1959,**92**,674.
63. A. V. R. Rao, S. P. Charvan and L. Sivadasan, *Tetrahedron,* 1986,**42**,5065.
64. Q. E. Thompson, *J. Am. Chem. Soc.,* 1961,**83**,845.
65. R. J. Kennedy and A. Hill, *Chem. Ber.,* 1900,**33**,533.
66. Y. Hayakawa, M. Ucheyame and R. Noyori, *Tetrahedron Lett,* 1986,**27**,4191.
67. A. Skowranska and E. Krawezyk, *Synthesis*, 1983,509.
68. R. L. Shriner and C. N. Wolf. *Org., Synth.* Collective Volume, 1963,**4**,910
69. K. A. Parker and J. G. Farmar, *J. Org. Chem.* 1986,**51**,4023.
70. E. P. Papadopovlos, A.Jarrar and C. H. Issidorides, *J. Org. Chem.,* 1966,**31**,615.
71. C. C. Price and C. W. Stacy, *Org. Synth.* Collective Volume, 1955,**3**,86.
72. L. Bauer and L. A. Gardella, Jr., *J. Org., Chem.,* 1961,**26**,82.
73. T. Nishimura, *Org. Synth.* Collective Volume, 1963,**4**,713.
74. C. G. Verer, T. G. Squires, Y. –Y. Chen, G. P. Hussmann, J. C. Shei and B. F. smit., *J. Org., Chem.,* 1982,**47**,3773.
75. D. Edwards and J. B. Stenlake, *J. Chem. Soc.,* 1954,3272.
76. F. Korte and K. H. Löhman, *Chem. Ber.,* 1957,**90**,1290.
77. B. M. Trost and R. Braslau, *J. Org. Chem.,* 1988,**53**,532.
78. K. Kondo, A. Oregishi and M. Fukuyama, *Tetrahedron Lett.* 1969,2461.
79. G. E. Hartzell and J. N. Paige, *J. Am. Chem. Soc.,* 1966,**88**,2616.
80. E. Santaniello, A. Manzocchi and C. Farachi, *Synthesis,* 1980,563.
81. D. S. Tarbell and C. Weaver, *J. Am. Chem. Soc.,* 1941,**63**,2939.
82. F. A. Carey, O.D. Dailey, Jr., O. Hernandez and J. R. Tucker, *J. Org. Chem.,* 1976,**41**,3975.
83. J. Trocha-Grimshaw and H. B. Henbest, *Chem. Commun.,* 1968,1035.
84. H. B. Henbest and S. A. Khan, *Chem. Commun.,* 1968,1036.
85. C. Y. Meyers, A. M. Malte and W. S. Matthews, *J. Am. Chem. Soc.,* 1969,**91**,7510.
86. L. D. Small, J. H. Bailey and C. J. Cavellito, *J. Am. Chem. Soc.,* 1947,**69**,1710.
87. T. J. Wallace and A. Schriesheim, *Tetrahedron Lett.,* 1963,1131.
88. E. Wertheim, *Org. Synth.* Collective Volume, 1943,**2**,471.
89. W. E. Truce and J. F. Lyons, J. Am. *Chem. Soc.,* 1951,**73**,126.
90. F. Ullmann and A. Lehner, *Chem., Ber.,* 1905, **38**, 729.

3

Enzymatic or Microbial Oxidations

Enzymes or microbes are useful in a large number of oxidations. The products obtained excel in regiospecificity, stereospecificity and enantioselectivity. Also, the optical purity (enantiomeric excess) is generally high and may frequently be 100 per cent. An added advantage of microbial oxidations is that they occur under very mild conditions, around 35–37°C and in dilute solutions and normally involve only one step.

Enzymes are also referred to as biocatalysts and the transformations are referred to as biocatalytic conversions. Enzymes are now readily available and are an important tool in organic synthesis.

Two enzymatic oxidations, known to mankind since early times are the conversion (oxidation) of alcohol into acetic acid by bacterium acetic in presence of air (the process is known as quick vinegar process), and the second is the conversion of sucrose into ethyl alcohol by yeast (this produss is used for the manufacture of ethyl alcohol). In a similar way, lactose is converted into lactic acid. The above enzymatic oxidations are referred to as fermentation (Scheme 3.1).

$$\underset{\text{Ethyl alcohol}}{CH_3CH_2OH} + O_2 \xrightarrow{\text{Bacterium acetic}} \underset{\text{Acetic acid}}{CH_3COOH} + H_2O$$

$$\underset{\text{Sucrose}}{C_{12}H_{22}O_{11}} + H_2O \xrightarrow[\text{Yeast}]{\text{Invertase}} \underset{\text{Glucose + fructose}}{2C_2H_{12}O_6}$$

$$\underset{\text{Glucose}}{C_6H_{12}O_6} \xrightarrow[\text{Yeast}]{\text{Invertase}} \underset{\text{Ethyl alcohol}}{2C_2H_5OH} + 2CO_2$$

$$\underset{\text{Lactose}}{C_{12}H_{22}O_{11}} + H_2O \xrightarrow[\text{Yeast}]{\text{Invertase}} \underset{\text{Lactic acid}}{4CH_3CH(OH)COOH}$$

(Scheme 3.1)

Considerable amount of work has been carried out in the hydroxylation of aromatic rings. Thus, benzene on oxidation with *Pseudomonas putida* in presence of oxygen gives the *cis*-diol[1]. However, Micrococuus spheriods like organism converts benzene into *trans*, *trans*-muconic acid[2] (Scheme 3.2).

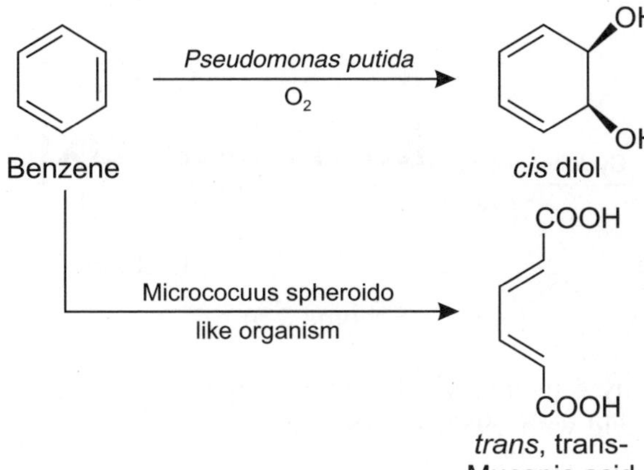

Benzene — *Pseudomonas putida*, O_2 → *cis* diol

Benzene — Micrococuus spheroido like organism → *trans*, trans-Muconic acid

(Scheme 3.2)

As in the case of benzene, other aromatic hydrocarbons like toluene, and halogenated toluenes also give the corresponding *cis*-diols[3] (Scheme 3.3).

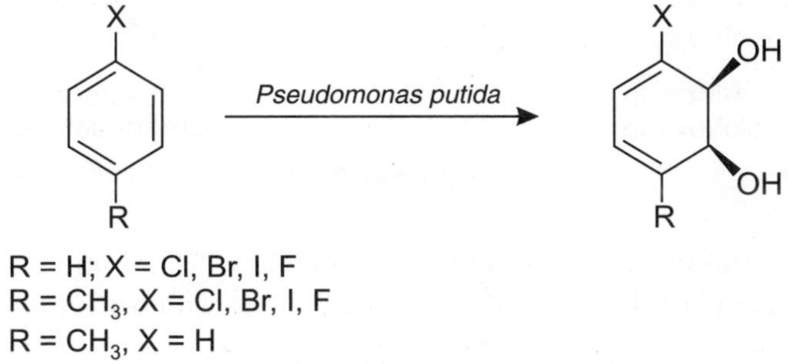

Pseudomonas putida

R = H; X = Cl, Br, I, F
R = CH_3, X = Cl, Br, I, F
R = CH_3, X = H

(Scheme 3.3)

Some of the biochemical transformations are not possible by conventional chemical means. Two such examples in heterocyclic compounds are given below[4] (Scheme 3.4.).

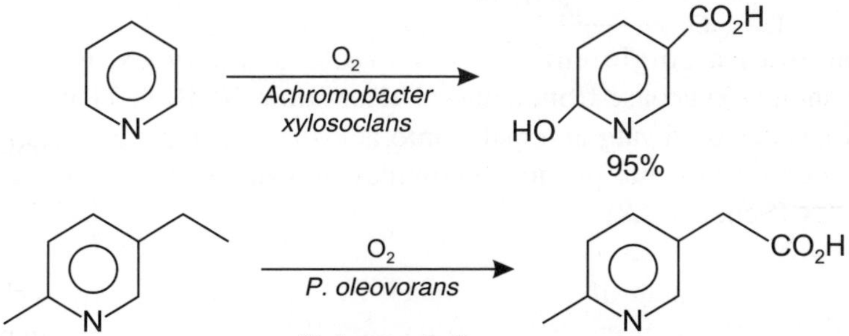

O_2, *Achromobacter xylosoclans* — 95%

O_2, *P. oleovorans*

(Scheme 3.4)

Enzymatic conversion of ketones to esters is very common in microbial degradations[5]. A typical transformation in the enzymatic **Baeyer-villiger oxidation** of cyclohexanone[6] using a purified cyclohexanone oxygenase enzyme[7] (Scheme 3.5).

Cyclohexanone —(Cyclohexanone oxygenase-FAD; NADPH, O_2)→ Lactone + $NADP^+$ + H_2O

(Scheme 3.5)

In a similar way, 4-methylcyclohexanone was converted into the corresponding lactone is 80% yield with >98% ee[8] (Scheme 3.6).

4-Methyl cyclohexanone —(Cyclohexanone oxygenase; NADPH, O_2; glucose 6-phosphate)→ Lactone 80% (> 98% ee)

(Scheme 3.6)

Enzymatic oxidation of phenyl acetaldehyde with enzyme cyclohexanone oxygenase gave phenylacetic acid (65% yield) along with benzyl formate (12%)[9] (Scheme 3.7).

$C_6H_5CH_2CHO$ —(Cyclohexanone oxygenase; NA DPH, O_2)→ $C_6H_5CH_2CO_2H$ + $C_6H_5CH_2OCHO$

Phenylacetaldehyde → Phenylacetic acid 65% + Benzyl formate 12%

(Scheme 3.7)

Sometimes a single enzyme is capable of many oxidations. Thus, cyclohexanone oxygenase from acinotobactor strain NCIB 9871 in presence of NADPH besides oxidising aldehydes into acids (*see* scheme 3.7) also oxidises phenyl boronic acids into phenols, sulphides into sulfoxides and selenides into selenoxides (Scheme 3.8).[9]

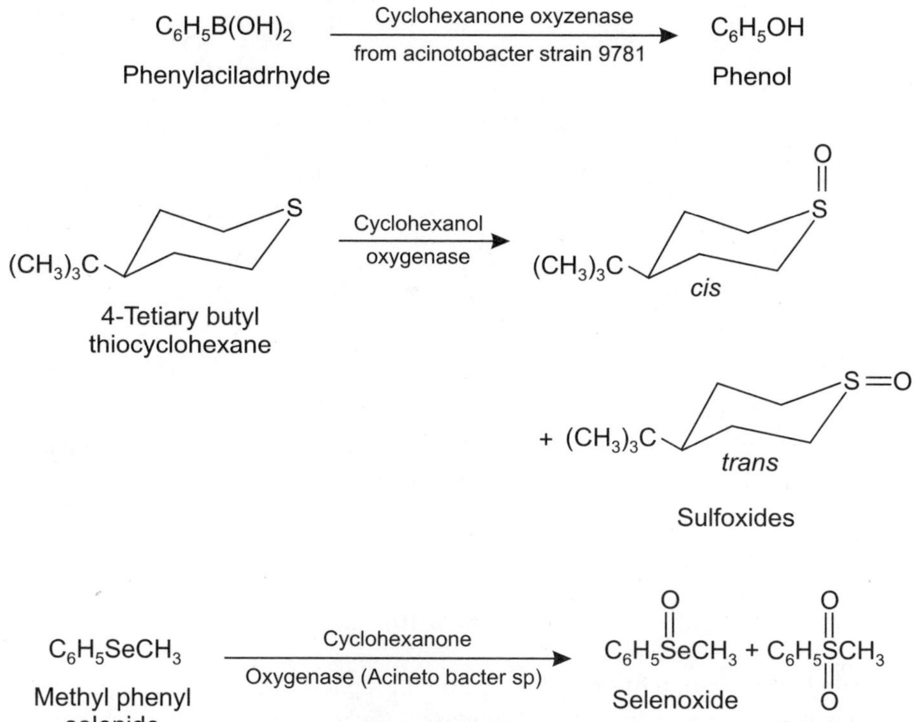

$C_6H_5B(OH)_2$ Phenylaciladrhyde —Cyclohexanone oxyzenase / from acinotobacter strain 9781→ C_6H_5OH Phenol

$(CH_3)_3C$– 4-Tetiary butyl thiocyclohexane —Cyclohexanol oxygenase→ *cis* + *trans*

Sulfoxides

$C_6H_5SeCH_3$ Methyl phenyl selenide —Cyclohexanone / Oxygenase (Acineto bacter sp)→ $C_6H_5Se(=O)CH_3$ Selenoxide + $C_6H_5S(=O)_2CH_3$ Selenone

(Scheme 3.8)

Another enzyme, horse liver dehydrogenase oxidises primary alcohols to acids (esters)[10] and secondary alcohols to ketones[11] (Scheme 3.9).

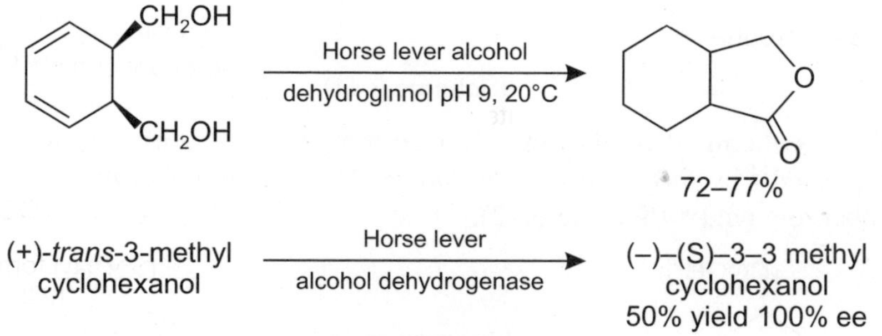

CH_2OH / CH_2OH —Horse lever alcohol / dehydroglnnol pH 9, 20°C→ 72–77%

(+)-*trans*-3-methyl cyclohexanol —Horse lever / alcohol dehydrogenase→ (–)-(S)-3-3 methyl cyclohexanol 50% yield 100% ee

(Scheme 3.9)

Oxidation of phenols to quinones can be accomplished[12] by the enzyme Horse radish peroxidase. Some examples are given below (Scheme 3.10).

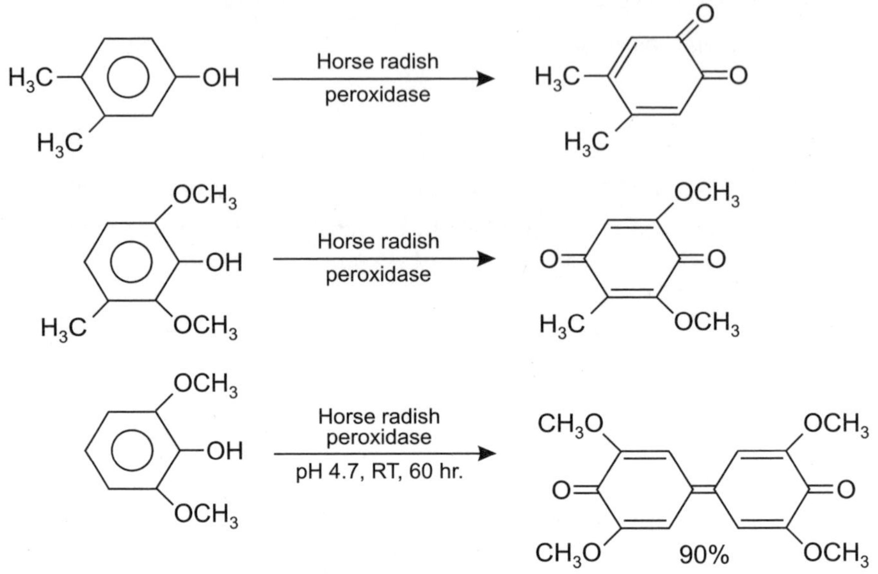

(Scheme 3.10)

Para hydroxylation of phenols (*e.g.*, o-chlorophenol) can be accomplished by the enzyme Mushroom Polyphenol oxidase[13] (Scheme 3.11)

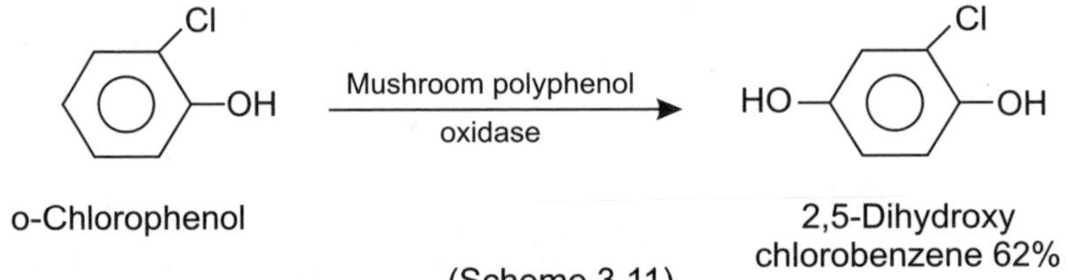

o-Chlorophenol

2,5-Dihydroxy chlorobenzene 62%

(Scheme 3.11)

The biochemical oxidation of an amino group to a nitrogroup can be accomplished[14] by *Streptomyces thioluteus*. Thus, p-aminobenzoic acid gives *p*-nitrobenzoic acid[14] (Scheme 3.12).

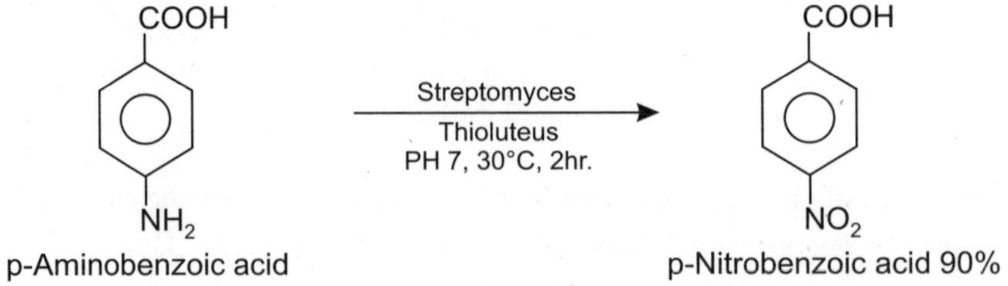

p-Aminobenzoic acid

p-Nitrobenzoic acid 90%

(Scheme 3.12)

The amino group in an side chain gets converted into carboxylic acid with *Hygrophorus conicus*[15]. Thus, tryptamine is converted into 2-indolone-3-acetic acid[15] (Scheme 3.13).

$CH_2CH_2NH_2$ (Tryptanine) —[*Hygrophorus canciss*, 27°C, 7 days]→ $CH_2CH_2NH_2$ (=O, N–H) 43% Indolone-3-acetic acid

(Scheme 3.13)

Terminal methyl groups as well as the halomethyl groups in aliphatic halides with 15–18 carbon atoms are oxidised[16, 17] biohemically to carboxyls. Some examples are given below (Scheme 3.14).

$$CH_3(CH_2)_4CH_2X \xrightarrow[\text{25°C, 24 hr.}]{\text{Torulopsis gropengiesseric}} bOOC(CH_2)_{14}COOH$$

X	% yield
Cl	33
Br	55
I	43

(Scheme 3.14)

In the above oxidation (Scheme 3.14) the formed products were isolated as methyl esters.

Considerable amount of work has been reported on enzymatic hydroxylation of steroids. Thus, progesterone on oxidation with *Rhizopus arrhizus*[18,19] yields 11α-hydroxyprogesterone alongwith 6β, 11α-dihydroxyprogesterone. However, *Calonectria decora* converts[20] progesterone into 12β, 15β-dihydroxyprogesterone (Scheme 3.15).

Progesterone —[*Rhizopus arrhizus*]→ 11 α-Hydroxyprogestrone + 6β, 11α-Dihydroxy progesterone

Progesterone —[*Calonectria decora*]→ 12 β, 15β-Dihydroxy progesterone

(Scheme 3.15)

In steroids, a number of Baeyor-villiger oxidations have been accomplished biochemically. Thus, progestrone and testosterone on treatment with *Cylindrocarpon radicicola* yielded[21] Δ^1-dehydrotestolactone. However, progesterone on oxidation with *Penicillium chrysogenum* yielded[21] testolactone, which is also obtained from 4-androstene–3,17–dione by treatment with *Penicillium lilacinum*[22] (scheme 3.16).

$COCH_3$ OH

4-Androstene-
-3,17-dione

Progesterone

Testosterone

Penicillium
lilacinum
79%

Pencillium
chrysogenum
70%

Cylindrocarpon
radicicola

Cylindrocarpon
radicicola
50%

Testololactane

Δ′-Dehydrotestolactone

(Scheme 3.16)

Besides the above, there are a number of other examples involing the use of enzymes in oxidations[23].

References

1. L.M. Shirley and S.C. Taylor. *J. Chem. Soc. Chem.,* Commun., 1983,954.
2. A. Kleinzeller and Z. Fenel, *Chem. Listy,* 1952, **46**, 300; *Chem. Abstr.,* 1953, **47**, 4290.
3. D.T. Gibson, J. R. Koch, C. L. Schuld and R. E. Kallio, *Biochemistry,* 1968, **7**, 3797; D.T. Gibson, M. Hansley, H. Yoshioka and T.J. Mabry, *Biochemistry,* 1970, **9**, 1626.
4. A. Kiener, CHEMTECH, September 1955, pp. 31–35.
5. R. A. Johnson, Chemical Oxidations with Microorganisms, Marcel Dekker, New York, 1972, pp. 157–164.
6. C.C. Ryerson, D.P. Ballou and C. Walsh, *Biochemistry*, 1982, **21**, 2644.

7. N.A. Donoghue, D.B. Norris and P.W. Trudgills Eur. J. Biochem., 1976, **63**, 175.
8. M.J. Taschner and D.J. Black, *J. Am. Chem. Soc.* 1988,**110**,6892.
9. B. P. Branchard and C. T. Walsh, *J. Am. Chem. Soc.* 1985,**107**,2153.
10. J.B. Jones and I. J. Jokovac., *Org. Synth.,* 1984,**63**,10.
11. J. Grunwald, B. Wirz, M. P. Scollar and A.M. Klibanov, *J. Am. Chem. Soc.,* 1986,**108**,6732.
12. B. C. Saunders and B. P. Stark, *Tetrahedron,* 1967,**23**,1867.
13. R. Z. Kazandjian and A.M. Klibanov. *J. Am. Chem. Soc.,* 1985,**107**,5448.
14. S. Kawai, T. Oshima and F. Egami, *Biochem. Biophys. Acta,* 1965,**97**,391.
15. D. J. Siehr, *J. Am. Chem. Soc.,* 1961, **83**, 2401.
16. D. F. Jones and R. Howe, *J. Chem., Soc.,* C., 1968,2816.
17. D. F. Jones and R. Howe, *J. Chem. Soc.,* C, 1968, 2801.
18. D. H. Peterson, H. C. Murray, S. H. Eppstein, L. M. Reineke, A. Weintraub, P. D. Meister and H. M. Leigh, *J. Am. Chem. Soc.,* 1952,**74**,5933.
19. D. H. Peterson, A. H. Nathan, P. D. Meister, S. H. Eppstein, H. C. Murray, A. Weintraub, L. M. Reineke and H. M. Leigh, *J. Am. Chem. Soc.,* 1953,**75**,419.
20. A. Schubert. G. Langbein and R. Siebert, *Chem. Ber.,* 1957,**90**,2576.
21. J. Fried, R. W. Thoma and A. Klingsberg, *J. Am. Chem. Soc.,* 1953, **75**, 5764.
22. R. L. Prairie and P. Talalay, *Biochemistry,* 1963, **2**, 203.
23. Milos Hudlicky, *Oxidations in Organic Chemistry*, ACS Monograph 186, American Chemical Society, Washingtion DC, 1990.

4

Some Oxidations Under Benign Conditions

4.1 INTRODUCTION

A large number of conventional oxidizing agents have been used for various types of oxidations (*see* chapters 1 and 2). These reagents have their own limitations in terms of toxicity, work up and associated waste disposal problems. Besides the catalysts used produce harmful effects. Even the solvents used for various oxidations being volatile have harmful effect on the environment. It is best to carry out the oxidations under benign conditions. Some of these conditions include use of phase transfer catalysts and other benin catalysts. Use of solvents like water, super critical carbon dioxide, ionic liquids, polyethlene glycol. Some of the oxidations have also been performed in solid state. Techniques like sonication, microwave irradiation besides being benign increases the yields of oxidation products. Besides these, enzymatic oxidations are one hundred percent benign (for details *see* Chapter 3).

4.2 OXIDATION OF ALKENES

The most common oxidation reactions of alkenes are epoxidation, hydroxylation and oxidation to carboxylic acids.

4.2.1 Epoxidation of Alkenes

Peracids are known to react with alkenes to give three-membered rings containing oxygen atom, called epoxides or oxiranes (Scheme 4.1).

Alkene + Peracid ⟶ Epoxide (Oxirane) + R—C(=O)—OH

(Scheme 4.1)

The above epoxidation takes place in non-polar solvents, such as dichlorobenzene or benzene. Simple alkenes can be epoxised with *m*-chloroperoxybenzoic acid (MCPBA) in aqueous $NaHCO_3$ at room temperature to yield epoxides in good yield. Alkenes like cyclopentene, cyclohexene, cycloheptene, cycloctene, methyl cyclohexene and (+)–3–carene have been epoxised with MCPBA at 20°C in 30 min to give 90–95% epoxide. Using this method, styrene, α-methyl styrene and trans-α-methylstyrene could also be epoxised. In aqueous medium, the reaction occurs in aqueous phase, but this does not affect the reactivity.

For direct epoxidation of simple alkenes, the peroxide must be activated in buffered aqueous THF, by stoichiometric amounts of organophosphorus anhydride (Scheme 4.2).

Alkane → Epoxidl 60–100%

$H_2O_2/(Ph_2PO)_2O$; K_2CO_3 / THF-H_2O, –5°C, 20 hr,

(Scheme 4.2)

Using this procedure a variety of alkenes could be epoxidized[2]. A safe and cheap method for the epoxidation has been developed[3]. It consists in using nacent oxygen generated by electrolysis of water at room temperature using Pd black as an anode. Using this procedure, cyclohexene ould be epoxised is good yield. In this procedures, water is used as a reaction medium as well as a reagent.

Allyl alcohols could be epoxidised regioselectively in presence of other C=C bonds by using monoperphthalic acid (MPPA) in presence of cetyltrimethyl ammonium hydroxide (CT AOH) (Scheme 4.3)[3].

OH → MPPA / H_2O/CTAOH → O OH

(Scheme 4.3)

Epoxidation of α, β, unsaturated carbonyl compounds could be accomplished[4,5]by using sodium perborate (SPB) in water at pH 8 (Scheme 4.4)

R1, COMe, R2, H → SPB, 57–80°C / H_2O–THF–N HC_2O-dioxane → O, R1, COMe, R2, H

Alkane 65–100%

R_1 = H, Me

R_2 = Me, Ph

(Scheme 4.4)

The epoxidation of α, β-unsaturated carbonyl compound with H_2O_2 under basic conditions, known as the **Weitz-scheffer epoxidation** (Scheme 4.5)[6] is a conveneint and efficient method of producing epoxides.

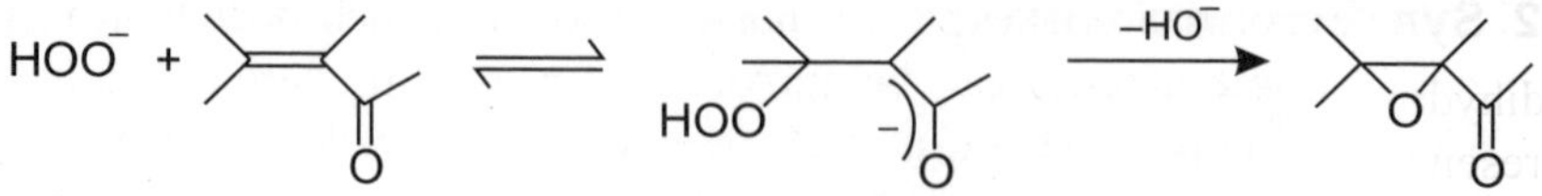

(Scheme 4.5)

The epoxidation of chalcones proceed very well with NaOCl (Commercially available) in water suspension[7] in presenec of a PTC hexadecyltrimethylammonium bromide in 50–100% yield (scheme 4.6).

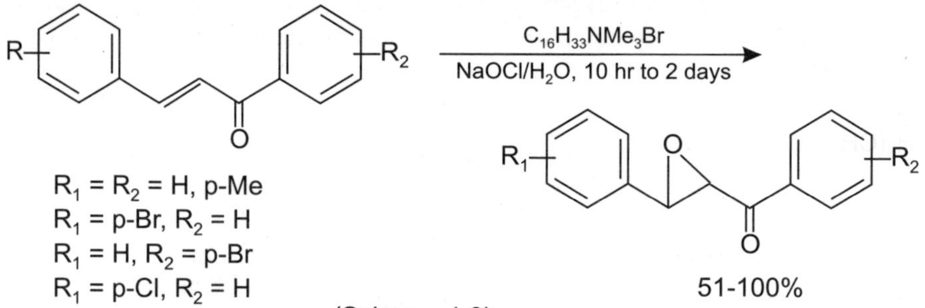

$R_1 = R_2 = H$, p-Me
R_1 = p-Br, $R_2 = H$
$R_1 = H$, R_2 = p-Br
R_1 = p-Cl, $R_2 = H$

$C_{16}H_{33}NMe_3Br$
$NaOCl/H_2O$, 10 hr to 2 days

51-100%

(Scheme 4.6)

For more details about epoxidation *see* section 2.1.2.2.

Epoxidation of internal double bond could be convenienlty accomplished[7a] with O_2 in SC-CO_2. In this case, an aldehyde was used as a sacrificial oxygen transfer agent. Thus, there is quantitative conversion of cis-cyclooctene into epoxycyclooctane (98% ee). In this case, case of 2-propinaldehyde gave better results. Under similar conditions, cyclohexene and *trans*-3-hexene were quantitatively converted to epoxy cyclohexane (91% selectivity) and *trans*-3-epoxyhexane (> 99% selectivity), respectivity (Scheme 4.6a).

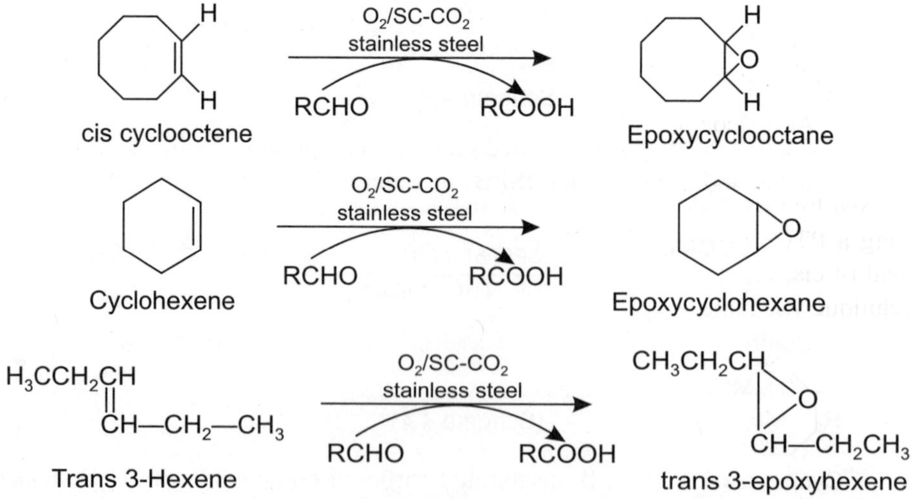

cis cyclooctene → Epoxycyclooctane (O_2/SC-CO_2, stainless steel; RCHO → RCOOH)

Cyclohexene → Epoxycyclohexane (O_2/SC-CO_2, stainless steel; RCHO → RCOOH)

Trans 3-Hexene → trans 3-epoxyhexene (O_2/SC-CO_2, stainless steel; RCHO → RCOOH)

(Scheme 4.6a)

4.2.2 Syn Dihydroxylation of Alkenes

Syn dihydroxylation of alkenes is achieved by treatment with dilute $KMnO_4$ solution in presence of alkali. In fact, the change in purple colour is the basis for the presence of double bond and this is known as **Bayers test** for unsaturation. This method is used for *Syn* hydroxylation of oleic acid and norbornene (Scheme 4.7).

dil. $KMnO_4$, NaOH

dil aq. $KMnO_4$, NaOH (aq)

1,2-diol (erythro)

Norbornene — aq. $KMnO_4$, dil. NaOH — Diol

(Scheme 4.7)

The *syn*-hydroxylation of alkenes can also be carried out by H_2O_2 in presence of catalytic amount of OsO_4[8]. By this method, allyl alcohol is quantitatively hydroxylated in water (Scheme 4.8).

Allyl alcohol — OsO_4, H_2O_2, H_2O — 1,1,3-Trihydroxypropane (glycerol)

(Scheme 4.8)

Syn-hydroxylation of olefins has also been carried out with $KMnO_4$ solution using a PTC catalyst under alkaline conditions. Thus, cyclooctene gives[9] 50% yield of cis, 1,2–cyclooctane diol, compard to an yield of about 7% by classical technique (Scheme 4.9).

Cyclooctene + aq $KMnO_4$ — $C_6H_5CH_2\overset{+}{N}Et_3Cl$, NaOH — *cis*-1,2-cyclooctane diol 50%

(Scheme 4.9)

4.2.3 Anti Dihydroxylation of Alkenes

Hydrogen peroxide in presence of tungsten oxide (WO_3) or selenium dioxide (SeO_2) gives anti-dihydroxylation product (Scheme 4.10.).

WO_3(50-70°C), H_2O_2

(Scheme 4.10)

For more details *see* section 2.1.2.6.

Asymmetric dihydroxylation of olefins has been carried out using catalytic amount of osmium tetraoxide and PEG–400[10a]. In this case, the ligand is efficiently recovered and recycled with good enantioselectivity. Thus, the dihydroxylation of styrene, α-methyl styrene and trans stilbene using OsO_4 (0.5 mol %) in PEG (400 MW) using N-methyl morpholine (NMO, 1.3 eq.) as the reoxidant gave good yields of the respective diols (Scheme 4.10 a).

OsO_4 (0.5 mol %), PEH (400 MW), 2 hr, NMO (1.3 eq)

Styrene, R = H
α-Methyl styrene, R = Me

Diol R = H 94%
Diol R = Me 97%

OsO_4 (0.5 mol %), PEH, 400 MW, 2 hr, NMO (1.3 eq)

Trans Stilbene

1,2-Diphenyl-1,2-ethanediol (95%)

(Scheme 4.10a)

4.2.4 Oxidation of Carbon-Carbon Double Bonds in Alkenes

The C–C double bonds in alkenes can be fissioned; the product obtained depends on the reagent and reaction conditions. Some typical examples are given below (Scheme 4.11).

[O], neutral 1-$KMnO_4$

Narbarnene

(Scheme 4.11) *cont...*

$$CH_3(CH_2)_5CH{=}CH_2 \xrightarrow[CH_3(CH_2)_{15}\bar{N}(CH_3)_3Cl^-]{C_6H_6,\ aq.\ KMnO_4} CH_3(CH_2)_4CH_2CO_2H$$

1-Octene → Heptanoic acid 95%

$$\alpha\text{-Pinene} \xrightarrow[\text{Dicyclohexeno-1}\delta\text{-crawn-6}]{C_6H_6,\ aq.\ KMnO_4} cis\text{-Pinonic aicd}$$

α-Pinene → *cis*-Pinonic aicd

(Scheme 4.11)

For more details *see* section 2.1.2.1.

Besides the reaction of alkenes, discussed above, alkenes react with singlet oxygen (formed by photolysis of oxygen in methylene chloride containing methylene blue) give dioxetanes (compound containing two atoms of oxygen in a four membered ring). For details *see* section 2.1.2.3.

4.3 OXIDATION OF ALKYNES

Alkynes on oxidation with $KMnO_4$ in aqueous medium give a mixture of carboxylic acids. Some examples are given below (Scheme 4.12).

$$R{-}C{\equiv}C{-}R' + 4[O] \xrightarrow{KMnO_4} RCOOH + R'COOH$$

$$CH_3(CH_2)_7C{\equiv}C(CH_2)_7COOH \xrightarrow[pH7.5]{aq.\ KMnO_4} CH_3(CH_2)_7\overset{O}{\overset{\|}{C}}{-}\overset{O}{\overset{\|}{C}}{-}(CH_2)_7{-}COOH$$

$$CH_3(CH_2)_7C{\equiv}C(CH_2)_7COOH \xrightarrow{alk\ KMnO_4} CH_3(CH_2)_7COOH + HOOC(CH_2)_7COOH$$

(Scheme 4.12)

4.4. OXIDATION OF SIDE CHAINS IN AROMATIC RING SYSTEM

Some examples are given below (Scheme 4.13)

$$C_6H_5CH_3 \xrightarrow[\text{Cetyltrimethyl ammonium chloride, 4 hr}]{aq.\ KMnO_4/\Delta/C_6H_6} C_6H_5COOH$$

Toluene → Benzoic acid > 78%

Quinoline → ($[O]/C_6H_6$; $KMnO_4$, reflux PTC) → Quinolinic acid

(Scheme 4.13)

Toluene can also be oxidised to benzoic acid by $KMnO_4$ under microwave irradiation (yield 40% after 5 min of microwave irradation)[9a].

4.5 OXIDATION OF ALCOHOLS

Normally alcohols are oxidised with aqueous potassium permanganate solution. Solid potassium permanganate can be used to oxidise alcohols in benzene or hexane suspension under sonication[11] (Scheme 4.14).

$R^1R^2C(H)OH$ → ($KMnO_4$/hexane or benzene; RT,))))) → $R^1R^2C{=}O$

(Scheme 4.14)

It has been shown that in case of oxidation of octan-2-ol ($R^1 = C_6H_{13}$; $R^2 = H$), the corresponding ketone is obtained in 92.8% yield (5 hr) compared to 2% yield by mechanical stirring. Similarly, cyclohexanol gives 53% yield of cyclohexanone by oxidation with $KMnO_4$ under sonication (5 hr) compared to 4% yield under usual conditions. In a simlar way, oxidation of cinnamaldehyde under sonication gave 82% yield of cinnamic acid (3 hr) compared to 4% under usual conditions.

Alcohols can also be oxidised to aldehydes with manganese dioxide on sonication[12], in an inert solvent like octane. It has been shown that ultrasonic treatment activates effcienlty crystalline manganase dioxide which is otherwise of low reactivity[12]. Such a reagent oxidises cinnamyl alcohol, geraniol or 1-phenyl ethanol to the corresponding aldehyde. The use of a less volatile solvent like octane accelerates the reaction[13] (Scheme 4.15).

Ph–CH=CH–CH₂OH (Cinnanyl alcohol) → (MnO_2, 30°C; Octane,)))) 20 min) → Ph–CH=CH–CHO 73% Cinnamaldehyde

Geraniol $\xrightarrow[\text{90 min}]{MnO_2,\ 30°C,\ \text{octane,))))}}$ Neral 80%

$PhCH_2CH_2OH \xrightarrow[\text{20 min}]{MnO_2,\ 30°C,\ \text{octane,))))}} PhCH_2CHO$

1-Phenylethanol → Phenylacetaldehyde

(Scheme 4.15)

Oxygen is highly soluble in SC-CO_2 and this medium is particularly useful for the oxidation of weakly polar, water insoluble alcohols, due to low polarity of SC-CO_2. Thus, oxidation of alchols to carbonyl compounds in SC-CO_2 has been shown to proceed with high selectivity and high rate[13]. The oxidations are performed using noble metal catalysts (4% Pd, 1%, Pt, 5% Bi/C). Using this procedure, primary alcohols could be converted to aldehydes. Benzyl alcohols and related substrates could be oxidised to the corresponding aldehydes with > 99% selectivity.

The oxidation of benzyl alcohol with $KMnO_4$ in 1-butyl–3–methylimidazolium tetra fluoroborate, [bmin] [BF_4] ionic liquid at room temperature gave benzaldehyde in 90% yield in 1 hr (Scheme 4.16)[14].

$C_6H_5CH_2OH \xrightarrow[\text{[bmin] [}BF_4\text{], 1 hr, RT}]{KMnO_4} C_6H_5CHO$

Benzyl alcohol → 90% Benzaldehyde

(Scheme 4.16)

Using the procedure, a large number of substituted benzyl alcohols could be oxidised[14].

Benzyl alcohol can also be oxidised to benzaldehyde by oxidation with potassium hypochlorite in PEG–6000 and ethylacetate (Scheme 4.17)[15].

$C_6H_5CH_2OH + KOCl \xrightarrow{\text{PEG-6000/}CH_3COOC_2H_5} C_6H_5CHO$

Benzylalcohol → Benzaldehyde

(Scheme 4.18)

Some of the reagents used for the oxidation of alcohols are toxic, have cumbersome properties and difficulties encountered in the isolation of the required products and waste disposal. However, with the introduction of metallic reagents on solid support have overcome some of the problems providing an attractive alternative in organic synthesis due to selectivity and ease of manuplation. Also by the immobilation of metals on the surface avoids their leaching into the environment. Following are given some the convenient procedures used for this purpose.

(i) Using Clayfen

A facile method for the oxidation of alcohols to carbonyl compounds has been developed. In this method, montmorillonite K clay-supported iron (III) nitrate (clay fen) is used under solvent free conditions. The process is acclerated by exposure to MW irradiation[16]. The reaction is believed to proceed via the intermediacy of nitrosonium ions. In this oxidation, no carboxylic acids are formed in the oxidation of primary alcohols (Scheme 4.19).

$$R_2(R_1)CH—OH \xrightarrow[\text{MW, 15-60s}]{\text{Clayfen}} R_2(R_1)C{=}O \quad 87\text{-}96\%$$

R_1 = Ph, p-MeC_6H_4, p-$MeOC_6H_4$, (tetrahydrofuran-2-yl), R_2 = H

R_1 = Ph, R_2 = Et, PHCO, $R_1 - R_2$ = (cyclohexane)

R_1 = p-$MeOC_6H_4$, R_2 = p-$MeOC_6H_4CO$

(Scheme 4.19)

In the above oxidation, the experimental procedure involves mixing of neat alcohols with clayfen and a brief irradiation of the reaction mixtures in a MW oven for 15–60s in the absence of solvent.

(ii) Using Activated Managanese dioxide-Silica

Benzyl alcohols are selectively oxidised to carbonyl compounds using 35% MnO_2 'doped' silica gel under MW irradiation (Scheme 4.20)[17].

$$R_2(R_1)CHOH \xrightarrow[\text{MW, 20-60s}]{MnO_2\text{-silica}} R_2(R_1)C{=}O \quad 67\text{–}96\%$$

R_1 = Ph, p-MeC_6H_4, p-$MeOC_6H_4$, PhCH = CH, R_2 = H

R_1 = Ph, R_2 = Et, Ph, PhCO, R_1 = R_2 = hydroquenone

R_1 = p-$MeOC_6H_4$, R_2 = p-$MeOC_6H_4CO$

(Scheme 4.20)

(iii) Using Chromium Trioxide Impregnated Wet Alumina

Chromium trioxide impregnated pre-moistened alumina is an efficient oxidising agent for the oxidation of benzyl alcohols to carbonyl compounds by simply admixing the substrates with the reagent at room temperature (Scheme 4.20)[18].

$$\underset{R_1}{\overset{R}{>}}CH{-}OH \xrightarrow[\text{MW, 40 S}]{\text{wet } CrO_3\text{-}Al_2O_3} \underset{R_1}{\overset{R}{>}}C{=}O \quad 68-90\%$$

R = Ph, p-MeC_6H_4, p-$MeOC_6H_4$, p–$NO_2C_6H_4$; R_1 = H

R = Ph, R_1 = Me, Ph, PhCO; R = R_1 = (cyclohexane), (tetralin)

(Scheme 4.20)

In the above oxidation (Scheme 4.20) no overoxidation of carboxylic acid takes place. Also the reaction is clean and there is no formation of tarry.

(*iv*) Oxidation with iodobenzene diacetate (IBD) 'doped' Alumina.

Oxidation of alcohols to carbonyl compound occurs rapidly with alumina-supported IBD under solvent free conditions with MW irradiation in quantative yield (Scheme 4.21)[19].

$$\underset{R_1}{\overset{R}{>}}CH{-}OH \xrightarrow[\text{IBD–Neutral Alumina}]{\text{MW, 1–3 min,}} \underset{R_1}{\overset{R}{>}}C{=}O$$

(Scheme 4.21)

4.6 OXIDATION OF ALDEHYDES AND KETONES

A number of procedures are available for the oxidation of aldehydes to the corresponding carboxylic acid in aqueous and organic media (*see* section 2.5).

Aromatic aldehydes can be oxidised by aqueous performic and (obtained *in situ* by the addition of H_2O_2 to HCOOH at low temperauture (0–4°C)[22]. Using this procedure, heteroaromatic aldehydes like formyl pyridines, formyl quinolines can also be oxidised to the corresponding carboxylic acids; in this procedure, the formation of N-oxides is avoided.

Oxidation of formyl group in presence of other oxidizable groups can be carried out in aqueous media in presence of a surfactant. Thus, 4-(methylthio) benzaldehyde is oxidised quantitatively to 4-(methylthio) benzoic acid with TBHP in a basic aqueous medium in presence of cetyltrimethyl ammonium sulphate[23].

It is known that Dakins oxidation[24] is used for the oxidation of aromatic alcohols in *ortho* or *para* position to the formyl group with alkaline H_2O_2. However, the yields are low. This reaction has been carried out in high yields using sodium percarbonate (SPC; Na_2CO_3 1.5H_2O) in H_2O- THF under ultrasonic irradiation[25]. Using this procedure a number of aldehydes could be oxidised in good yield (Scheme 4.22).

R^1	R^2
H	OH
H	OCH_3
OCH_3	H
OCH_3	OH
	H

(Scheme 4.22)

Ketones can be converted into the corresponding esters by the well-known **Baeyer-Villiger oxidaton**[26]. This conversion is usually affected by peracids in organic solvents. It is now possible to carry out this conversion in aqueous heterogeneous medium using MCPBA at room temperature[27]. Some examples are given below (Scheme 4.23.).

Acetophenone → ($PhCOOH, CHCl_3$, 25°C) → Phenyl acetate

R-ml, *t*-Bu → (MCPBA, 20°C, H_2O, 1hr.) → 95%

R = H, Cl, OMe → (MCPBA, 80°C, H_2O, 0.5 .15 hr) → 70–90%

(Scheme 4.23)

Balyer-Villiger oxidations of ketons with MCPA proceed much faster at room temperature in the solid state[28]. The yields obtained in the solid state are much better than in $CHCl_3$. Some typical examples are given below (Scheme 4.24).

$$\text{MeOC}-C_6H_4-\text{Br} + m\text{-CPBA} \xrightarrow[\text{solid state}]{\text{RT. 5 days}} \text{MeOCO}-C_6H_4-\text{Br}$$

64% (50% in $CHCl_3$)

$$PhCOCH_2Ph + \text{mePBA} \xrightarrow[\text{solid state}]{\text{RT. 24 hr.}} PhCOOCH_2Ph$$

Benzylphenylketone

97% (46% in $CHCl_3$)

$$PhCOPh + m\text{-CPBA} \xrightarrow[\text{solid state}]{\text{RT. 24 hr.}} PhCOOPh$$

85% (13% in $CHCl_3$)

$$\text{PhOC}-C_6H_4-\text{Me} + m\text{-CPBA} \xrightarrow[\text{solid state}]{\text{RT. 24 hr.}} \text{PhOCO}-C_6H_4-\text{Me}$$

50% (72% in $CHCl_3$)

(Scheme 4.24)

4.7 MISCELLANEOUS OXIDATIONS

(*i*) Oxidation of alkanes, bromides, carboxylic acids, cyanides and amines to carbonyl compounds: Substrates like alkanes, bromides, carboxylic acids, cyanides and amides can be oxidised to carbonyl compounds by copper (II) nitrate impregnated on K10 clay (claycop)-hydrogen peroxide (Scheme 4.25)

$$R_2R_1CH-R_3 \xrightarrow[\text{claycop-}H_2O_2]{\text{MW}} R_2R_1C=O$$

R_1 = Ph, p-$NO_2C_6H_4$

R_2 = H, Ph

R_3 = H, Br, COOH, CN, NH_2

(Scheme 4.25)

In the above oxidation (Scheme 4.25), the reaction of copper (II) nitrate and H_2O_2 produces a hydroxy copper Cu(II) compound (which is also obtained from Cu (II) acetate and H_2O_2); it is the hydroxy copper (II) compound which oxidises the organic substrates to carbonyl compound.

$$2Cu(NO_3)_2 + 2H_2O + H_2O_2 \longrightarrow 2CuO_2H + 4HNO_3$$

The resulting nitric acid is neutralised with $KHCO_3$ to maintain pH ≈ 5.

***(ii)* Oxidation of α-hydroxyketones to 1,2-diketones:** α-Hydroxyketones (benzoins) can be easily oxidised to 1,2–diketones (benzils) in high yield using solid reagent systems, copper (II) sulphate–alumina[30] or oxone-wet alumina[31] under the influence of microwaves (Scheme 4.26).

Note: Oxone is a registered trade mark of E. I du Pont de Nemours and company. It has the composition $2KHSO_5.KHSO_4.K_2SO_4$.

$$R-CH(OH)-CO-R_1 \xrightarrow[\text{MW, 2–3.5 min}]{CuSO_4\text{-}Al_2O_3 \text{ or oxone-}Al_2O_3} R-CO-CO-R_1$$

$R = R_1 = C_6H_5$, *p*-MeC_6H_4, *p*-$MeOC_6H_4$, *p*-ClC_6H_4,
$R = C_6H_5$; R_1 = *p*-MeC_6H_4, *p*-$MeOC_6H_4$
and R = Me; $R_1 = C_6H_5$

(Scheme 4.26)

***(iii)* Oxidation of sulphides to sulfoxides and sulfones:** Oxidation of sulphides to sulfoxides and sulfones can be achieved with the desired selectivity (to either sulfoxides or sulfones) by using silica 'doped' with 10% sodium periodate on microwave irradiation under reduced power and reaction time (Scheme 4.27)[32].

$$R-S-R_1 \text{ (Sulphides)} \xrightarrow[\text{M.W. 1–3 min}]{20\%\ NaIO_4\text{-Silica (3.0 eq)}} R-SO_2R_1 \text{ (72–93\% Sulfones)}$$

$$R-S-R_1 \text{ (Sulphides)} \xrightarrow[\text{M.W. 0.5–2.5 min}]{20\%\ NaIO_4\text{-Silica (1.7 eq)}} R-S(=O)-R_1 \text{ (76–85\% Sulfoxides)}$$

$R = R_1$ = Ph, *p*-Me C_6H_4, *p*-$MeOC_6H_4$, P-ClC_6H_4
$R = C_6H_5$, R_1 = *p*-MeC_6H_4, *p*-$MeOC_6H_4$
and R = Me; $R_1 = C_6H_5$

(Scheme 4.27)

Selective oxidation of alkyl, aryl and cyclic sulphides to the corresponding sulfoxides can also be achieved by the solid reagent system 1BD-alumina upon microwave irradiation (Scheme 4.28)[33].

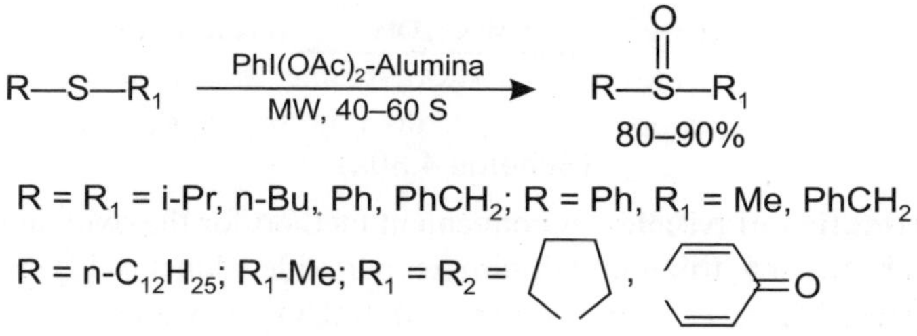

$$R—S—R_1 \xrightarrow[\text{MW, 40–60 S}]{\text{PhI(OAc)}_2\text{-Alumina}} R—\overset{O}{\overset{\|}{S}}—R_1 \quad 80–90\%$$

$R = R_1$ = i-Pr, n-Bu, Ph, $PhCH_2$; R = Ph, R_1 = Me, $PhCH_2$

R = n-$C_{12}H_{25}$; R_1-Me; $R_1 = R_2$ = (cyclopentane), (cyclohexadienone)

(Scheme 4.28)

In a solid state reaction with clayfen, a variety of alkyl, aryl and cyclic sulphides are readily oxidised to the corresponding sulfoxides in high yield on microwave thermolysis[34].

(iv) Oxidation of Enamines: β,β-Disubstituted enamines on oxidation with $KMnO_4$–Al_2O_3 in domestic MW oven (225W, 82°C) as well as in focussed (330W, 140°C) microwave oven under solvent free conditions give[35] the carbonyl compounds. Use of focussed MW oven gave better yields (Scheme 4.29). In the reactions conducted in an oil bath at 140°C, no ketone formation was observed.

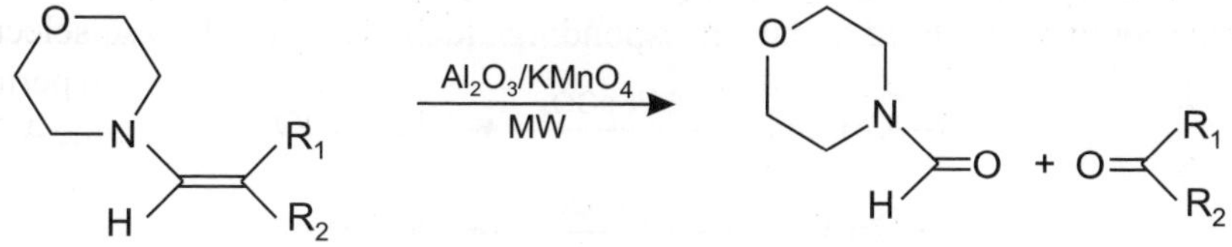

$Al_2O_3/KMnO_4$, MW

(Scheme 4.29)

(v) **Oxidation of Arenes:** $KMnO_4$ impregrated alumina oxidises arenes to ketone within 10 min. in solvent free conditions using focussed microwaves (Scheme 4.30)[36].

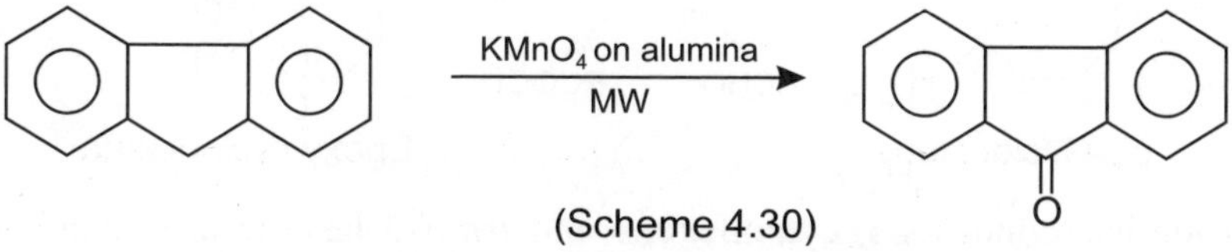

$KMnO_4$ on alumina, MW

(Scheme 4.30)

(vi) **Oxidation of Nitro compounds into Carbonyl compounds:** Primary and secondary nitro compounds are oxidised into the corresponding carbonyl compounds by alkaline $KMnO_4$ (scheme 4.30 a)[37]

$$R—CH_2NO_2 \xrightarrow[H_2O,\ 0–5°C]{KMnO_4,\ OH^-} R—CHO \quad 85\%$$

$$\text{Cyclobutyl-}NO_2 \xrightarrow[H_2O,]{KMnO_4,\ OH^-} \text{Cyclobutanone}\ (95\%)$$

(Scheme 4.30a)

***(vii)* Oxidation of Nitriles:** A convenient method for the oxidation of nitriles into amides is possible using urea-hydrogen peroxide adduct (UPH, H_2NCONH_2, H_2O_2) in presence of catalytic amount of K_2CO_3 in water-acetone at room temperature (Scheme 4.31)[38]

$$\underset{R = Ph,\ Me,}{R\text{—}CN} \xrightarrow[H_2O\text{-acetone, RT}]{UPH,\ K_2CO_3} \underset{85\text{–}95\%}{R\text{—}CONH_2}$$

(Scheme 4.31)

4.8 OXIDATIONS IN SUPER CRITICAL CARBON DIOXIDE

The system supercritical carbon dioxide-carbondioxide-oxygen ($ScCO_2$–O_2) is particularly useful for the oxidation of water insoluble alcohols to carbonyl compounds. This procedure gives high selectivity and excellent yields[39]. Benzyl alcohols could be oxidised to the corresponding aldehyde with >99% selectivity.[39]

$$R\text{—}CH_2OH \xrightarrow{O_2/ScCO_2} R\text{—}CHO$$

$$ArCH_2OH \xrightarrow{O_2/ScCO_2} ArCHO$$

This system is also useful for the conversion of *cis*-cyclooctene into epoxycyclooctane (99% cc). In this case, an aldehyde (*e.g.*, 2-propinaldehyde) is used as a sacrifical oxygen transfer agents[39].

$$\textit{cis}\text{-Cyclooctene} \xrightarrow[\text{RCHO} \rightarrow \text{RCOOH}]{O_2/SCCO_2,\ \text{Stainless steel}} \text{Epoxy cyclooctane}$$

Under similar conditions, cyclohexene and *trans*-3-hexene are quantitatively converted to epoxycyclohexane (91% selectivity) and *trans*-3-epoxyhexane (>98% selectivity) respectively[39].

The system $SCCO_2$ has also been used for sharpless asymmetric epoxidation and diasteroselective sulfoxidations[39].

4.9 OXIDATIONS IN IONIC LIQUIDS

Epoxidations of alkenes and allylic alcohols, asymmetric dihydroxylation of olefines, oxidative carbonylation of amines and Wacker type oxidations have been satisfactorily performed in ionic liquids[35]. It is also possible to oxidise benzylic alcohols to the corresponding aldehydes (in 90% yield) using $KMnO_4$ in ionic liquids[36].

4.10 OXIDATIONS IN POLYETHYLENE GLYCOL

Benzyl bromide is oxidised to benzaldehyde using $K_2Cr_2O_7$ in Polyethylene glycol (PEG-400)[42].

$$\underset{\text{Benzyl bromide}}{C_6H_5CH_2Br} \xrightarrow[\text{oxidation}]{K_2Cr_2O_7/\text{PEG-400}} \underset{\text{Benzaldehyde}}{C_6H_5CHO}$$

Internal olefins could be dihydroxylated using PEG as solvent and OsO_4 as catalyst[42]

$$\underset{\text{Styrene}}{C_6H_5CH = CHC_6H_5} \xrightarrow[\substack{OsO_4 \\ \text{PEG-400 (2-3 hr.)}}]{\text{N-methyl morpholine (NMO)}} \underset{\text{Styrene (95\%) diol}}{C_6H_5CH(OH)\,CH(OH)C_6H_5}$$

Trans-Ethylcinnamate, R = H
4-Methoxy trans ethylcinnamate
R = OMe

$\xrightarrow[\text{PEG=400, 2 hr (or Mo)}]{OsO_4\text{ (05 mol \%)}}$ 92–93%

An environmentally benign synthesis of adipic acid in PEG based aqueous biphasic system using sodium tungstate and H_2O has been reported[43].

$$\underset{\text{Cyclohexene}}{C_6H_{10}} + 4H_2O \xrightarrow[\text{PEG/NaHSO}_4]{Na_2WO_4} \underset{\text{Adipic acid}}{HOOC(CH_2)_4COOH} + 4H_2O$$

4.11 OXIDATIONS IN SOLID STATE

Using urea hydrogen peroxide adduct (uPH), following oxidations have been performed:[44]

(*i*) Oxidation of hydroxy aldehydes and ketones to the corresponding phenols:

$\xrightarrow[\substack{50\text{–}80°C \\ \text{20 min to 72 min}}]{\text{uHP}}$

R = OH, NO_2, OMe, H

80–90%

COCH$_3$ ... OH, R = H, OH $\xrightarrow[85°C,\ 60min]{uPH}$ catechol (OH, OH, R) 80–86%

R = H, OH

(*ii*) Oxidation of nitriles to amides

Eco-friendly oxidation of nitriles (*e.g.*, C_6H_5CN, $C_6H_5CH_2CN$) has been achieved using uPH at 85°C (1–1.5 hr) to yield the corresponding amides ($C_6H_5CONH_2$,$C_6H_5CH_2CONH_2$, respectively).

(*iii*) Selective oxidation of sulphides to sulfoxides or sulfones.

Urea-formaldehyde complex (UPH) oxidises sulphides to sulfoxides or sulfones. Some examples are given below.

$$C_6H_5SMe \xrightarrow[15\ min]{UPH,\ 85°C} C_6H_5SOMe \quad 80\%$$

$$[CH_3(CH_2)_3]_2S \xrightarrow[10\ min]{UPH,\ 85°C} [CH_3(CH_2)_3]_2SO \quad 85\%$$

Tetrahydrothiophene $\xrightarrow[60\ min]{UPH,\ 85°C}$ sulfolane (S with two O) 88%

$C_6H_5(CH_2)_2S \xrightarrow[2\ hr]{UPH,\ 85°C} C_6H_5(CH_2)_2SO_2$ 90%

(*iv*) Oxidation of nitrogen heterocycles to N-oxides.

Nitrogen heterocycles could be oxidised to the corresponding N-oxides using UPH. Some exampels are given below:

Pyridine (N) $\xrightarrow[45\ min]{UPH,\ 85°C}$ pyridine N-oxide (N^+–O^-) 87%

Tetramethylpyrazine (N, N) $\xrightarrow[45\ min]{UPH,\ 85°C}$ tetramethylpyrazine N-oxide (N, N^+–O^-) 92%

4.12 OXIDATIONS USING POLYMER SUPPORTED REAGENTS

In these oxidations, the polymer supported reagents are used. Following are the advantages of using such reagents:

1. The isolation of the product in pure state easy and the reaction can be easily carried

Substrate + Polymer supported regent ⟶ Product (can be isolated by filtration) + Excess of the polymer supported reagent

2. The polymer supported reagents can be recycled after use.
3. Most of the oxidations can be carried out rapidly and in high yields under mild conditions.

A number of polymers can be used as polymer support. However, the most widely used polymer is polystyrene.

Following is given some of the polymer supported reagents, which can be used for various oxidations.

Polymer supported peracid (P)–C_6H_4–CO_3H

Polymer supported chromic acid, Amberlyst A–26, $HCrO_4^-$ form is commercially available.

Polymer supported phase transfer catalysts and crown ethers

(P)——PTC

Poly-styrene boronic acid (P)–C_6H_4–$B(OH)_2$

For a discussion on Polymer supported reagents, their preparation and applications *see* V. K. Ahluwalia and R. Aggarwal, Organic Synthesis, Special Techniques, Narosa Publishing House, 2005, Chapter 5.

References

1. F. Fringuelli, F. Germani and G. Sovelli, *Tetrahedron, Lett.,* 1989, **30**, 1427.
2. A. S. Kende, P. Delair and B. E. Blass, *Tetrahedron Lett.,* 1994, **44**, 8123.
3. B. Notari., Stud. Surf. Sci., Catal, 1961, **67**, 243; R. A. Sheldon, *J. Mol. Catal.,* 1980, **7**, 107.
4. H. Hudlicky, *Oxidations in Organic Chemistry,* ACS Monograph, American Chemical Society, Washington DC, 1990.
5. J. Muzart, *Synthesis,* 1955,1325; K. LO. Read, J. T. Gupton and T. L. Solarz, *Synth. Commun.,* 1989, **19**, 3579.
6. G. Beri, *Topics in Stereochemistry,* N. L. Allinger, E. L. Eliel, Eds., vol 7, p–93, John Wiley, 1967.
7. F. Toda, H. Takumi, M. Nagami and K. Tanaka, *Heterocycles*, 1998, **47**, 469.

7a. G. Jener., D. Sueur, T. Mallat and A. Baiker, *Chem. Commun* 2000,2247.

8. N. A. Milas and S. Sussman, *J. Am. Chem. Soc.,* 1936, **58**, 1302; R. Daniels and J. L. Fischer, *J. Org. Chem.,* 1963, **28**, 320.

9. W. B. Webber and J. P. Shephard, *Tetrahedron Lett.,* 1972, 4907.

9a. R. Gedye, F. Smith, K. Westaway, A. Humera, L. Baldisera and L. R. Laberge, *Tetrahedron Lett.,* 1986, **26**, 279.

10. M. Mugdan and D. P. Young, *J. Chem. Soc.,* 1949,2488.

10a. S. ChandraSekhar, Ch. Narshihmulu, S. S. Sultana and N. R. Reddy, *Chem., Commun.,* 2003,1716.

11. J. Yamawaki, S. Sumi, T. Ando and T. Hanafusa, *Chemistry Lett.,* 1983,379.

12. T. Kimura, M. Fujita and T. Ando, *Chemistry Lett.,* 1988, 137:

13. G. Jerner, D. Sueur, T. Mallat and A. Baiker, *Chem. Commun.,* 2000,2247.

14. A. Kumar, N. Jain and S. M. S. Chauhan, *Synthetic Communication,* 2004, **34**, 2835.

15. P. Balasubramian, P. Sakumar and R. Chandani, *Tetrahedron Lett.,* 1979, **20**, 3543.

16. R. S. Varma and R. Dahiya, *Tetrahedron Lett.,* 1997, **38**, 2043.

17. R. S. Varma, R. K. Saini and R. Dahiya, *Tetrahedron Lett.,* 1997, **38**, 7823.

18. R. S. Varma and K. K. Saini. *Tetrahedron Lett.,* 1998, **39**, 1481.

19. R. S. Varma, R. Dahiya and R. K. Saini, *Tetrahedron Lett.,* 1997, **38**, 7029.

22. R. H. Dodd and M. LeHyaric, *Synthesis,* 1993, 295.

23. F. Fringuelli, R. Pellegino, O. Piematti and F. Pizzo, *Synth. Commun.,* 1994, **24**, 2665.

24. H. D. Dakin, OS, 1941, **1**, 149; J. E. Lettler, *Chem. Rev.,* 1949, **45**, 385; H. D. Dakin, *J. Am. Chem., Soc.,* 1909, **42**, 477.

25. G. W. Kabalka, N. K. Reddy and C. Narayana, *Tetrahedron Lett.,* 1992, **33**, 865.

26. A. V. Baeyer and V. Villiger. Ber., 1899, **32**, 3625.

27. F. Froriguelli, F. Germani; F. Pizzo and G. Savelli, Gazz. *Chem., Ital.,* 1989, **119**, 249.

28. K. Tanaka and F. Toda, *Chem. Rev.,* 2000,**100**,1028,29.

29. R. S. Varma and R. K. Saini, *Tetrahedron Lett.,* 1998, **39**, 1307.

30. R. S. Varma, D. Kumar and R. Dahiya, *J. Chem.,* Res (S), 1998, 324.

31. R. S. Varma, R. Dahiya and D. Kurmar, *Molecules on Line,* 1998, **2**, 82.

32. R. S. Varma, R. K. Saini and H. M. Meshram, *Tetrahedron Lett.,* 1997, **38**, 6525.

33. R. S. Varma, R. K. Saini and R. Dahiya, *J. Chem.,* Res. (S), 1998, 120.

34. R. S., Varma and R. Dahiya, *Synth. Commun.,* 1998, **28**, 4087.

35. H. Benhaliliba, A. Derdour, J.–P. Bazureau, F. Texier–Boullet and J. Hamelin, *Tetrahedron Lett.,* 1998, **39**, 541.

36. A. Oussaid and A. Loupy, *J. Chem. Res* (S), 1997, 342.

37. V. K. Ahluwalion and R. S. Varma, Green Solvents for Organic Synthesis, Narosa, 2009, Page 1.41.

38. R. S. Verma and K. P. Naicker, *Org., Lett.,* 1999, **1**, 189.

39. D. Clarke, M. A. Ali, A. A. Clifford, A. Parratt, P. Rose, D. Schwinn, W. Bannwarth and C. M. Rayner, *Current Topics in. Medicinal Chemistry* 2004, **4**, 753–754 and the references cited there in.

40. For a discussion on oxidations in ionic liquids see V. K. Ahluwalia and R. S. Varma, *Green solvents for organic synthesis*, Narosa Publishing House, 2009, page 7.17 to 7.19 and the references cited there in.

41. See Reference 40 page 7.42 and the referencis cited therein.

42. See Reference 40 page 8.13 and the references cited therein.

43. See Reference 40 page 8.23 and the references cited therein.

44. H. Tanka, S. Kishigami and F. Toda, *J. Chem. Soc.,* 1991, **56**, 4333.

Suggested Readings

1. Oxidation in Organic Chemistry, Milos Hudlický, ACS Manograph 186, American Chemical Society, Washington, DC, 1990.
2. Oxidation and Reductions in Organic Synthesis, Timothy J. Donohol Oxford Science Publications, UK.
3. Oxidation in Organic Syntheses Michael B. Smith, McGraw Hill International Editions, 1994, Page 213–342.
4. Oxidation in Organic Reaction Mechanism V.K. Ahluwalia and Rakesh Kumar Parashar, Narosa Publishing House, New Delhi, 2007, Page 137-220 and the reference cited therein.
5. Oxidation in Some Modern Methods of Organic Synthesis, W. Carruthers, Cambridge University Bess, 1886, Page 344-410.
6. Oxidations in Enzymes for Green Organic Synthesis V.K. Ahluwalla, Narosa Publishing House, New Delhi, 2010, Page 3.6 and 4.9 to 4.21 and the references cited therein.
7. Oxidations in Green Solvents for Organic Synthesis V.K. Ahluwalia and Rajender S. Varmaa, Narosa Publishing House, New Delhi 2009, Page 1.30, 5.19 to 5.21, 5.4, 7.19, 7.42, 8.4 and 8.8 and the references cited therein.
8. Oxidations in Organic Synthesis, Special Techniques, V.K. Ahluwalia and Renu Aggarwal, Narosa Publishing House, New Delhi, 2006, Page 51 to 56, 129, 177 to 180, 187 to 189 and 207 to 208 and the references cited thesein.
9. Oxidations in Alternate Energy Processes in Chemical Synthesis, V.K., Ahluwalia and R.S. Varma, Narosa Publishing House, New Delhi, 2008, page 4.9 to 4.13, 7.4 and 9.3 to 9.5 and the references cited therein.
10. Oxidations, Thomas N. Sorrell, in Organic Chemistry, Viva Books, 2010, page 332 to 356.
11. Oxidations, Francis A. Carey and Richard J. Sundberg in Advanced Organic chemistry, part B Springer 2001, page 747-808.
12. Oxidations, Marye Anne Fox and James K. Whitesell, in Organic Chemistry, Jones and Bartlett Publishers, 2004, page 449-451.

Index

Q

R

S

T